Marine Oil Spills 2018

Marine Oil Spills 2018

Special Issue Editor
Merv Fingas

MDPI • Basel • Beijing • Wuhan • Barcelona • Belgrade

MDPI

Special Issue Editor
Merv Fingas
Spill Science
Canada

Editorial Office
MDPI
St. Alban-Anlage 66
4052 Basel, Switzerland

This is a reprint of articles from the Special Issue published online in the open access journal *Journal of Marine Science and Engineering* (ISSN 2077-1312) from 2018 to 2019 (available at: https://www.mdpi.com/journal/jmse/special_issues/bz_oil_spills_2018)

For citation purposes, cite each article independently as indicated on the article page online and as indicated below:

LastName, A.A.; LastName, B.B.; LastName, C.C. Article Title. *Journal Name* **Year**, *Article Number*, Page Range.

ISBN 978-3-03897-854-1 (Pbk)
ISBN 978-3-03897-855-8 (PDF)

Cover image courtesy of Elastec, Carmi, Illinois.

Contents

About the Special Issue Editor

Merv Fingas is a scientist working on oil and chemical spills. He was Chief of the Emergencies Science Division of Environment Canada for over 30 years and is currently working on research in Western Canada. Dr. Fingas has a PhD in environmental physics from McGill University and three Master's degrees, in chemistry, business, and mathematics, all from the University of Ottawa. He also has a Bachelor of Science majoring in Chemistry from Alberta and a Bachelor of Arts from Indiana. He has published more than 940 papers and publications in the field. Dr. Fingas has prepared 10 books on spill topics. He has served on two oil spill committees on the National Academy of Sciences of the United States, including the Committee on Oil in the Sea. He is chairman of several ASTM and intergovernmental committees on spill matters. Importantly, he was the founding chairman of the ASTM subcommittee on in situ burning, as well as the chairman of a subcommittee on oil spill-treating agents and of another on oil spill detection and remote sensing, positions he holds to date. In the summer of 2010, he testified before a US congressional committee on oil spills in conjunction with the Gulf spill. He was one of three scientists working with US NOAA to examine the mass balances of the Gulf spill.

Journal of
Marine Science and Engineering

MDPI

Editorial

Marine Oil Spills 2018

Merv Fingas

Spill Science, Edmonton, Alberta T6W 1J6, Canada; fingasmerv@shaw.ca

Received: 26 March 2019; Accepted: 26 March 2019; Published: 27 March 2019

check for
updates

Major oil spills can attract the attention of the public and the media. In past years, this attention had created a global awareness of the risks of oil spills and the damage they do to the environment. In recent years, major spill incidents have been fewer in number; still, the public is aware of very major spills, but generally is unaware that spills are a daily fact of life.

Oil is a necessity in our industrial society and a major element of our lifestyle. Most of the energy used in much of the developed world is for transportation which runs on oil and petroleum products. According to trends in energy usage, this is not likely to decrease much in the future. Industry uses oil and petroleum derivatives to manufacture such vital products as plastics, fertilizers, and chemical feedstocks, which will still be required in the future. In fact, the production and consumption of oil and petroleum products is increasing worldwide, and the risk of oil pollution is increasing accordingly. The movement of petroleum from oil fields to the consumer involves as many as 10 to 15 transfers between many different modes of transportation, including tankers, pipelines, railcars, and tank trucks. Oil is stored at transfer points and at terminals and refineries along the route. Accidents can happen during any of these transportation steps or storage times. Fortunately, in the past few years, the actual number of spills has decreased, but oil spills will still continue to form part of our industrial fabric.

Obviously, an important part of protecting the environment is ensuring that there are as few spills as possible. Both government and industry are working to reduce the risk of oil spills, with the introduction of strict new legislation and stringent operating codes. Industry has invoked new operating and maintenance procedures to reduce accidents that lead to spills. Intensive training programs have been developed to reduce the potential for human error.

Oil spills necessitate a multiplicity of talents to deal with them and a multiplicity of disciplines to study and research them. This special edition provides a glimpse into these multiple facets of oil spills. Topics include oil spill modeling, risk analysis and preparation for oil spills. We hope that this special edition will be helpful and enlightening to those in the ever-changing and advancing field of oil spills.

Conflicts of Interest: The author declares no conflict of interest.

Journal of
Marine Science and Engineering

MDPI

Article

Oil Droplet Transport under Non-Breaking Waves: An Eulerian RANS Approach Combined with a Lagrangian Particle Dispersion Model

Roozbeh Golshan [1], Michel C. Boufadel [1,*], Victor A. Rodriguez [1], Xiaolong Geng [1], Feng Gao [1], Thomas King [2], Brian Robinson [2] and Andrés E. Tejada-Martínez [3]

[1] Center for Natural Resources, Department of Civil and Environmental Engineering, New Jersey Institute of Technology, Newark, NJ 07102, USA; rgolshan@njit.edu (R.G.); var5@njit.edu (V.A.R.); gengxiaolong@gmail.com (X.G.); gaofeng0414@gmail.com (F.G.)

[2] Bedford Institute of Oceanography, Department of Fisheries and Oceans, Dartmouth, NS B2Y 4A2, Canada; Tom.King@dfo-mpo.gc.ca (T.K.); Brian.Robinson@dfo-mpo.gc.ca (B.R.)

[3] Department of Civil and Environmental Engineering, University of South Florida, 4202 East Fowler Avenue ENB 118, Tampa, FL 33620, USA; aetejada@usf.edu

* Correspondence: boufadel@gmail.com; Tel.: +1-973-596-6079

Received: 5 December 2017; Accepted: 8 January 2018; Published: 15 January 2018

Abstract: Oil droplet transport under a non-breaking deep water wave field is investigated herein using Computational Fluid dynamics (CFD). The Reynolds-averaged Navier–Stokes (RANS) equations were solved to simulate regular waves in the absence of wind stress, and the resulting water velocities agreed with Stokes theory for waves. The RANS velocity field was then used to predict the transport of buoyant particles representing oil droplets under the effect of non-locally generated turbulence. The RANS eddy viscosity exhibited an increase with depth until reaching a maximum at approximately a wave height below the mean water level. This was followed by a gradual decrease with depth. The impact of the turbulence was modeled using the local value of eddy diffusivity in a random walk framework with the added effects of the gradient of eddy diffusivity. The vertical gradient of eddy viscosity increased the residence time of droplets in the water column region of high diffusivity; neglecting the gradient of eddy diffusivity resulted in a deviation of the oil plume centroid by more than a half a wave height after 10 wave periods.

Keywords: RANS; non-breaking ocean waves; random walk method; Lagrangian particle dispersion; oil spill model

1. Introduction

Waves play an important role in the transport and fate of oil spills [1,2]. Waves at sea are accompanied by breakers of various magnitudes due to the interaction of various waves and the presence of wind. These breakers result in shear stress that breaks the oil slick into droplets [3] that get injected into the water column. The subsequent motion of waves, even regular waves, results in further downward spreading of the small oil droplets [4]. The "rule of thumb" [5] is that droplets smaller than 100 μm remain below the surface whereas those larger than 100 μm return to the water surface. Waves combined with wind generate Langmuir turbulence characterized by Langmuir cells which also play a significant role in the vertical and horizontal distribution of oil slicks [6].

Extensive studies have been conducted for predicting tracer transport, and relations between tracer properties, fluid motion, and the spatial distribution of bubbles, solute or droplets were sought and developed [7–9]. In [10] the direct effect of waves on transport were addressed using an Eulerian formulation. Boufadel and co-workers [4,11,12] assumed second order waves and used a Lagrangian formulation to investigate the effects of regular and irregular waves on dispersed oil and explained, among other, the comet shape of spills based on the droplet sizes and the Stokes drift; as the large droplets stay closer to the surface, they get entrained forward by the Stokes drift, which is maximum at the surface. The smaller droplets, thus, trail behind, giving the appearance of a comet. The velocity field above the mean water level was obtained by using either Taylor expansion from the surface [4,11] or Wheeler stretching [12]. The impact of turbulence was assessed using an empirical eddy viscosity expression from [13].

The aforementioned studies treated the droplets as passive tracers with a term that accounts for buoyancy and another term to account for turbulent diffusion through a random walk. However, a whole category of studies focused on the dynamics at the droplet scale, and solved the equation of motion for each droplet accounting for inertia and added mass. For example, it was demonstrated that, under certain conditions (related to the droplet Stokes number, defined below) inertia and added mass play an important role in moving droplets [14]. It was further noted the presence of a vertical Stokes drift due to droplet's inertial effects [15]. Bakhoday-Paskyabi expanded on these works to consider various types of waves (regular, conidial, and solitary), in which theoretical arguments were given for the need to include the added mass for the conditions considered [16]. These studies relied on the irrotational theory of waves, and neglected the impact of turbulence on droplet transport.

In a study of flow in a tidal channel, the authors argues that the gradient of eddy diffusivity needs to be explicitly accounted for when evaluating the transport of particles in the water column [17]. In that study, the eddy diffusivity corresponded to a wind and tidal driven flow characterized by surface and bottom boundary layers. It was shown that by ignoring the gradient of diffusivity, particles accumulate in the areas of lower diffusivity. Thus, there is a need to account for the gradient of eddy diffusivity [18–21], which is the work pursued herein in conducting random walk simulations.

We focus on the particular problem of oil droplets spreading under non-breaking deep-water waves with turbulence advected into the domain by the waves and assess the effect of turbulence-engendered diffusion on the transport of the droplets. The turbulence may be considered nonlocal, generated elsewhere perhaps by the action of winds, and transported by the Stokes drift to a zone where the winds have desisted. The combined effect of the non-breaking waves and the turbulence on the droplet motions is of focus here.

In the present RANS (Reynolds-averaged Navier-Stokes) simulation, the turbulence is injected at the left boundary of the domain and is advected into the domain by the motion of the surface waves. The present simulation configuration is considered as an idealization of turbulence generated over a finite horizontal span in the open ocean and advected to an adjacent region where the turbulence source is no longer present. For example, in the upper ocean, Langmuir turbulence [22] associated with Langmuir circulations occurs over a limited horizontal span where winds and waves are sufficiently aligned allowing for the generation of the turbulence. The turbulence may then be transported by the waves beyond (outside) of the region of production where the wind-driven shear and waves are misaligned and thus where the source of the turbulence is no longer present.

Our chosen numerical framework to simulate the movement of water waves and evolution of the turbulence consists of the RANS equations with turbulence closure provided by the RNG (Re-Normalisation Group) k-ε model. We used for this purpose the commercial software Fluent 15.0 [23]. The free surface was modeled using the Volume of Fluid (VOF) module within Fluent. In terms of capturing the hydrodynamics [24–26] (especially due to turbulence), the RANS approach may be viewed as a compromise between the potential flow theory solutions [27] and the highly resolved Large Eddy Simulation (LES) [28] approach.

The motion induced by the regular waves studied is taken as two-dimensional, thus the present simulations are two-dimensional spanning horizontal (along wave) and vertical directions (see Figure 1). Turbulence is three-dimensional and thus its resolution would require a three-dimensional simulation approach such as a large eddy simulation or direct numerical simulation. In the current approach based on Reynolds-averaging, the motion induced by the waves is resolved while the turbulence is not simulated (resolved), but rather modeled through the k-ε turbulence model. In this approach, the turbulence intensity (i.e., the turbulence kinetic energy (TKE)), the TKE dissipation rate and the ultimate eddy viscosity and eddy diffusivity are predicted through the governing k-ε turbulence model transport equations subject to the resolved two-dimensional flow field induced by the waves. Thus, the k-ε model equations can be taken as two-dimensional given that the flow field forcing these equations (e.g., through turbulence production by mean flow shear) is also two-dimensional. The turbulence being modeled can be alternatively considered as an ensemble-average prediction, averaged over the third dimension not resolved by the simulation.

The generated waves for the investigation were regular waves with a period of $T = 1.0$ s and a wave height $H = 0.15$ m in a domain whose water depth is 1.2 m (see Figure 1). This ensured deep-wave conditions as it is larger than half of the wave length which is ~1.56 m.

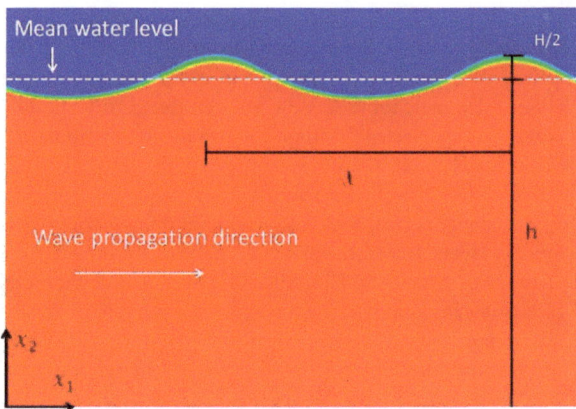

Figure 1. Details of multi-phase wave simulation. The areas colored in red represent cells with water phase (i.e., $\alpha_w = 1$ in Equation (7a)) and areas colored in blue represent cells with air phase (i.e., $\alpha_w = 0$).

2. Materials and Methods

2.1. Governing Equations: Eulerian RANS Framework

For an unsteady, viscous incompressible, two dimensional flow, the Reynolds-averaged governing equations are

$$\frac{\partial u_j}{\partial x_j} = 0 \tag{1}$$

$$\rho\left(\frac{\partial u_j}{\partial t} + u_j\frac{\partial u_i}{\partial x_j}\right) = -\frac{\partial p}{\partial x_i} + \frac{\partial}{\partial x_i}\left[(\mu + \mu_t)\frac{\partial u_j}{\partial x_j}\right] \tag{2}$$

$$\rho\left(\frac{\partial K}{\partial t} + u_j\frac{\partial K}{\partial x_j}\right) = \frac{\partial}{\partial x_j}\left[\alpha_k(\mu + \mu_t)\frac{\partial K}{\partial x_j}\right] + P_k - \rho\varepsilon \tag{3}$$

$$\rho\left(\frac{\partial \varepsilon}{\partial t} + u_j\frac{\partial \varepsilon}{\partial x_j}\right) = \frac{\partial}{\partial x_j}[\alpha_\varepsilon(\mu + \mu_t)\frac{\partial \varepsilon}{\partial x_j}] + C_{1\varepsilon}\frac{\varepsilon}{K}P_K - C_{2\varepsilon}p\frac{\varepsilon^2}{K} - R_\varepsilon \tag{4}$$

where Equations (1) and (2) represent the conservation of mass and conservation of momentum in two spatial directions, respectively. Equations (3) and (4) represent the turbulence k-ε model consisting of a transport equation for turbulent kinetic energy K (TKE) (Equation (3)) and a transport equation for TKE dissipation rate ε (Equation (4)) [29]. In the equations above, $i = 1, 2$ with index 1 corresponding to horizontal and 2 to vertical directions, respectively. A repeated index indicates summation over the index. Here, t is time, u_i is the Reynolds-averaged fluid velocity vector, p is the Reynolds-averaged pressure, ρ is constant density of the fluid, μ is dynamic viscosity, P_k is TKE production rate by mean velocity shear, α_k and α_ε are the inverse Prandtl numbers for K and ε, respectively, $C_{1\varepsilon} = 1.42$ and $C_{2\varepsilon} = 1.68$ are model constants, and μ_t is eddy viscosity. Note that the eddy viscosity tensor is taken to be diagonal with equal diagonal entries, thus we consider an isotropic eddy diffusivity. The k-ε model isotropic eddy viscosity is given by:

$$\mu_t = \rho C_\mu \frac{K^2}{\varepsilon} \tag{5}$$

with $C_\mu = 0.0845$. The eddy viscosity in Equation (5) is described as isotropic in the sense that at a fixed point in space, the same value of the eddy viscosity is used for the vertical and horizontal RANS momentum equations. However, the eddy viscosity is spatially dependent thereby possessing a non-zero spatial gradient.

The last term on right hand side of Equation (4) is specific to the RNG k-ε model [30] and is given as:

$$R_\varepsilon = \frac{C_\mu \rho \eta^3 (1 - \eta/\eta_0)}{1 + \beta \eta^3} \frac{\varepsilon^2}{K} \tag{6}$$

with $\eta \equiv SK/\varepsilon$ and $\eta_0 = 4.38$ and $\beta = 0.012$ being model constants. Note that $S = (2S_{ij}S_{ji})^{1/2}$ with $S_{ij} = \frac{1}{2}\left(\frac{\partial u_i}{\partial x_j} + \frac{\partial u_j}{\partial x_i}\right)$.

The fluid is composed of two phases, air and water. While both phases share the same governing equations (described above), the density, dynamic viscosity, and eddy viscosity vary depending on the phase in the local cell. The density and dynamic viscosity are calculated with the following equations:

$$\rho = \alpha_W \rho_W + (1 - \alpha_W)\rho_\alpha \tag{7a}$$

$$\mu = \alpha_w \mu_w + (1 - \alpha_w)\mu_\alpha \tag{7b}$$

where α_w is a scalar value representing volume fraction of water with value of 1 corresponding to a full water cell and 0 corresponding to an air cell. A transport equation for α_w is solved to track the interface during the simulation.

2.2. Governing Equations: Lagrangian Particle Dispersion Framework

By assuming that oil particle dynamics have no feed-back effect on the dynamics of the carrier water phase (i.e., passive tracers), we treat oil as a discrete phase being dispersed by the flow. The response of each particle to the advection and diffusion induced by the turbulent flow field is studied using the random walk method. The method consists of time integration of the Lagrangian velocity equation for individual particles. To track a particle located at the position $x^{(n)}$ and at the starting time $t^{(n)}$, the location at the time $t^{(n)} + \Delta t$ is found by the following stochastic equation [12,31]:

$$x_i^{(n+1)} = x_i^n + u_i^{(n)}\Delta t + \delta_{i,2}w_b\Delta t + \frac{\partial D_i}{\partial x_i}\Delta t + R\sqrt{2D_i\Delta t} \tag{8}$$

The second term on the right hand side of Equation (8) represents the advection induced by the carrier velocity field. In the third term on the right hand side of Equation (8), w_b is the particle upward rising velocity induced by buoyancy and given by [32,33] as

$$w_b = \sqrt{\frac{4g|\rho_d - \rho|D_d}{3C_D\rho}} \tag{9}$$

where g is gravitational acceleration, D_d is oil particle diameter, ρ_d is oil particle density and C_D is a particle drag coefficient. The fourth term is the gradient of the eddy diffusivity which for a depth-dependent eddy viscosity, as will be the case here, serves to induce a vertical velocity or vertical transport in the direction of increasing diffusivity. The last term is a stochastic model representing the fluctuating turbulent field constructed from the RANS simulation data, where R is a random number with Gaussian distribution. Using the Boussinesq hypothesis and assuming that the eddy diffusivity is isotropic [29,34,35], the eddy diffusion coefficient is

$$D = \nu + \nu_t \approx \nu_t = C_\mu \frac{K^2}{\varepsilon} \tag{10}$$

where $v = \mu/\rho$ is the kinematic viscosity of the fluid. Note that for a high Reynolds number flow, the eddy viscosity is orders of magnitude larger than the dynamic viscosity and hence the latter can be neglected.

The particle tracking Equation (8) does not include inertial effects, which is a shortcoming considering the studies [14,15]. However, a criterion for deviation from the sound theories of these works is through the Stokes number defined as $S_t = \tau_p/\tau_c$ where τ_c is the time scale of the turbulence which may be taken as K/ε and thus around 1.0 s in this work. The term τ_p is the particle inertial (or Stokes) response time, defined as $\tau_p = \beta D^2/18v$, where β is the particle density-to-water ratio, D is particle diameter, and v is water kinematic viscosity. As the oil density is taken herein as 866 kg/m^3, and considering the largest droplet sizes herein, which is 1000 microns, the term τ_p is around 0.05s. This gives a Stokes number $S_t < 0.05$, which is a relatively small value (note that $S_t = 0$ for neutrally buoyant particles) indicating that inertial effects can be neglected.

In the present investigation, we build on our earlier work [4,11,12] through the usage of depth-dependent eddy diffusivity, but we calculate it using the k-ε closure model for turbulence. We believe this is more realistic than imposing a generic value. Thus, neither water motion nor the eddy diffusivity are imposed in the interior of the domain, and they are directly calculated by the simulation.

2.3. Problem Setup

2.3.1. Wave Formulation

The waves simulated here are non-breaking deep water waves. The setup is depicted in Figure 1 with wavelength λ being the horizontal distance between the crests (or troughs), wave height H is the vertical distance between the crest and trough of the wave, and the wave period is T. The wave number is defined as $k = 2\pi/\lambda$ and the angular frequency is given by $\sigma = 2\pi/T$. For deep water waves, the wavelength is linked to the frequency using the dispersion relation [27] as

$$\sigma = \sqrt{kg} \tag{11}$$

where g is gravitational acceleration. The analytical solution for velocities using the first order linear wave theory is given as

$$u_i(x_1, x_2, t) = \delta_{1,i}\left(\frac{kgH}{2\sigma}\frac{\cosh kx_2}{\cosh kh}\cos(kx_1 - \sigma t) + \frac{3H^2\sigma k}{16}\frac{\cosh 2kx_2}{\sinh^4(kh)}\cos 2(kx_1 - \sigma t)\right)$$
$$+\delta_{2,i}\left(\frac{kgH}{2\sigma}\frac{\sinh kx_2}{\cosh kx}\cos(kx_1 - \sigma t) + \frac{3H^2\sigma k}{16}\frac{\sinh 2kx_2}{\sinh^4(kh)}\sin 2(kx_1 - \sigma t)\right) \tag{12}$$

where h is the mean water depth.

The water density was taken as 998.2 kg/m^3, air density as 1.225 kg/m^3, water dynamic viscosity as 1.003×10^{-3} kg/m^3, air dynamic viscosity as 1.7894×10^{-5} kg/m^3, and oil density as 866 kg/m^3, representing Alaskan North Slope oil [36].

2.3.2. Flow Simulation

The domain of simulation consists of a rectangular box of 20 m $\times$ 3 m in the x_1 (horizontal) and x_2 (vertical) directions, respectively (see Figure 2). A mesh comprising 835,229 nodes and 834,239 mixed quadrilateral cells was used for the simulation. The grid resolution has been chosen to properly resolve the motion induced by the wave field as predicted by linear wave theory. Given the sensitivity of the VOF method to coarse meshes near the interface area, the mesh was refined significantly near the free surface.

The refinement was performed within a rectangular region. The size of the rectangular region was 8 m $\times$ 0.3 m in the x_1 and x_2 directions, respectively, with the middle of the region corresponding to the mean water level. Within this rectangular region the mesh size varied from 0.001 m and 0.005 m. Outside of this region the mesh size varied from 0.005 m and 0.01 m.

A second-order upwind scheme within Fluent was used to discretize the equations in space. Time discretization consisted of the SIMPLE method given its good stability and convergence attributes [23]. The resulting time integration was implicit and second order. A time step of 0.0005 s was used. The time step was particularly small, given the CFL number restrictions, and was chosen to capture the unsteadiness of the flow field due to the forcing conditions resulting from the wave train [37].

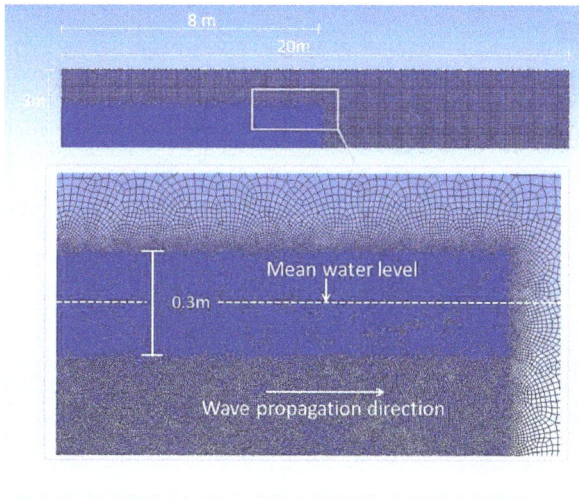

Figure 2. Details of the problem domain (**top**) and mesh refinement near the surface (**bottom**). The darker blue represents areas with more refined mesh.

The solution started from rest. At the left edge of the rectangular domain in Figure 2, a Dirichlet velocity boundary condition was imposed using the first order velocity terms in Equation (12). TKE and TKE dissipation values were prescribed at the left boundary. Note that this boundary condition represents turbulence generated elsewhere being advected into the computational domain by the Stokes drift of the waves. The right and bottom sides of the domain were modeled as solid walls with standard wall models being imposed. At the top edge, a pressure outlet boundary condition was applied. The mesh downstream of the domain was left coarse to dissipate the wave energy before reaching the end wall at $x_1 = 20$ m.

2.3.3. Particle Tracking

The Lagrangian particle tracking approach taken in our study is to predict the position of the particles via Equation (8). After simulating the hydrodynamics of water through solution of the Reynolds-averaged governing equations for the hydrodynamics of the flow underneath the regular waves, the results were imported to our in-house particle tracking code NEMO3D [38–41]. The imported results included velocity components, eddy viscosity, and the water volume fraction α_w. The NEMO3D code is capable of constructing a triangular unstructured mesh over any set of points and using linear interpolation to calculate the in-between values of the variables. The particle search algorithm locates the particle and links it to the corresponding triangular element in which the particle has traveled to. Moreover, the transient input data are updated at each tracking time step. Using an unstructured linear mesh provided the capability of constructing the spatial gradients of eddy diffusivity.

Particle tracking was performed for two groups of particles. The first group of 500 had a diameter of 100 micron, while the second group of 500 had a diameter of 1000 micron. The tracking was performed over a 10-wave period duration after the flow had become fully developed. Particles were released at two depths. The first particle group was uniformly distributed in the spatial span of $x_1 \in [2.5, 3]$ m and fixed elevation of $x_2 = 1.1$ m. The second particle group shared the same horizontal distribution as the first group while its elevation was at $x_2 = 0.8$ m. Here and in what follows, particles in the first group shall be referred to as "near surface particles" while particles in the second group shall be referred to as "deep particles".

3. Results

3.1. Numerical Simulation Validation

The simulation was started from rest and was continued until the resolved flow was fully developed. Once a fully developed flow was attained, the simulation was continued for another 10 wave periods and the outputted data over this time span was used to perform particle tracking. Figure 3 shows instantaneous snapshots of the contours of horizontal and vertical water velocity components. Figure 4 shows a comparison of the time series of velocity components with the analytical solution from the second order linear wave theory (Equation (12)) at the point $x_1 = 2.7$ m, $x_2 = 1.1$ m corresponding to a location where particle tracking was performed. Figure 5 shows a comparison of depth profiles of the horizontal and vertical components of the velocity with the analytical solution from second order wave theory (Equation (12)) at $x_1 = 2.7$ m at two different times corresponding to the crest and trough of a wave. Overall, a very good agreement is noted between the numerical and analytical solutions and in particular Figures 3 and 4 demonstrate that no significant spatial and temporal damping in the numerical solution existed.

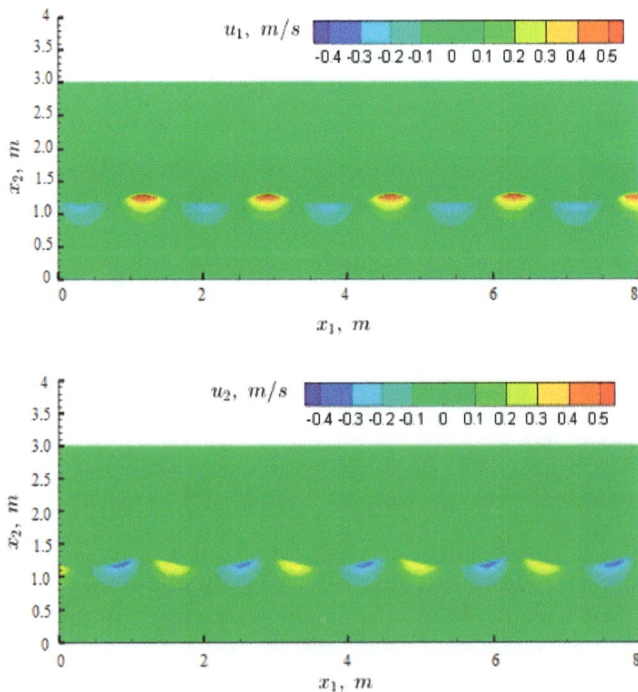

Figure 3. Instantaneous contours of vertical (**top**) and horizontal (**bottom**) velocities after the flow is fully developed. Horizontal velocity is positive to the right and vertical velocity is positive upward.

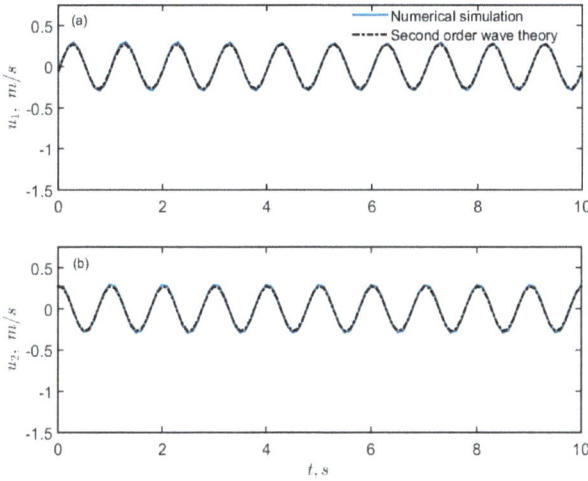

Figure 4. Comparison of the numerical solution with the analytical solution in terms of time series of horizontal (**a**) and vertical (**b**) velocities at the point corresponding to $x_1 = 2.7$ m and $x_2 = 1.1$ m during 10 periods of the wave for the fully developed flow.

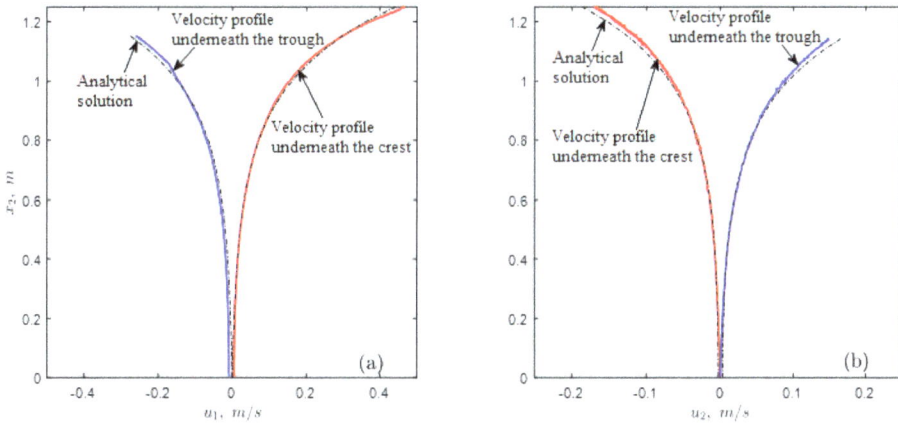

Figure 5. Comparison of horizontal (**a**) and vertical (**b**) velocities between numerical (solid line) and analytical solutions at $x_1 = 2.7$ m at two different times corresponding to the crest and trough of a wave. In the left panel, the left curves are under the trough while the right curves are under the crest. In the right panel, the left curves occurred under the crest while the right curves occurred under the trough.

3.2. Vertical Profiles of Turbulent Quantities

Figure 6 shows vertical profiles of TKE K, TKE dissipation ε, and eddy diffusivity, D, underneath the crest and trough of the waves. The values are close to 0 at the free surface and grow sharply to their maxima within one wave height of the MWL. After reaching the maximum, each value decreased to almost zero at an elevation approximately equal to 0.7 m. At this elevation, the vertical and horizontal velocity components, shown in Figure 5, also drop sharply to near 0 at lower elevations. The sharp decrease of TKE and ε is associated with a decrease in the velocity (shear) itself. The profile of eddy diffusivity is similar to what was observed in field studies [17,42].

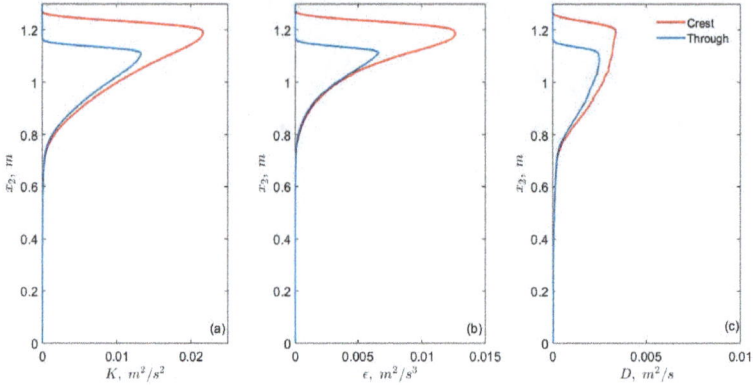

Figure 6. Vertical variation of turbulent kinetic energy ($m^2\ s^{-2}$) (**a**), turbulence kinetic energy (TKE) dissipation rate ($m^2\ s^{-3}$) (**b**), and eddy diffusivity ($m^2\ s^{-1}$) (**c**) underneath a crest ($x_1 = 1.56$ m) and a trough ($x_1 = 2.34$ m) of the wave.

The sharp vertical gradient of eddy diffusivity near the free surface observed in Figure 6 is expected to provide an important contribution to the vertical mass transport in that region, as the gradient of eddy diffusivity acts as advection in the random walk method (Equation (8)). The maximum value of D was attained underneath the wave crest and the smallest value at the same elevation was under the trough. This implies that mass transport due to turbulent diffusion is higher underneath the crest. Moreover, there is also a horizontal gradient of eddy viscosity which acts as a horizontal advective velocity, ultimately affecting the Stokes drift velocity. However, such a transport is negligible in comparison to the Stokes drift.

3.3. Particle Trajectories

Figure 7 presents instantaneous positions of 100 and 1000 μm particles released near the water surface. This experiment was conducted with 500 particles of each diameter, but only 50 of each diameter are shown in Figure 7 for clarity. Particle tracking was performed with Equation (8) with inputs derived from the previously described flow field underneath the surface wave. As can be seen, particles of 1000 μm diameter tend to drift forward faster than the 100 μm diameter particles. This causes the comet shape of oil plumes described in previous studies [11,12]. The 100 micron diameter particle tended to submerge and disperse deep in the water column while the particles of 1000 μm diameter remained close to the surface, thereby experiencing a greater Stokes drift.

Ensemble-averaged particle trajectories were analyzed to understand the combined effect of advective velocities, diffusion, and buoyancy. Results were obtained by tracking 500 individual droplets and calculating the plume trajectory by averaging the trajectories over all of the particles. The number of particles was chosen sufficiently large so as to ensure that the result is not affected by the number of particles.

Figures 8 and 9 present the averaged plume trajectory for particles released near the water surface at the elevation $x_2 = 1.1$ m and for particles released at $x_2 = 0.8$ m, respectively. As can be observed, the particle motion orbits are not closed paths for both sets of particles of 100 and 1000 micron diameters. This results in a net forward motion of mass which is the well-known Stokes drift [27], and it probably factors in the Lagrangian drift [12]. We also conducted particle tracking for neutrally buoyant particles (i.e., we made the diameter equal to 0). The averaged trajectories in this case were nearly identical to those of 100-micron particles in Figures 8 and 9 and hence the results are not presented.

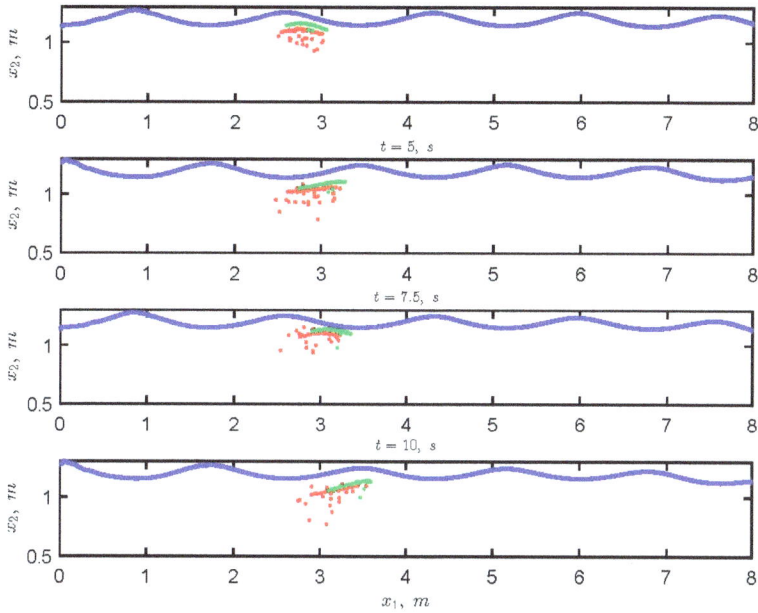

Figure 7. Particle positions at different times. Particles of diameter 1000 μm are represented with green points and the red points represent particles of diameter equal to 100 μm. Each group has 50 particles that are tracked during 10 wave periods. Starting point of all particles was at $x_2 = 1.10$ m.

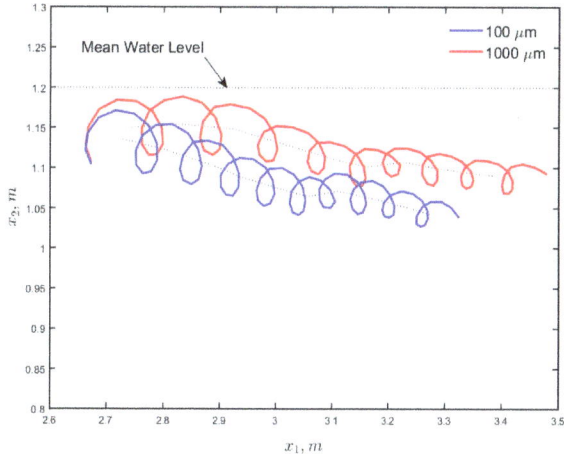

Figure 8. Ensemble averaged plume trajectory (over 500 particles of each size) of particles of diameter 100 and 1000 μm. The particles were released at $x_1 \in [2.5, 3]$ m, $x_2 = 1.10$ m. The dotted line through the plume trajectory was obtained by window averaging over each trajectory loop corresponding to the wave period.

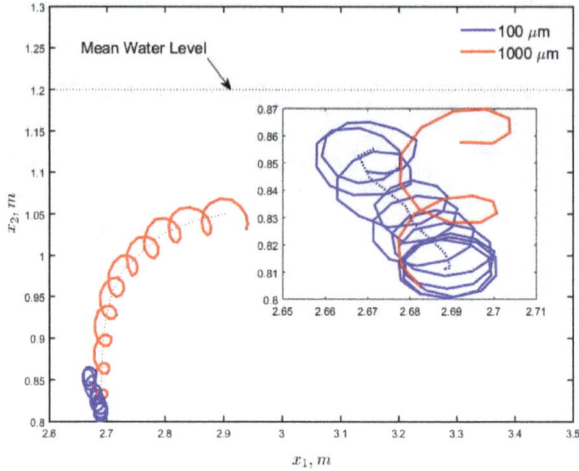

Figure 9. Ensemble averaged plume trajectory of particles of diameters 100 and 1000 μm (500 particles were used to obtain the averages). The particles were released at $x_1 \in [2.5, 3]$ m, $x_2 = 0.8$ m. Dotted line through the plume trajectory was obtained by window averaging over each trajectory loop corresponding to the wave period.

3.4. Effect of Turbulent Diffusion

Considering Figure 8, a gradual downward shift of the 1000 μm and 100 μm near surface particles is observed as particles drift with the flow. This is due to the combined effect of the wave kinematics, buoyancy and turbulence. More specifically, particles diffuse from high concentration to low concentration, and initially there are more oil droplets on the surface. In addition, the boundary (the free surface) prevents diffusion upward of the water surface, and thus the net diffusion of particles is downward. In particular, imagine particles at the mean water level (MWL), the trough brings all of them down (as they cannot stay above the water surface), but the crest brings only some of them up (approximately 50 percent based on randomness), and thus the net motion is downward.

The larger downward shift of the near surface 100 μm particle plume trajectory observed when compared to that of the 1000 μm particles is consistent with the field observations of [5], who noted that due to the lesser buoyancy of smaller droplets, turbulence disperses these smaller droplets further down into the water column. Looking at Figures 6 and 8, the 100 μm particles submerge despite being in regions of positive vertical gradient of diffusivity, as the downward motion induced by the wave kinematics and turbulent dispersion is able to overcome the upward motion induced by buoyancy and the positive vertical gradient of diffusivity. Thus, the vertical gradient of diffusivity does not stop the downward migration of particles, but only slows it down as the particle pass through the high diffusivity region.

Figure 9 shows the trajectory of 1000 μm particles released at the elevation $x_2 = 0.80$ m. It is apparent that the movement is dominated by buoyancy, which is due to the large buoyancy of the droplets. Particles initially travel upward until reaching an equilibrium depth modulated by the action of the wave motion and the turbulence by the end of the simulation. The trajectory tends to be more horizontal at $x_2 \approx 1.05$ m corresponding to the region of highest diffusivity, which is consistent with the slowing down of 100 μm droplets as they travel downward. It is also consistent with the results mentioned earlier [17].

In order to better highlight the effect of gradient of eddy diffusivity, Figure 10 presents the ensemble averaged trajectory of near surface 100 and 1000 μm particles with and without inclusion of the gradient of the eddy diffusivity in the particle tracking Equation (8). It can be observed that inclusion of the gradient of diffusivity causes particles to move towards the region of high eddy diffusivity values, which is in agreement with [17]. For example, 1000 μm particles remained close to the surface when the eddy diffusivity gradient is excluded from the particle tracking equation. When the gradient of diffusion is included, the 1000 μm particles submerge to the region of greatest diffusivity at about $x_2 = 1.1$ m (see Figure 6c). Overall, neglecting the gradient of eddy diffusivity causes a deviation of particle trajectories elevations by more than half a wave height after 10 wave periods, which is a non-trivial amount.

In Figure 10, the 100 μm particles continuously descend in the water column. However, the rate of descent is not as pronounced when the gradient of diffusion is included in the particle tracking equation. This further confirms that the eddy diffusivity gradient tends to keep these particles closer to the region of greater eddy diffusivity at about $x_2 = 1.1$ m.

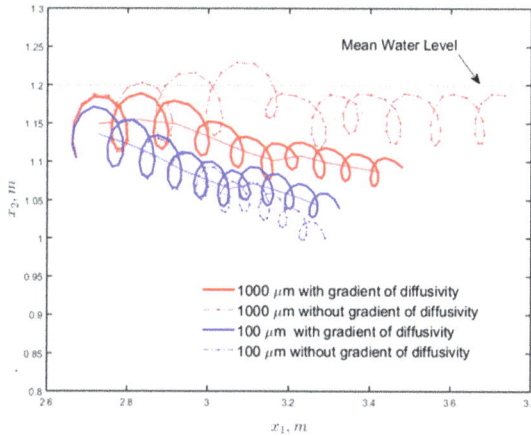

Figure 10. Ensemble averaged plume trajectory (over 500 particles of each size) of particles of diameter 100 and 1000 μm. Release location was at $x_1 \in [2.5, 3]$ m, $x_2 = 1.1$ m. Dotted line through the plume trajectory was obtained by window averaging over each trajectory loop corresponding to the wave period.

3.5. Stokes Drift Calculations

Using the averaged particle trajectories in Figure 10, we calculated the Stokes drift velocities of ensemble data. We used for this purpose neutrally buoyant particles. Figure 11 shows the average speed of 500 near surface particles, when no turbulence effect was included (i.e., by setting the eddy diffusivity and its gradients equal to 0). Due to the proximity of particles to the free surface, the plume exhibited a weak oscillation around the starting elevation of $x_2 = 1.1$ m. The calculated Stokes drift was 0.072 m s^{-1} while the theoretical value [27] was 0.075 m s^{-1}. The fact that the two values are close suggests that the RANS simulation is capturing, at least, second-order accurate kinematics.

$$U_s = \frac{H^2 k}{4} \sigma e^{(2kx_2)}$$

(13)

The calculated values of averaged Stokes drift velocity with turbulence effect for near surface 100 μm particles with and without gradient of diffusivity were 0.065 m s^{-1} and 0.062 m s^{-1}, respectively. The smaller Stokes velocity compared to the case without turbulence effect is because turbulence (i.e., randomness) results in a net downward movement of the plume (discussed earlier), where the Stokes velocity is smaller. The gradient of eddy diffusivity in this case appears to have only a slight impact on the Stokes drift, increasing the velocity by 5% (from 0.062 to 0.065 m s^{-1}). The greater Stokes drift induced by the gradient of eddy diffusivity is consistent with the fact that the plume remains closer to the surface when the gradient is included (Figure 10).

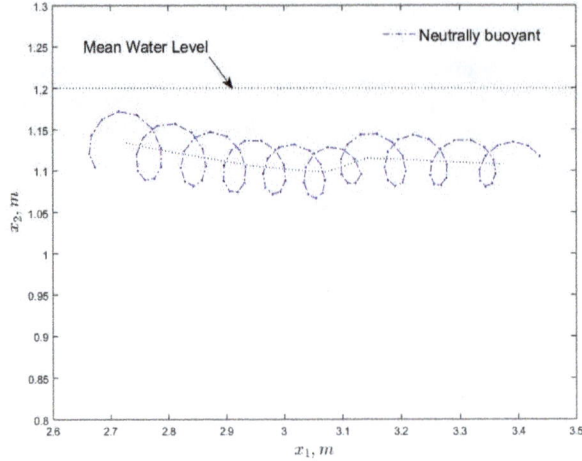

Figure 11. Ensemble averaged plume trajectory (over 500 neutrally buoyant particles of each size) for the case with no turbulence effect and thus with pure advection. Release location was at $x_1 \in [2.5, 3]$ m, $x_2 = 1.10$ m.

Note that in Figure 11 the orbital motion of the 1000 μm particles is larger. This is due to the fact that greater buoyancy of the 1000 μm particles pushes these particles closer to the surface thereby exposing them to greater Stokes drift.

4. Discussion

Transport of oil droplets due to non-breaking wave and buoyancy effects was investigated. The present study focused on smaller spatial and temporal scales compared to our previous works [4,11,12] which were hundreds of meters and on the order of hours. We considered the transport of oil droplets due to wave motion, buoyancy, and non-local turbulence, and we found that other forces, such as inertia are negligible for the scenarios that we considered. A key issue explored here was the effect of inclusion of the gradient of eddy viscosity (i.e., diffusivity) in random walk particle dispersion, and it was observed that the gradient of eddy viscosity tends to advect particles from low diffusivity regions to high diffusivity regions. This is in agreement with the earlier work of [17] for wind and tidal boundary layers.

The vertical profile of eddy diffusivity shows that it increases rapidly from zero at the free surface to maximum value at a depth below the mean water level comparable to the wave height (elevation $x_2 \approx 1.1$ m, see Figure 6). The resulting steep negative vertical gradient above $x_2 = 1.1$ m induces a downward advective velocity in the random walk method Equation (8). Also, the positive vertical gradient below $x_2 \approx 1.1$ m causes an upward advective velocity. Therefore, accounting for the turbulent diffusivity slows down the predominantly downward movement of neutral (and low) buoyancy droplets (or particles).

It is likely that the gradient of eddy viscosity is commonly neglected in numerical works due to the uncertainty in estimating it. But excluding the gradient causes a systematic bias in the results, and thus, might not be justified in all situations. To provide more insight on the origin of the gradient of the diffusion coefficient in Equation (8) we expand the diffusion term in the following diffusion equation (used also by [17]):

$$\frac{\partial C}{\partial t} = -\frac{\partial}{\partial x_j}\left(D\frac{\partial C}{\partial x_j}\right) = -\frac{\partial D}{\partial x_j}\frac{\partial C}{\partial x_j} - D\frac{\partial^2 C}{\partial x_j^2} \tag{14}$$

where C is the Reynolds-averaged tracer concentration (e.g., oil with C being oil mass per unit volume of water). The first term on the right hand side of Equation (14) is of advective form with the horizontal and vertical derivatives of eddy diffusivity acting as horizontal and vertical advective velocities, respectively. This is particularly important because as it was seen in the results section, there are sharp vertical eddy diffusivity gradients near the surface. The importance of the gradient of eddy viscosity for Eulerian approaches of the form in Equation (14) has been observed in RANS modeling of wind and wave forced oceanic turbulent boundary layers [17]. We believe it is important to consider it in Lagrangian approaches, as done herein.

Unsurprisingly, buoyancy was found to play an important role in oil droplet movement. For submerged particles, buoyancy causes upward transport of particles as seen in Figure 9. A weak upward buoyancy force allows for wave motion and the turbulence to submerge the 100 μm particles deeper than the 1000 μm particles which is in agreement with field observations [5]. These overall effects of buoyancy were observed to be modulated or tempered by the gradient of the eddy diffusivity tending to move the particle closer to the region of higher diffusivity.

5. Conclusions

In this manuscript the effect of non-breaking wave motion on oil droplet transport was evaluated. Numerical simulations were conducted using the RANS equations with regular waves of period 1.0 s and height 0.10 m. The current work is an extension of the earlier work of [11] in which Lagrangian tracking of oil particles was performed under regular surface waves in order to understand the combined effects of waves, turbulent diffusion, and buoyancy on the transport of oil droplets at sea. Turbulent diffusion was represented via a random walk model similar to that implemented for the current work, but with constant eddy diffusivity. In the present work, we have extended the study in [11] to spatially dependent eddy diffusivity as calculated by the k-ε closure for turbulence. Unlike in the case of [11] which imposed wave-induced motion via second order wave theory, in the present work the waves and thus the wave induced velocity are directly resolved by the simulation. Furthermore, we have investigated the role of the gradient of the eddy diffusivity acting as an advective component in the random walk model driving vertical transport towards the zone in the water column characterized by high diffusivity. The latter behavior has been observed for the first time under the action of waves in the present study. This behavior induced by the gradient of eddy diffusivity has also been observed/explained by [17] in simulations of vertical distributions of particles in a water column subject to wind and tidal forcing and moderate stratification; orbital motions induced by surface waves were not considered in the work of [17].

We found that the resulting RANS velocities underneath the regular waves closely compare to the velocity values obtained based on the second order theory (Stokes theory), giving credence to the numerical results. The turbulence advected into the computational by the waves was characterized by an eddy viscosity (taken equal to eddy diffusivity) increasing sharply from the surface until reaching a maximum value at depth comparable to double the wave height and then decreasing gradually with depth. We used the RANS velocities and the diffusivity to track the movement of oil droplets of size 100 μm and 1000 μm in a Lagrangian framework using the random walk method.

It was found that when the particles are released at the water surface, the 100 μm droplets migrated downward in the water column while the 1000 μm droplets remained close to the water surface, which can be explained based on the buoyancy. We showed that including the diffusivity gradient would increase the rate of descent of particles in the water column until reaching the zone of maximum diffusivity. Conversely, large buoyancy (1000 μm) droplets below the maximum value of diffusivity would migrate faster upward to reach the zone of high diffusivity.

In the future we will explore the effect of turbophoresis [17]. Turbophoresis drives particles away from regions of large turbulent kinetic energy, thus away from regions of high eddy diffusivity, in contrast to the effect of the gradient of eddy diffusivity investigated here.

Acknowledgments: This research paper was made possible, in part, by a grant from The Gulf of Mexico Research Initiative to the Consortium CARTHE II and by funding from the Department of Fisheries and Ocean Canada (DFO). Data are publicly available through the Gulf of Mexico Research Initiative Information and Data Cooperative (GRIIDC) at https://data.gulfresearchinitiative.org (doi:10.7266/N7KK999Q). However, it does not reflect the views of these entities, and no official endorsement should be implied.

Author Contributions: Michel C. Boufadel and Andrés E. Tejada-Martínez conceived and designed the experiments; Roozbeh Golshan performed the experiments; Michel C. Boufadel, Andrés E. Tejada-Martínez and Roozbeh Golshan analyzed the data; Feng Gao contributed materials and analysis tools; Roozbeh Golshan wrote the paper. The remaining authors contributed the conception of the work and to improving the overall quality of the work and the manuscript.

Conflicts of Interest: The authors declare no conflict of interest.

References

1. Sobey, R.; Barker, C. Wave-driven transport of surface oil. *J. Coast. Res.* **1997**, *13*, 490–496.
2. Korotenko, K.A.; Bowman, M.J.; Dietrich, D.E. High-resolution numerical model for predicting the transport and dispersal of oil spilled in the Black Sea. *Atmos. Ocean. Sci.* **2010**, *21*, 123–136. [CrossRef]
3. Zhao, L.; Torlapati, J.; Boufadel, M.C.; King, T.; Robinson, B.; Lee, K. VDROP: A comprehensive model for droplet formation of oils and gases in liquids-incorporation of the interfacial tension and droplet viscosity. *Chem. Eng. J.* **2014**, *253*, 93–106. [CrossRef]
4. Boufadel, M.C.; Bechtel, R.D.; Weaver, J. The movement of oil under non-breaking waves. *Mar. Pollut. Bull.* **2006**, *52*, 1056–1065. [CrossRef] [PubMed]
5. Lunel, T. Dispersion: Oil droplet size measurements at sea. In Proceedings of the 16th Arctic and Marine Oil Spill Program (AMOP), Calgary, AB, Canada, 7–9 June 1993; Technical Seminar; Environment Canada: Fredericton, NB, Canada, 1993; pp. 1023–1055.
6. Yang, D.; Chamecki, M.; Meneveau, C. Inhibition of oil plume dilution in Langmuir ocean circulation. *Geophys. Res. Lett.* **2014**, *41*, 1632–1634. [CrossRef]
7. Pohorecki, R.; Moniuk, W.; Bielski, P.; Zdrjkowski, A. Modelling of the coalescence/redispersion processes in bubble columns. *J. Geophys. Res.* **1988**, *56*, 572–586. [CrossRef]
8. Mewes, D.; Loser, T.; Millies, M. Modelling of two-phase flow in packings and monolith. *Chem. Eng. Sci.* **1999**, *54*, 4729–4747. [CrossRef]
9. Monismith, S.G.; Cowen, E.A.; Nepf, H.M.; Magnaudet, J.; Thais, I. Laboratory observations of mean flows under surface gravity waves. *J. Fluid Mech.* **2007**, *573*, 131–147. [CrossRef]
10. Elliott, A.J.; Wallace, D.C. Dispersion of surface plumes in the southern north sea. *Ocean Dyn.* **1989**, *42*, 1–16. [CrossRef]
11. Boufadel, M.C.; Du, K.; Kaku, V.; Weaver, J. Lagrangian simulation of oil droplets transport due to regular waves. *Environ. Model. Softw.* **2007**, *22*, 978–986. [CrossRef]
12. Geng, X.; Boufadel, M.C.; Ozgokmen, T.; King, T.; Lee, K.; Lu, Y.; Zhao, L. Oil droplets transport due to irregular waves: Development of large-scale spreading coefficients. *Mar. Pollut. Bull.* **2016**, *104*, 279–289. [CrossRef] [PubMed]
13. Ichiye, T. Upper ocean boundary-layer flow determined by dye diffusion. *Phys. Fluids* **1967**, *10*, 270–279. [CrossRef]
14. Eames, I. Settling of particles beneath water waves. *J. Phys. Oceanogr.* **2008**, *38*, 2846–2853. [CrossRef]
15. Santamaria, F.; Boffetta, G.; Afonso, M.; Mazzino, A.; Onorato, M.; Pugliese, D. Stokes drift for inertial particles transported by water waves. *EPL* **2013**, *102*, 1023–1055. [CrossRef]
16. Bakhoday-Paskyabi, M. Particle motions beneath irrotational water waves. *Ocean Dyn.* **2015**, *65*, 1063–1078. [CrossRef]
17. Visser, A.W. Using random walk models to simulate the vertical distribution of particles in a turbulent water column. *Mar. Ecol. Prog. Ser.* **1997**, *158*, 275–281. [CrossRef]
18. Dimou, K. *3-D Hybrid Eulerian-Lagrangian/Particle Tracking Model for Simulating Mass Transport in Coastal Water Bodies*; Department of Civil and Environmental Engineering, Massachusetts Institute of Technology: Cambridge, MA, USA, 1992.
19. Gouesbet, G.; Berlemont, A. Eulerian and lagrangian approaches for predicting the behaviour of discrete particles in turbulent flows. *Prog. Energy Combust. Sci.* **1999**, *25*, 133–159. [CrossRef]

20. Ermak, D.; Nasstrom, J. A lagrangian stochastic diffusion method for inhomogeneous turbulence. *Atmos. Environ.* **2000**, *34*, 1059–1068. [CrossRef]

21. Hunter, J.; Craig, P.; Phillips, H. On the use of random walk models with spatially variable diffusivity. *J. Comput. Phys.* **1993**, *106*, 366–376. [CrossRef]

22. Thorpe, S.A. Langmuir circulation. *Annu. Rev. Fluid Mech.* **2004**, *36*, 55–79. [CrossRef]

23. ANSYS. *Fluent 12.0 User's Guide*; ANSYS, Inc.: Canonsburg, PA, USA, 2009.

24. Zhang, B.; He, P.; Zhu, C. Modeling on Hydrodynamic Coupled FCC Reaction in Gas-Solid Riser Reactor. In Proceedings of the ASME 2014 4th Joint US-European Fluids Engineering Division Summer Meeting Collocated with the ASME 2014 12th International Conference on Nanochannels, Microchannels, and Minichannels, Chicago, IL, USA, 3–7 August 2014; p. V01DT32A005.

25. Gao, F.; Zhao, L.; Boufadel, M.C.; King, T.; Robinson, B.; Conmy, R.; Miller, R. Hydrodynamics of oil jets without and with dispersant: Experimental and numerical characterization. *Appl. Ocean Res.* **2017**, *68*, 77–90. [CrossRef]

26. Geng, X.; Boufadel, M.C.; Cui, F. Numerical modeling of subsurface release and fate of benzene and toluene in coastal aquifers subjected to tides. *J. Hydrol.* **2017**, *551*, 793–803. [CrossRef]

27. Dean, R.; Dalrympl, R. *Water Wave Mechanics for Engineers and Scientists*; World Scientific: Singapore, 1991.

28. Pope, S. *Turbulent Flows*; Cambridge University Press: Cambridge, UK, 2000.

29. Chen, Y.S.; Kim, S.W. *Computations of Turbulent flows Using an Extended k-ε Turbulence Closure Model*; Report NASA CR-179204; NASA: Greenbelt, MD, USA, 1987.

30. Yakhot, V.; Orszag, S.; Thangam, S.; Gatski, T.; Speziale, C. Development of turbulence models for shear flows by a double expansion technique. *Phys. Fluids* **1992**, *4*, 1510–1520. [CrossRef]

31. Gardiner, A. *Handbook of Stochastic Methods: For Physics, Chemistry, and the Natural Sciences*; Springer Series in Synergetics; Springer: New York, NY, USA, 1990.

32. Clift, R.; Grace, J.; Weber, M. *Bubbles, Drops and Particles*; Academic Press: New York, NY, USA, 1978.

33. Zhao, L.; Boufadel, M.; Adams, E.; Socolofsky, S.; King, T.; Lee, K.; Nedwed, T. Simulation of scenarios of oil droplet formation from the deepwater horizon blowout. *Mar. Pollut. Bull.* **2015**, *101*, 93–106. [CrossRef] [PubMed]

34. Wilcox, A. Reassessment of the scale determining equation for advanced turbulence models. *AIAA J.* **1988**, *26*, 1299–1310. [CrossRef]

35. Lai, A.C.K.; Nazaroff, W.W. Modeling indoor particle deposition from turbulent flow onto smooth surfaces. *J. Aerosol Sci.* **2000**, *31*, 463–476. [CrossRef]

36. Personna, Y.; King, T.; Boufadel, M.; Zhang, S.; Kustka, A. Assessing weathered endicott oil biodegradation in brackish water. *Mar. Pollut. Bull.* **2014**, *86*, 102–110. [CrossRef] [PubMed]

37. Laccarino, G.; Ooi, A.; Durbin, P.; Behnia, M. Reynolds averaged simulation of unsteady separated flow. *Int. J. Heat Fluid Flow* **2003**, *24*, 147–156. [CrossRef]

38. Geng, X.; Boufadel, M.; Xia, Y.; Li, H.; Zhao, L.; Jackson, N.; Miller, R.S. Numerical study of wave effects on groundwater flow and solute transport in a laboratory beach. *J. Contam. Hydrol.* **2014**, *165*, 37–52. [CrossRef] [PubMed]

39. Geng, X.; Boufadel, M.C. Numerical study of solute transport in shallow beach aquifers subjected to waves and tides. *J. Geophys. Res.* **2015**, *120*, 1409–1428. [CrossRef]

40. Zhao, L.; Gao, F.; Boufadel, M.C.; King, T.; Robinson, B.; Lee, K.; Conmy, R. Oil jet with dispersant: Macro-scale hydrodynamics and tip streaming. *AIChE J.* **2017**, *63*, 5222–5234. [CrossRef]

41. Gao, F.; Zhao, L.; Shaffer, F.; Golshan, R.; Boufadel, M.; King, T.; Robinson, B.; Lee, K. Prediction of oil droplet movement and size distribution: Lagrangian method and VDROP-J model. *Int. Oil Spill Conf. Proc.* **2017**, *2017*, 1194–1211. [CrossRef]

42. McPhee, M.G.; Smith, J.D. Measurements of the turbulent boundary layer under pack ice. *J. Phys. Oceanogr.* **1976**, *6*, 696–711. [CrossRef]

Journal of
*Marine Science
and Engineering*

MDPI

Article

The Implications of Oil Exploration off the Gulf Coast of Florida

Jake R. Nelson * and Tony H. Grubesic

Center for Spatial Reasoning & Policy Analytics, College of Public Service & Community Solutions,
Arizona State University, 411 N Central Avenue, Suite 400, Phoenix, AZ 85004, USA; grubesic@asu.edu
* Correspondence: jrnels20@asu.edu; Tel.: +1-303-931-5263

Received: 6 March 2018; Accepted: 30 March 2018; Published: 2 April 2018

check for
updates

Abstract: In the United States (U.S.), oil exploration and production remain critical economic engines for local, state, and federal economies. Recently, the U.S. Department of the Interior expressed interest in expanding offshore oil production by making available lease areas in the U.S. Gulf of Mexico, the U.S. West Coast and East Coast, as well as offshore Alaska. With the promise of aiding in energy independence, these new lease areas could help solidify the U.S. as one of the world's largest oil-producing countries, while at the same time bolstering the local and regional energy job sectors. Of all the newly proposed lease areas, the Gulf Coast of Florida is particularly contentious. Opponents of drilling in the area cite the sensitive ecosystems and the local and state tourism economy that depends heavily on the numerous beaches lining Florida's coast. In this analysis, we use a data-driven spatial analytic approach combined with advanced oil spill modeling to determine the potential impact of oil exploration off of Florida's Gulf Coast given a loss-of-control event. It is determined that plume behavior varies drastically depending on the location of the spill but that overall impacts are comparable across all spill scenario sites, highlighting the necessity of contingency-type analyses. Implications for spill response are also discussed.

Keywords: oil spill; impact modeling; simulation; contingency planning

1. Introduction

The Gulf of Mexico (GOM) is home to several large and rich oil reservoirs. As a result, for many decades, the GOM has been a primary production site for U.S. oil. Recent reports show that almost all of the offshore oil production in the United States takes place in the GOM (~97%) and accounts for about 17% of the total oil and gas produced in the United States [1]. The oil-based energy sector associated with the GOM also employs thousands of people in the U.S. As of January 2016, the two largest offshore-oil-producing states, Texas and Louisiana, had roughly 260,000 and 44,000 workers employed in the oil-based energy sector, respectively [2].

The importance of oil to the communities along the GOM cannot be overstated and it continues to grow. In a 2015 report by the Bureau of Ocean Energy Management (BOEM), oil reserves in the GOM were estimated to be upwards of 3.67 billion barrels, with contingent reserves estimated to be about 3.29 billion barrels [3]. Given the already large economic footprint of the oil industry in the GOM region, combined with future production potential, the continued exploration and development of oil reserves in the GOM is cited as one of the most important pathways to establishing U.S. energy independence [4]. In short, oil exploration, extraction, and production are critically important to the U.S. economy, but there are risks. Oil extraction can, and does, exact a significant toll on the environment—disrupting complex ecosystems and the overall environmental vitality of the region [5,6].

Consider, for example, one of the most catastrophic environmental disasters in recent history. The 2010 Deepwater Horizon blowout, which ultimately released between 4.5 and 4.9 million barrels of oil into the GOM waters over a four-month period [7], devastated coastal environments and economies. Estimates of economic loss to the entire region range from $8.7 billion [8] to upwards of $37 billion [9]. In Florida alone, recent estimates suggest that 4.1 million recreational trips were cancelled as a direct result of the spill, totaling an estimated loss of $2.04 billion [10]. Since the 2010 blowout of the Deepwater Horizon there have been a handful of disruptive accidents and events that have taken place in the GOM. The underwater infrastructure of the Shell Brutus platform failed in May 2016, eventually releasing 2100 barrels of oil into the offshore environment [11]. More recently, the 1 October 2017 pipeline rupture 40 miles south of Venice, Louisiana, resulted in the release of 9350 barrels of oil into the water column [12]. Neither spill made landfall but they serve as stark reminders of the risks and potential consequences of oil production in the offshore environment.

Recent efforts by the U.S. federal government and its administration to develop and solidify energy independence have resulted in proposals to open vast swaths of U.S. coastal waters to offshore oil exploration and production [13]. The proposed areas include the Atlantic seaboard, coastal Alaska, the Eastern GOM planning area, and much of the U.S. west coast (ibid.). Quickly following this announcement, Florida governor Rick Scott negotiated with the administration to exempt the Florida coast from offshore oil production, citing the potential harm to the region's tourism industry, worth 60 billion dollars per year [14]. More importantly, the coastal waters of Florida have been off limits to drilling for many years as a result of the GOM Energy Security Act of 2007 [15]. In fact, most of the Eastern GOM Planning Area was placed under a drilling moratorium until 2022, with recent calls to make the moratorium permanent [16]. Regardless of the final decision concerning Florida's waters, it is important to revisit and reevaluate the potential outcomes associated with oil extraction efforts in the region, especially given recent interest in possibly allowing drilling to take place.

To be sure, while the number of oil spill impact assessments has increased dramatically since the Deepwater Horizon catastrophe [17], none have quantified or directly addressed the potential outcomes of a disaster in the Eastern GOM Planning Area. This is not to say that site-specific research is nonexistent. On the contrary, there is a growing body of research concerning site-specific impact and risk quantification in Europe [18–21], coastal Asia [22,23], and some parts of the United States [24,25]. However, the major substantive foci of these studies are on methodological development. With recent proposals to open the GOM Eastern Planning Area to oil exploration and production, the purpose of this paper is to develop a broad understanding of the risks and impacts associated with drilling in the Eastern GOM. We apply the core methodological procedures found across many oil spill risk and impact methodologies (e.g., [17]) to the Eastern GOM. This process of tracking the final fate of oil, from blowout locations to the shoreline, will provide some insight into the behavior of oil spills in the area along with their potential impacts to the state of Florida and beyond.

2. Background

On 4 January 2018, the United States Secretary of the Interior, Ryan Zinke, announced plans to open almost all of the U.S. Outer Continental Shelf (OCS) area to oil exploration and production to support U.S. energy independence [13]. Although portions of U.S. coastal waters, such as those found in California and Alaska, have historically functioned as areas of active oil production, many others, including the Atlantic Seaboard and the Eastern GOM, have not. Thus, it was not surprising to see that all Pacific and Atlantic states, with the exception of Maine, formally voiced objections to offshore oil extraction activities [26]. Thus far, only Florida has been granted an exception (ibid.).

Many of the calls to remove Florida (and other states) from consideration for offshore oil drilling cite the potential impact that an oil spill may have on the environment and tourism. Tourism is the largest economic driver in the state of Florida, and much of it depends on a pristine coastal environment [27]. Specifically, tourism generates about 23% of the state's sales tax revenue and employs over one million individuals [28]. Threats to the tourism industry in Florida, whether real or perceived, can have a dramatic impact on the state economy as the perception that a beach might be oiled can alter the vacation plans of individuals who fear a health risk or a contaminated shoreline [29,30]. This was exactly the case following the Deepwater Horizon. Although the majority of the oil from the Deepwater Horizon made landfall on the shorelines of Louisiana, Mississippi, and Alabama [31], Florida experienced significant economic losses from cancelled recreation trips due to the perception that the beaches and ocean in/around Florida were oiled [10].

Clearly, the economic loss from just the perception of an oiled coastline can be significant for any community, state, or region with an economy rooted in coastal tourism. However, the concern regarding drilling off the Florida coast also comes from the potential for actual damage caused by physical oiling of the ecosystems and allied coastal assets. For some perspective on this issue, one only needs to reexamine the consequences of the Deepwater Horizon spill. Consider the nature of the spill, which formed a large, deep sea and surface oil plume, along with massive amounts of sinking oil [32,33], all of which contributed to a significant loss of the nearshore and deep-sea benthic fauna in the GOM region [34–36]. Furthermore, related studies found evidence of harm to shallow water coral communities [37], some coastal fish species [38], seabirds [39,40], sea turtles [41], and possibly (but not fully confirmed) marine mammals [42,43]. From an economic perspective, it was estimated that the closure of fishable waters cost local economies several billion dollars [8,44]. Add to that the environmental impacts [45] and harm to the tourism industry [10] and economic costs could easily be in the tens of billions.

2.1. Implications for Florida

Florida is situated in a unique geographical position in relation to ocean currents. Recent work suggests that Florida's western shelf is isolated from cross-shelf "squeezelines" or current velocity fields that tend to attract nearby particle trajectories [46]. In other words, the western coastline of Florida may not be particularly susceptible to oiling. The construction of Lagrangian coherent structures (LCS) from twelve-year-long ocean surface circulation data confirms the absence of squeezelines from the shore to about the 50 m isobath off Florida's west shelf [46] making the 50 m isobath an important boundary. From the shore to the 50 m isobath one can expect very little ocean current activity, meaning that oil within this area is likely to stagnate or move very slowly. From the 50 m isobath and beyond, oil is much more likely to be pulled and transported by the squeezelines which can rapidly increase the plume extent. Important to keep in mind is that the distance between Florida's western shoreline and the 50 m isobath displays significant geographic variation. For portions of the Florida Panhandle, it is only 20 miles offshore. In other locales, including the areas west of Tampa, it is found over 100 miles off the coast. In short, oil within the area encompassed by the 50 m isobath will remain fairly stagnant, possibly making response and cleanup operations for a spill more effective. This recent work on squeezelines, however, is based on a 12 year average [46]. Averages often mask natural variations in currents, wind speeds, and direction. Thus, a closer look into these natural variations may reveal scenarios where spills can generate significant impacts to the Florida coast regardless of spill origination.

It is also important to acknowledge that the evaporation of oil, and the rate at which evaporation occurs, is partly dependent on movement of oil. If oil remains stagnant, the natural degradation rates are slowed [47]. As a result, the stagnation of oil over a particular area may result in an increased amount of deposition. As oil mixes with the sea water during emulsification, some amount will sink, coating the benthic communities below. At the same time, portions of the petroleum derivatives remain closer to the surface, affecting species that call this part of the water column, home. This overall oiling process was clearly evident following the Deepwater Horizon spill [36,48].

Given the complex web of interaction between oil, sensitive ecosystems, weathering processes, and the environment, as well as the sensitive facilities and economic compositions of proximal communities, it is critically important to have a strong sense of what could happen in the event of a spill off of Florida's Gulf coast before decisions are made on drilling. This is true, not just for Florida, but for any state on the Atlantic or Pacific coasts of the U.S. That said, Florida offers a particularly interesting case because of the location of the 50 m isobath, prevailing currents, and associated weather patterns. Again, Florida may be protected from oil spill intrusions further offshore, reducing the potential for harm. If so, oil production could bring additional jobs and revenue to the state, as well as royalties to the Federal Government. On the other hand, a catastrophic spill could also result in billions of dollars in damage to the state, depressing tourism and causing severe harm to local ecosystems.

2.2. Oil Spill Impact Modeling

One way to evaluate the potential outcomes of a spill is through the use of contingency analysis, where hypothetical spills and their impacts are modeled. This type of "what if" modeling generates valuable geospatial intelligence, helping to both visually and quantitatively depict how oil spills may behave during a catastrophic event at a particular time and in a specific place. More importantly, the ability to explore the implications of such spills on the environment can help first responders prepare tactical intervention efforts [49] and help communities to develop strategies to reduce their vulnerabilities to extreme spill events [24].

As mentioned previously, in the years following the Deepwater Horizon oil spill there was a significant increase in the amount of work being done in the oil spill risk and impact modeling area [17]. There are three major families of analysis: (1) vulnerability analysis, (2) risk analysis, and (3) normative impact modeling. Vulnerability analysis primarily focuses on characterizing the susceptibility of shorelines to damage following an oiling event. Major vulnerability studies include environmental sensitivity index mapping and development [50–52], industry-specific vulnerability metrics [17], or creating composite vulnerability scores based on economic, social, and environmental assets [21,53]. Risk analysis is primarily concerned with estimating the probability that an oil spill will impact a specific geographic area [17]. This can be done with an ensemble-type approach, using hundreds of oil spill simulations to determine probability [54,55], or risk can be inferred based on historical accounts of oil spills in a particular area [56]. Although these approaches often yield similar results, simulation approaches are growing in popularity due to their higher levels of accuracy and the robustness of the associated modeling techniques.

Lastly, the use of normative impact modeling represents a hybrid approach, combining the best of both vulnerability and risk analysis [17]. For example, Azevedo and colleagues [57] determined risk via oil transport and shoreline exposure which is combined with a vulnerability metric based on biologic and physical indicators related to their sensitivity to oil. The combination of these two metrics into a spatially explicit normalized impact index is used to characterize different segments of coastline. Olita et al. [18] and Canu et al. [20] perform similar analyses that result in a normalized index of total impact for a number of coastal environments. In short, the way in which these two important metrics are combined varies from study to study, but their combination helps account for the probability of occurrence, the degree of oiling, and how susceptible to damage the surrounding communities (social and environmental) are to the effects of oil [24].

3. Study Area, Methods, and Data

The GOM is divided into three drilling districts: the Western, Central, and Eastern Planning Areas. As of 1 January 2018, there were 2795 active leases in the GOM with 815 of those actively producing [58]. Only 37 leases are active in the Eastern GOM—none of which are currently producing oil. The Eastern Planning Area is home to 13 complete protraction areas, and 6 partial areas. Across the GOM, oil production depth varies between active platforms, with 1937 in water depths from 0 to 200 m, 20 active in water depths of 201–400 m, 10 active in water depths from 401 to 800, 9 active in depths of 801–1000 m, and 32 active in water depths greater than 1000 m. Since 2015, seven new platforms have been installed, three of which are in water depths greater than 1000 m, indicating that oil exploration remains active in the GOM region.

For the purposes of this research, a handful of locations were identified in the GOM, proximal to Florida, to simulate potential oil spills. Using the protraction diagram provided by the Bureau of Ocean Energy Management [59], ten locations were selected, ensuring geographic diversity and variety in offshore environmental conditions (Figure 1). Because the Florida shelf is so large (in terms of its extension into the GOM), seven of the selected spill locations fall within the area between the coastline and the 50 m isobath. Recall that this is the area without the presence of "squeezelines" and theoretically an area where very little particle movement will take place. The other four locations are in deeper waters and further from the Florida shoreline. It is important to acknowledge that locations off the Florida shelf tend to intersect the GOM loop current [60] for at least part of the year.

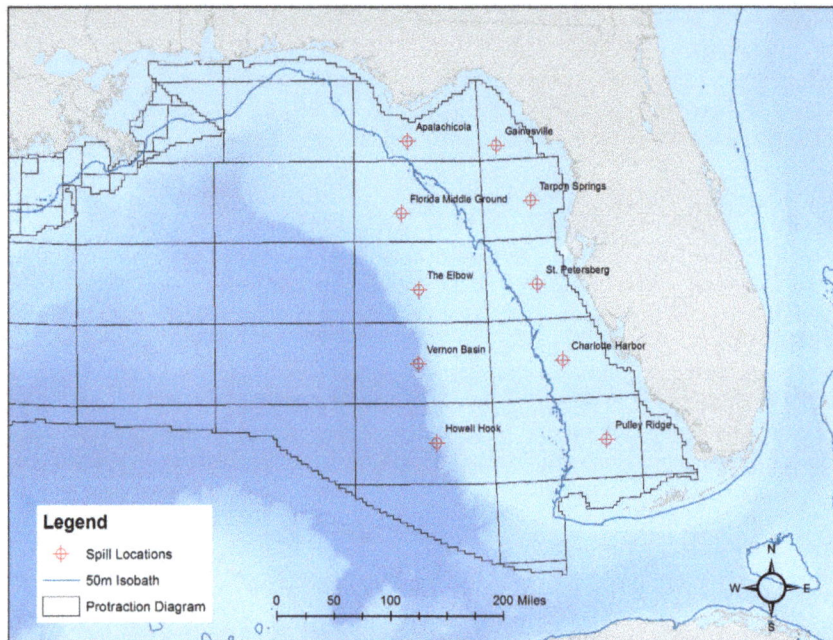

Figure 1. Study area and the set of locations used for spill simulation. Scenario names are derived from the name of the protraction block in which they occur.

3.1. Ambient Data and Spill Model

The spill model used for this research is the Blowout and Spill Occurrence Model (BLOSOM) [24,61,62], which is combined with the 2017 Navy Coastal Ocean Model (NCOM) American Seas (AmSeas) data to model the blowout and subsequent oil transport [63]. The NCOM AmSeas ocean model has a temporal resolution of 3 h and a spatial resolution of 3.3 km. The NCOM comes in the NetCDF data format with 40 different depth levels [64]. For each level, information on water salinity, temperature, velocity, and direction are available. At the highest (shallowest) layer, ambient information includes the surface atmospheric pressure, surface roughness, surface temperature, and wind stress in the x and y directions.

BLOSOM is a four dimensional spill modeling suite designed for simulating offshore spills in deepwater and ultra-deepwater environments. Most of the equations and associated functions within BLOSOM are designed for high-pressure environments; however, with slight modification, BLOSOM has the ability to handle the simulation of offshore surface spills as well. BLOSOM comprises several individual models that, in conjunction, make up the integrated modeling suite. It begins with the Jet/Plume model and progresses through the conversion model which handles the oil as it is converted from behaving under jet-like influences to buoyant forces. From there the transport model simulates the long-term final fate of the oil within the water column and on the ocean surface. While these three models are operating, there are several other models handling the physics of the oil and ocean environment. These include the crude oil model, the gas/hydrates model, the weathering model, and the hydrodynamic handler. The individual models as well as BLOSOM as a whole are described in more detail in [62] and the model itself can be found at https://edx.netl.doe.gov/blosom/.

3.2. Spill Locations

The timing of simulated blowouts was carefully modeled to try and ensure that the velocity and direction vectors would favor conditions where oil would move quickly toward the Florida coast. With a temporal resolution of three hours, each modeled day consists of eight individual current files beginning with hour 0 and ending with hour 21 (on a 24 h time scale). The northward and eastward velocity values were averaged over each day and then over each month. These calculations yielded two sets of 12 raster files (one file for each month in the year)—one set for the northward direction and another set for the eastward direction. We then created a 25 mile buffer around each blowout location and, using the two sets of monthly current files, calculated the average direction and velocity values of the ocean currents within the buffer (Table 1). Based on velocity strength and directions trending toward the shore, two months were selected (June and July) as the worst possible time to have a spill in the eastern GOM region. However, there is a caveat to the selection of current direction. For the blowout locations in the southern portion of the study area, it must be acknowledged that a large southward velocity means strong loop current activity. This increases the potential for oil-related problems in the Florida Keys, Cuba, and the Atlantic Coast and was taken into account when choosing which months to simulate the spills. The locations and starting months for each spill, as well as the ambient environmental conditions, are detailed in Table 1.

Table 1. Geographic location, depth, and the average monthly current speed within a 25 mile radius around the spill location. Negative numbers represent current directions opposite to the direction noted in the column headers. For example, negative numbers for the June and July average current velocity north would mean the dominant current direction is south.

Spill Scenario	Latitude	Longitude	Depth (m)	June Average Current Velocity (North, m/s)	July Average Current Velocity (North, m/s)	June Average Current Velocity (East, m/s)	July Average Current Velocity (East, m/s)
				Eastern Locations			
Gainesville	29.12 N	−83.79 W	62.33	0.037	−0.013	−0.021	0.02
Tarpon Springs	28.43 N	−83.32 W	62.43	0.049	−0.001	−0.015	−0.0009
St. Petersburg	27.39 N	−83.26 W	111.54	0.092	−0.001	−0.03	0.001
Charlotte Harbor	26.44 N	−82.93 W	121.39	0.088	0.008	−0.017	−0.007
Pulley Ridge	25.43 N	−82.36 W	91.86	0.047	0.011	−0.017	−0.016
				Western Locations			
Apalachicola	29.21 N	−85.10 W	104.98	0.033	−0.024	−0.05	0.059
Florida Middle Ground	28.32 N	−85.2 W	600.39	0.015	−0.004	−0.014	−0.002
The Elbow	27.36 N	−84.97 W	1417.32	−0.049	−0.058	0.049	0.113
Vernon Basin	26.45 N	−85.00 W	10833.33	0.039	−0.186	−0.062	0.1
Howell Hook	25.46 N	−84.76 W	5754.59	−0.22	−0.764	0.494	0.208

3.3. Spill Scenarios

All spill scenarios were structured equally. For the purposes of this analysis, we are more interested in the final fate of the oil, rather than the many nuances and variations associated with spill locations.[1] Spill duration and discharge rate were 10 days and 500 barrels per day, respectively. These

---

[1] We readily acknowledge that spill locations and their local environmental conditions are important factors to consider, but these details will be explored in future research efforts.

are relatively modest spill settings and do not reflect a catastrophic blowout such as the Deepwater Horizon. However, there is enough oil released into the environment for it to be a concern. The resulting plume is tracked for a total of 60 days, providing ample time to characterize the behavior and transport of the plume in the water column. Additional settings for BLOSOM reflect widely accepted standards in spill modeling. Horizontal diffusion was modeled as a random walk, with a Smagorinsky coefficient of 0.15 [46]. Default models for spreading [65], emulsion [66], mass transfer [67], and evaporation [68] were also selected to improve repeatability.

3.4. Impact Calculation

As detailed previously, the Deepwater Horizon event triggered a substantial resurgence in the development and testing of methods to quantify actual and potential impact [57,69]. This process is necessarily data driven, relying on the incorporation of data sets to represent the vulnerability of coastal communities which can be distinguished by sector (biologic, economic, social) and then grouped together for a composite vulnerability measure (Table A1). More data yields a richer analysis and provides the ability to better express the local spatial and temporal aspects of vulnerability [17].

For the purposes of this research, a 2 km × 2 km impact grid [24] was generated for the study area. This grid helps to aggregate and represent the environmental and socio-economic assets in the region, as well as their exposure to the effects of oil along the coast. Specifically, for each of the individual grid cells in the impact grid, the number of assets present across both the environmental and socio-economic sectors was calculated and used to inform the total vulnerability measure for each cell. If more assets are present within a grid cell, that area is more susceptible to harm from oil. No weighting scheme was used, although scalars could be easily implemented.

Overall impacts were determined on a cell-by-cell basis reflecting the amount of oil within a grid cell and the vulnerability of the cell (and its assets) for each of the spill scenarios. Impacts for the coastal environment and the water column were calculated separately and a derived final impact score for each scenario (based on the combined coastal and water column) was generated. A simple standardization process was also used to keep the impact values comparable across all scenarios. We scaled the amount of oil within the grid cells along the coast to range between 0 and 1, with 0 being no oil present and 1 being the highest amount of oil that accumulated within one grid cell *during all scenarios* for that area (Oil_{mod}). Basically, the oil modifier for the coastal areas was based on the maximum amount of oil that had beached within one grid cell.[2]

The open water modifier was derived in a slightly different manner. In this case, the total amount of oil that passed through each grid cell over the course of the simulation period was tracked. Then, similar to the coastal oil modifier, the maximum amount of oil that had passed through any one grid cell was used to create the range of oil modifier values. After generating the oil modifier for each grid cell, the modifier value was multiplied by the vulnerability score for each grid cell:

$$\text{Grid cell impact } (G_{cell}) = Oil_{mod} \times Vulnerability \tag{1}$$

where Oil_{mod} is the modifier scaled to the amount of oil within an individual grid cell and Vulnerability is the number of assets present within the boundaries of an individual grid cell.

Finally, overall impact for each scenario was determined by the sum total of the impact values for all grid cells in the water column (Equation (2)) and coastal areas (Equation (3)), with values scaled to range between 0 and 10. To account for the oil remaining within the water column, an additional value was added to the water column impact score reflecting the amount of oil that could potentially cause impacts at a later time (Equation (4)). This value was derived in a manner similar to the oil modifier

---

[2] Although this modifier improves comparability between scenarios, it also unintentionally obfuscates the overall impacts of oil in coastal locations. Any amount of oil that comes into contact with the shoreline is a problem.

where the total amount of oil to pass through a grid cell (OW_{oil}) was divided by the maximum amount of oil $(Max\ OW_{oil})$ that passed through any one grid cell for all spill scenarios. The value was then transformed to range from 0 to 5.

$$\text{Total Open Water Impact } (OW_{imp}) = \sum G_{cell1} + G_{cell2}, \ldots, + G_{celln} \tag{2}$$

$$\text{Total Coastal Impact } (C_{imp}) = \sum G_{cell1} + G_{cell2}, \ldots, + G_{celln} \tag{3}$$

$$\text{Oil Remaining Impact } (OR_{imp}) = \frac{OW_{oil}}{Max\ OW_{oil}} \tag{4}$$

$$\text{Total Scenario Impact } = (OW_{imp} + OR_{imp}) + C_{imp} \tag{5}$$

4. Results

Upon simulating the spills at each of the locations outlined in Figure 1, the results suggest stark spatial and temporal differences in spill behavior between the eastern and western locations. In an effort to decompose and explain these differences in behavior, the following subsections are framed geographically, where the spills for the eastern and western blocks are detailed separately. Finally, we detail how the behavior of these spills translates into overall impact for coastal and offshore Florida.

4.1. Spill Behavior—Eastern Locations

The model results show that spills beginning in the eastern locations were generally smaller in spatial extent than their western counterparts, with the exception of Gainesville, a clear outlier. At the end of the 60 day simulation period, the Gainesville spill has the largest spatial extent at 3272 square miles, followed by Tarpon Springs (616 sq. mi.), Charlotte Harbor (634), St. Petersburg (447 sq. mi.), and Pulley Ridge (304 sq. mi.) (Figure 2). Again, the lack of movement and geographic spread of these spills was somewhat expected given the minimal number of squeezelines in this area and lower east–west current velocity. In fact, for all locations except Gainesville, the plume remains relatively compact and moves slowly, in a northerly direction (Figure 3). In terms of plume dynamics, the lack of ocean current activity on the Florida shelf prevents the stretching and deformation of the plume as one typically sees further off shore (detailed in Section 4.2). At the end of the 60-day simulation, plumes are roughly 20 miles or more from the shore regardless of where the spill originated and *none* of the oil has reached the shores of Florida. However, as displayed in Figure 3, the plumes are on a northward trajectory toward the Florida panhandle.

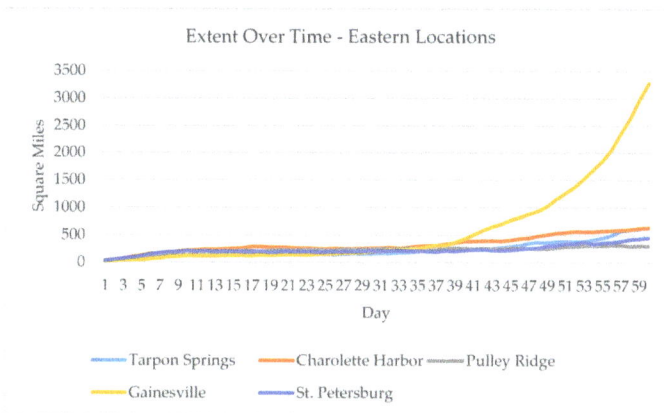

Figure 2. Plume extent as a function of time for the eastern locations.

Although much larger in geographic extent, the Gainesville spill scenario remains fairly compact until Day 39 when it begins to rapidly expand (Figure 2). It is at this time that a portion of the plume breaches the 50 m isobath boundary which is followed by stretching and pulling of the plume further from shore. Again, there are very few squeezelines of high particle attraction within the region from 0 to 50 m depth but once that line is passed, the plume rapidly expands. By Day 60, the plume begins its entry into the loop current and will eventually be transported into the Atlantic Ocean (Figure 3a). More importantly, because of the lack of east–west current activity on the Florida shelf, none of the oil from the Gainesville scenario makes landfall by the 60 day mark.

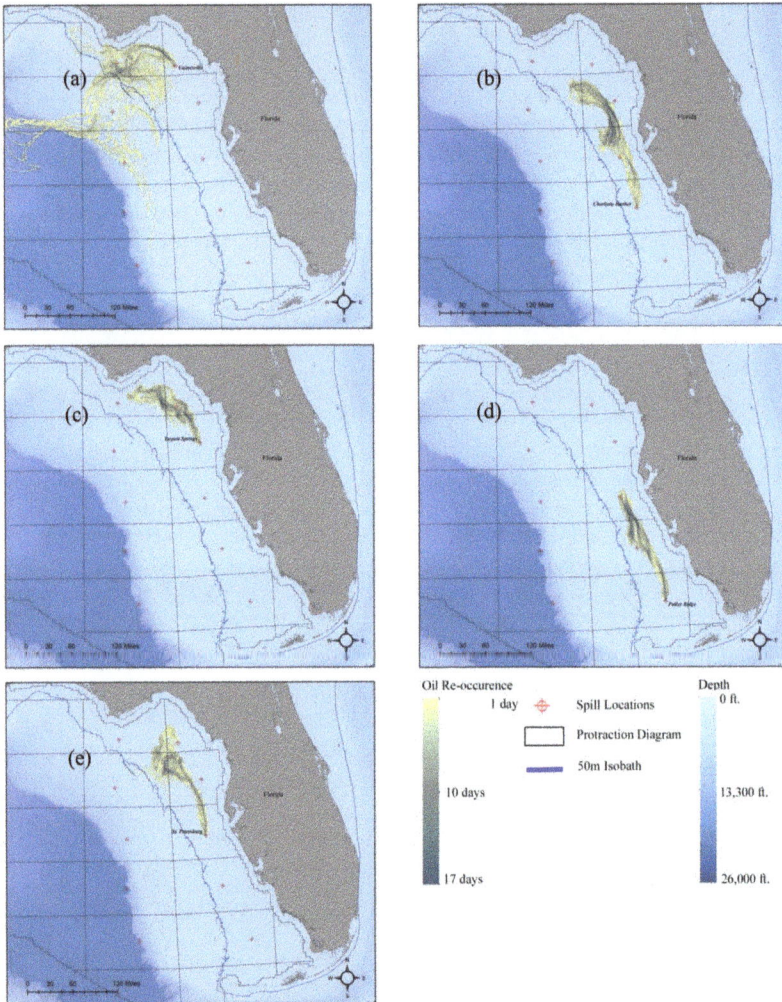

Figure 3. Eastern scenarios plume extent and shape over the course of the 60 day simulation period. (a) (top left) depicts the plume from the Gainesville scenario. (b) (top right) depicts the plume from Charlotte Harbor. (c) (middle left) depicts the Tarpon Springs plume. (d) (middle right) depicts the Pulley Ridge plume and (e) (bottom left) depicts the St. Petersburg plume. The plume is represented by the number of days that new oil particles moved through a specific location.

4.2. Spill Behavior—Western Locations

As expected, spills beginning in the western locations and outside of the 50 m isobath are significantly more dynamic than their eastern counterparts. For all western locations, at least some of the oil reaches the loop current and is transported south, past the Florida Keys and into the Atlantic Ocean. Even the northernmost western spill location, Apalachicola, had particles making the turn and heading into the Atlantic Ocean by the end of the 60-day simulation period. Interestingly, the origin of the Apalachicola scenario falls inside the 50 m isobath. From Figure 4 one can see that the plume behavior of Apalachicola is similar to the Gainesville scenario. The plume was relatively modest in extent until it breached the 50 m isobath on Day 29. After that, the plume grows rapidly in extent and also impacts the northern Florida shore by Day 30.

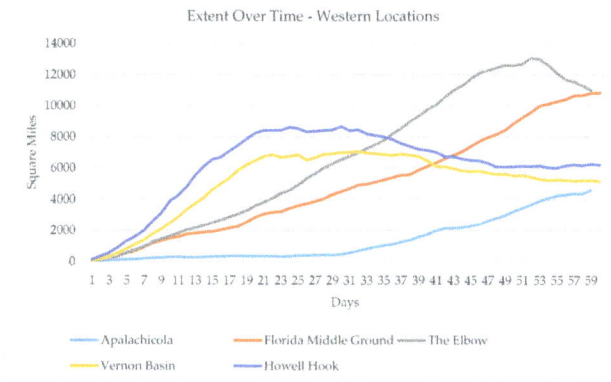

Figure 4. Plume extent as a function of time for the western locations.

The remaining western spill locations all experience strong and extremely rapid movement of oil as it is pulled by the loop current. As Figure 4 illustrates, the western spills tended to be much larger in geographic extent when compared with the eastern spills.[3] The Elbow was largest in extent at 10,963 square miles followed by Florida Middle Ground with a final-day extent of 10,815 square miles. Howell Hook had the next-largest extent at 6196 square miles followed by Vernon Basin (5144 sq. mi.) and Apalachicola (4583 sq. mi.). It is also important to acknowledge that some of these spills may be smaller in geographic extent than one might expect. This is due in part to data limitations, but it is also a function of the strength of the loop current. Once the plume enters the loop current, it is pulled and stretched and accelerates into a quickly moving (yet thin) train of oil that follows the loop current path. To be sure, spills beginning in the western locations are highly dynamic, never stagnating in a single geographic location for more than a day (Figure 5). As soon as these spills enter the water column, they begin to move rapidly. Thus, getting a handle on the plume in both time and space is difficult—making tactical interdiction efforts (e.g., application of dispersants or allocation of response equipment) more difficult.

Consider, for example, the speed at which plumes begin to be influenced by the loop current (Figure A1). In the case of Howell Hook, by the time the oil reaches the surface, it is already being pulled south (Day 2). The Vernon Basin spill exhibits a similar temporal profile, being pulled south by Day 3. The Elbow spill takes somewhat longer to connect with the loop current, but by Day 15, a clear southern trajectory begins to take shape. The Florida Middle Ground location serves as a small

---

[3] It is likely that the spills would be much larger if the simulated scenarios were not restricted to the boundary of the current files used for analysis.

exception. Its northern location means that it is not directly in line with the loop current. Although the spill stagnates for a number of days, it is eventually pulled south by the loop current on Day 32.

Similar to the spills emanating in eastern locations, the Gulf Coast of Florida is largely spared from oiling with spills that emanate from the western locations. In fact, a large majority of the oil never makes landfall when beginning from a western locale. Instead, the remaining oil is left swirling in the Gulf or launched into the Atlantic Ocean. The minimal amount of oil that does make landfall is largely concentrated on the shores of the Florida Keys, Cuba, and the Bahamas.

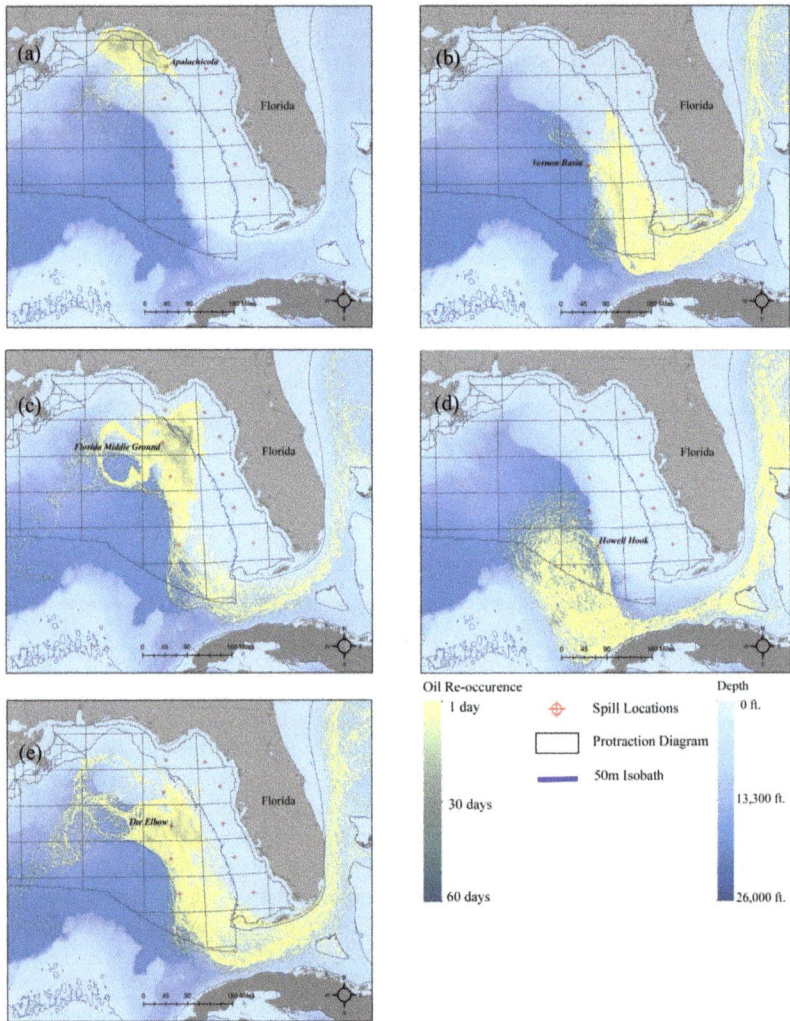

Figure 5. Western scenario oil plume extent and behavior over the 60-day simulation period. Plumes originating from the western locations were more dynamic and larger in extent, wrapping around Florida and out into the Atlantic Ocean. (**a**) (top left) depicts the 60 day plume from Apalachicola. (**b**) (top right) depicts the 60-day plume of Vernon Basin. (**c**) (middle left) depicts the 60-day plume from Florida Middle Ground. (**d**) (middle right) depicts the plume of Howell Hook and (**e**) (bottom left) depicts the plume of The Elbow. The plume is represented by the number of days that new oil particles moved through a specific location.

4.3. Impacts—Eastern Locations

On-shore impacts resulting from spills at the eastern locations were negligible. By the end of the 60 day simulation period, none of the plumes had reached the shoreline. As a result, the majority of impacts for the eastern locations are representative of the potential threat that the plume poses to the shore in the future. Figure 3 displays the propensity for plumes from eastern spill locations to move north, heading toward the Florida Panhandle, which could be a problem. However, given the plume behavior exhibited in the Gainesville scenario, it is questionable whether these plumes will ever hit the shore. Instead, they may turn and make their way further into the GOM. That being said, the potential remains for this oil to threaten the coastal environment because a significant amount of oil still remains in the water column (Table 2).

Table 2. Total oil and number of assets impacted from each of the spill scenarios and the total amount of oil not beached following 60 days of simulation. Values denoted here are used in the calculation of total scaled impacts.

Spill Scenario	Total Coastal Oil (Gallons)	Total Coastal Assets Impacted	Total Oil in Open Water Remaining (Gallons)	Total Open Water Assets Impacted
Eastern Scenarios				
Gainesville	0	0	90,932.89	23,307
Tarpon Springs	0	0	84,304.82	11,277
St. Petersburg	0	0	88,429.15	12,407
Charlotte Harbor	0	0	87,356.18	14,687
Pulley Ridge	0	0	84,439.23	9056
Western Scenarios				
Vernon Basin	7243	1752	23,073.15	49,498
Howell Hook	10,953.14	0	26,873.38	12,517
Florida Middle Ground	252.31	307	78,702.33	51,676
Apalachicola	190.92	142	85,861.80	32,123
The Elbow	5234.70	1878	51,711.00	67,948

In rank order, the eastern scenarios with the largest impact were St. Petersburg followed closely by Tarpon Springs, Charlotte Harbor, Pulley Ridge, and Gainesville (Table 3). Gainesville is an interesting case. It has the highest amount of oil left within the environment and the highest number of assets impacted (Table 2), yet it does *not* have the highest total water column impact. Since the impacts depend on the degree (i.e., intensity) of oiling, the geographically dispersed nature of the spill means that the total impact ends up being the smallest. Functionally, this is the core idea behind the application of dispersants to a plume. The dispersant makes the oil droplets smaller, encourages their spread, and reduces the potential for geographically intense oiling. Smaller droplets and a more dispersed plume also enhances natural oil degradation due to weathering and other processes. In sum, however, the eastern spill scenarios leave large quantities of oil in a fairly small area. Without some type of tactical response effort for cleaning up the oil it will continue to mix with water and sediment and eventually sink, coating the benthic communities below.

4.4. Impacts—Western Locations

Similar to the eastern locations, a substantial (but in general smaller) amount of oil remains within the water column at the end of the 60-day simulation period for western spills (Table 2). With the exception of Apalachicola, the majority of the oil is transported out around the southern coast of Florida and on into the Atlantic Ocean (Figure 5). This trajectory sets up the Florida Keys for experiencing the highest rates of oiling and subsequent impact. Although the oiling extents of the western scenarios are far greater than those of the eastern scenarios, the total open water impacts are much smaller (Table 3). Accelerated spill movements and dispersed plumes create a situation where the majority of oil does not collect in a single, open water location. Of course, with impacts being directly related to the degree of oiling, total impacts for western spill scenarios in open water were much smaller in comparison with those for the eastern spills.

A key differentiating factor between the outcomes of eastern and western spills is the *potential* for coastal impacts. Because of the behavior and strength of the GOM loop current, the Florida Keys, Florida's Atlantic shoreline, Bahamas, and Cuba are all at risk of coastal oiling, but this is dependent on the spill location. That said, regardless of the initial starting location, a substantial amount of oil makes its way into the Atlantic Ocean.[4] It is likely that the impacts reported for western locations are conservative and the effects of oiling would accrue if the analysis was extended both spatially and temporally.

Table 3. Total scaled impacts for each impact area of the spill scenarios. Overall impacts are denoted in the last column and are the sum total of coastal impacts, water column impacts, and remaining oil impacts.

Spill Scenario	Total Open Water Impact	Scaled Open Water Impact	Total Coastal Impact	Coastal Impact Scaled	Remaining Water Column Oil Impact	Overall Impact
			Eastern Locations			
Gainesville	680.72	8.05	0	0	5.00	13.05
Tarpon Springs	807.56	9.47	0	0	4.61	14.08
St. Petersburg	855.4	10.00	0	0	4.85	14.85
Charlotte Harbor	767.27	9.02	0	0	4.79	13.81
Pulley Ridge	704.98	8.33	0	0	4.62	12.94
			Western Locations			
Vernon Basin	164.63	2.31	143.51	10.00	1.00	13.31
Howell Hook	47.159	1.00	0	1.00	1.22	3.22
Florida Middle Ground	398.012	4.91	0.15	1.01	4.28	10.20
Apalachicola	737.57	8.69	2.23	1.14	4.70	14.53
The Elbow	303.36	3.85	92.63	6.81	2.69	13.35

4.5. Overall Impacts

Area-specific impacts varied greatly between the eastern and western locations. Due to the absence of oil beaching during the eastern scenarios, total impact came from residual oil remaining after the 60-day simulation period and the intersection of open-water assets with the oil plume as it moved north. Because the open water impact was relative to the largest amount of oil occurring within one grid cell, the western spill scenarios with their highly dynamic and rapidly moving plumes had much smaller open water impacts when compared with the eastern locations (Table 3). However, all western spills but Howell Hook had coastal impacts, increasing their overall impact scores (Figure 6). The takeaway here is that no matter where a spill begins, and regardless of the dynamics of the plume, the overall impacts of most spills are similar. In fact, the most detrimental spill is St. Petersburg which has no coastal impacts and has the second-smallest extent (Table 3). The resulting high concentration of oil over sensitive offshore environments makes it the most impactful of all scenarios modeled here. Following St. Petersburg is Apalachicola, with the second-highest total impact, and then Tarpon Springs, at 14.53 and 14.08, respectively.

Howell Hook is an outlier. It has few immediate impacts to the open water, no coastal impacts, and negligible effects from oil remaining in the water column. It is important to remember that Howell Hook reflects a western spill scenario, but is the most southerly location in the group. As a

---

[4]   Again, data restrictions and project scope limit our ability to gain a comprehensive picture of what that means in terms of final impacts.

result, the oil is immediately caught in the loop current. The residence time of the plume in any one location within the study area is approximately 10 h, and because of its rapid movement, the plume is widely dispersed. This helps explain the minimal impact of the oil in open water. However, this rapid movement also moves the oil beyond our study area and into the Atlantic Ocean, where data are sparse. So although the effects of the loop current pull the oil plume far enough south that coastal Florida is spared from oiling, it is important to reiterate that these results could be conservative, although we cannot say for certain.

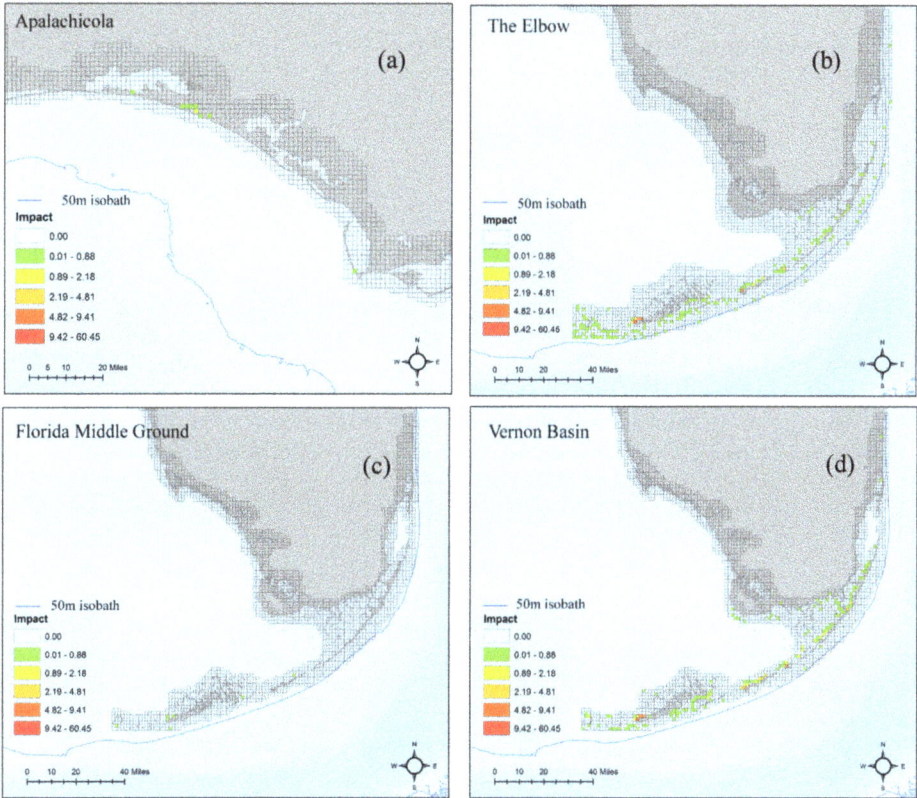

Figure 6. Spatial distribution of the impacts from the scenarios that had a coastal impact. (**a**) (top left) indicates the coastal impact from Apalachicola. (**b**) (top right) indicates coastal impacts from The Elbow. (**c**) (bottom left) indicates coastal impacts from Florida Middle Ground and (**d**) (bottom right) indicates the coastal impacts from Vernon Basin. With the exception of Apalachicola where the coastal impacts occurred on the Florida Panhandle, the majority of impacts were seen along the Florida Keys and Atlantic Coast of Florida.

5. Discussion and Conclusion

The current U.S. government administration has showed considerable interest in leveraging offshore oil resources as a means to support efforts towards energy independence. This paper provided site-specific contingency analyses or "what if" scenarios for evaluating the potential effects of drilling-related oil spills near coastal Florida. These types of analyses are critical for providing stakeholders with the geospatial intelligence necessary for evaluating the potential spatiotemporal impacts of spills for a region. In particular, when spill simulations are combined with data-driven, quantitative spatial analysis, the magnitude of impacts caused by a spill can also be obtained. This information can be used to identify locations for oil production that may pose less risk for development and operation, especially if a loss-of-control event was to occur.

As detailed previously, many of these newly opened lease areas are dense with ecosystems, environments, and economic clusters that are highly vulnerable to the effects of oiling. This analysis has taken these factors into account and provides a glimpse into what the potential implications of offshore oil exploration and production may be if a spill was to occur near the Florida Gulf Coast. There are several key findings concerning plume behavior and impacts worth further discussion.

When it comes to plume behavior, there were several substantial differences in plume evolution over time. For eastern locations, closer to the shoreline, plumes largely moved as a single, cohesive unit. This is meaningful for two reasons. First, the compact shape and slow movement of the plume will make tactical response and cleanup efforts less complex. The thickness of the plume on top of the water column lends itself well to in situ burning while dispersant applications could be concentrated over a relatively small area. Also, where response efficiency is concerned, plume behavior of this type is a benefit because it allows responders time to coordinate the myriad resources required to combat the spill. Second, the results of this paper confirm the absence of squeezelines within the 50 m isobath and highlight the dearth of cross-shelf current activity. This translates into plumes that remain offshore, keeping the Gulf Coast of Florida oil-free. This same outcome was also true for the western spill locations—with the Gulf Coast of Florida spared from oiling.

The western spill locations, however, yielded plumes with a more dynamic and unique suite of behaviors. The most obvious differences manifested in the geographic extents of the spills. Regardless of where the western spills originated, the oil plumes were eventually pulled into the loop current, undergoing a rapid increase in extent. Response efforts for these spills would need to occur quickly (i.e., a day or two), before the plume reaches the loop current. If, however, the plume reaches the loop current, response teams would be stretched from the GOM to the Atlantic Ocean. Regardless of the tactical precision associated with a response for this type of spill, the associated interdiction efforts would be daunting. The inevitable (vast) extent of a plume reaching the loop current also means that cleanup efforts would become an international affair, involving the United States, Cuba, and the Bahamas—drastically increasing the complexities of coordination.

Regarding the overall impacts of the simulated spills, there were some small surprises. Coastal impacts were negligible, but impacts in the open water have the potential to be severe, especially for eastern spill scenarios. As noted in the results section, impact is necessarily a function of degree of oiling. The more oil in any given area, the more likely (and severe) the damage will be. Plumes from the eastern scenarios were highly concentrated off the coast of Florida, meaning that assets in that area were likely exposed to a significant amount of oiling. More importantly, the simulations suggest that a large amount of oil remains present after 60 days, and will continue to impact assets as time goes on. On the contrary, although the western scenarios displayed the largest spill extents, the oil was dispersed such that the impacts to both coastal locations and the water column were smaller. Again, the geographic spread of these spills can be attributed to the GOM loop current. Relatedly, it is also important to acknowledge the "blind spots" associated with these simulations. The results reported here are conservative, at best, and because of data limitations cannot accurately reflect what happens to the oil once it reaches the Atlantic Ocean. Significantly more work is required to develop a fully comprehensive understanding of spill impacts on coastal ecosystems, especially when the plumes interact with the GOM loop current.

Perhaps the most surprising (but important) finding of this analysis concerns the derived impact scores for the oil spills. Given the differences in spill location and extent, as well as coastal and water column impacts for each scenario, the overall impact results were remarkably similar. Spills proximal to the Florida coast will have significant impacts to the water column, as well as local benthic and coral communities which are particularly sensitive to oil exposure. Given Florida's reliance on tourism and related activities, any decay in coastal water quality will likely have an impact on this sector. Further,

J. Mar. Sci. Eng. **2018**, *6*, 30

it is important to mention the thriving pelagic zones in/around coastal Florida, which underpin a significant portion of the U.S. seafood industry. If oiled, the economic impacts would be severe and far-reaching.

One final consideration worth noting concerns the aesthetic implications [70] of offshore oil operations for coastal states, including Florida. Both California and Florida have strongly resisted efforts to renew offshore drilling operations, arguing that environmental, tourism, and aesthetic values would be negatively impacted [71]. Given the importance of tourism to many coastal states, this is a legitimate concern worth acknowledging.

To conclude, this paper addresses one of the core issues associated with opening up lease areas in the eastern GOM for oil exploration and drilling, focusing on the potential impacts of oil spill events for coastal Florida. Catastrophic spills are rare, but smaller spills are frequent enough that both their immediate and long-term additive effects represent real concerns to proximal coastal communities and their associated economies. Oiling has a detrimental effect on ecosystems, the environment, and all industries tied to these natural resources. Any consideration to reopen protected waters to oil production must be informed by rigorous, empirically driven scientific research to evaluate the potential impacts of these plans, *prior* to implementation. Simply put, there are too many jobs and too many communities reliant on sensitive ecosystem services and natural capital to make an uninformed decision [72].

Acknowledgments: This work was supported by the National Academies of Science Gulf Research Program (# 2000007349). The authors would also like to thank the BLOSOM team at the National Energy Technology Laboratory for helping with oil simulations and offering guidance where needed.

Author Contributions: Nelson and Grubesic conceived and designed the study; Nelson performed the spill simulations and analyses; Nelson and Grubesic analyzed the data and wrote the paper. All authors read and approved the final manuscript.

Conflicts of Interest: The authors declare no conflict of interest.

Appendix A

Table A1. Data sets used to determine the vulnerability of each of the grid cells in the analysis. Data was marked as present or absent within a grid cell and summed to determine the vulnerability of the grid cells.

Data Set	Sector
Beach Access	
Marinas	
Boat Ramps	Recreation/Tourism
Drinking Water Intake	
Parks	
Piers	
Essential Fish Habitat	
Migratory Pelagic	
Red Drum	
Reef Fish	
Spiny Lobster	
Albacore Tuna	
Sharpnose Shark	
Big Eye Tuna	
Blacknose Shark	
Blacktip Shark	
Tiger Shark	
White Marlin	Ecologic/Environment
Yellowfin Tuna	
Bluefin Tuna	

Table A1. *Cont.*

Data Set	Sector
Coral Reef/Hardbottom Habitat	
Artificial Reef Locations	
Critical Wildlife Areas	
Sea Turtle Nesting Beaches	
Wildlife Refuge	
Oyster Habitat	
Environmental Sensitivity Index	
Marine Protected Areas	
Recreational/Tourism Businesses	
Seafood Processing Plant	
Airports	
Coastal Roads	
Refineries	Socio-Economic
Liquefied Natural Gas facilities	
Platforms	
Pipelines	
Wells	

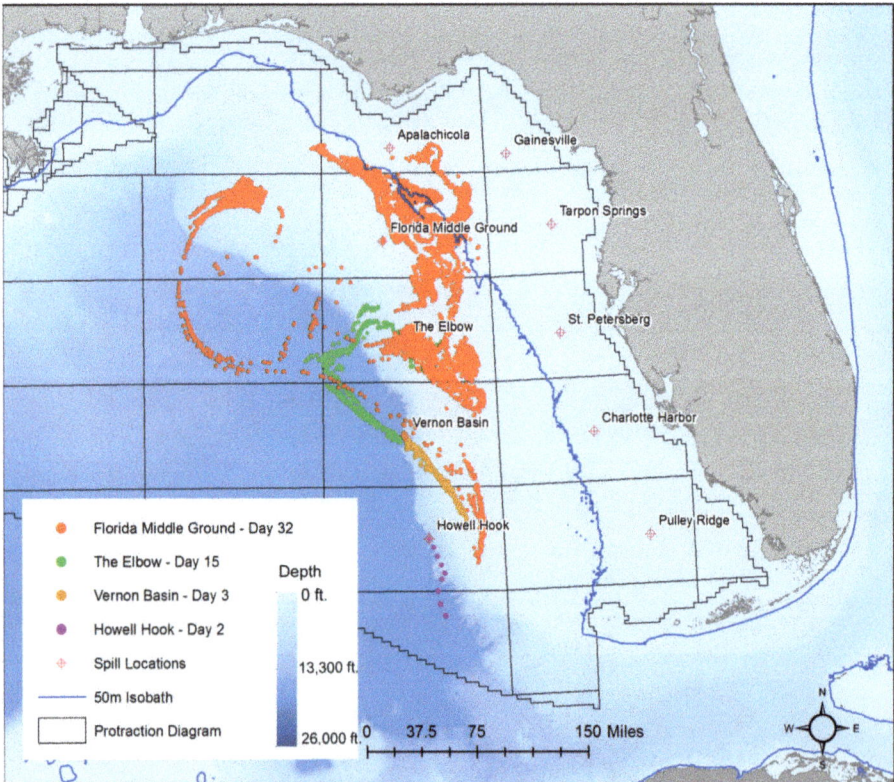

Figure A1. The day in which the oil particles begin to be pulled into the loop current for four out of the five western scenarios.

References

1. Energy Information Administration (EIA). Energy Information Administration Gulf of Mexico Fact Sheet. Available online: https://www.eia.gov/special/gulf_of_mexico/ (accessed on 15 February 2018).
2. Louisiana State University (LSU). Gulf Coast Energy Outlook. Available online: https://www.lsu.edu/ces/publications/2017/GCEO2017.pdf (accessed on 25 February 2018).
3. Kazanis, E.; Maclay, D.; Shepard, N. *Estimated Oil and Gas Reserves Gulf of Mexico OCS Region December 31, 2013*; Bureau of Ocean Energy Management (BOEM) Report; BOEM: New Orleans, LA, USA, 2015.
4. Sovacool, B.K. Solving the oil independence problem: Is it possible? *Energy Policy* **2007**, *35*, 5505–5514. [CrossRef]
5. Peterson, C.H.; Rice, S.D.; Short, J.W.; Esler, D.; Bodkin, J.L.; Ballachey, B.E.; Irons, D.B. Long-term ecosystem response to the Exxon Valdez oil spill. *Science* **2003**, *302*, 2082–2086. [CrossRef] [PubMed]
6. Silliman, B.R.; van de Koppel, J.; McCoy, M.W.; Diller, J.; Kasozi, G.N.; Earl, K.; Adams, P.N.; Zimmerman, A.R. Degradation and resilience in Louisiana salt marshes after the BP-Deepwater Horizon oil spill. *Proc. Natl. Acad. Sci. USA* **2012**, *109*, 11234–11239. [CrossRef] [PubMed]
7. Graham, B.; Reilly, W.K.; Beinecke, F.; Boesch, D.F.; Garcia, T.D.; Murray, C.A.; Ulmer, F. *Deep Water: The Gulf Oil Disaster and the Future of Offshore Drilling: Report to the President*; National Commission on the BP Deepwater Horizon Oil Spill and Offshore Drilling; U.S. Government Publishing Office: Washington, DC, USA, 2011.
8. Sumaila, U.R.; Cisneros-Montemayor, A.M.; Dyck, A.; Huang, L.; Cheung, W.; Jacquet, J.; Kleisner, K.; Lam, V.; McCrea-Strub, A.; Swartz, W.; et al. Impact of the Deepwater Horizon well blowout on the economics of US Gulf fisheries. *Can. J. Fish. Aquat. Sci.* **2012**, *69*, 499–510. [CrossRef]
9. Smith, L.C.; Smith, M.; Ashcroft, P. Analysis of environmental and economic damages from British Petroleum's Deepwater Horizon oil spill. *Albany Law Rev.* **2011**, *74*, 563–585. [CrossRef]
10. Court, C.D.; Hodges, A.W.; Clouser, R.L.; Larkin, S.L. Economic impacts of cancelled recreational trips to Northwest Florida after the Deepwater Horizon oil spill. *Reg. Sci. Policy Pract.* **2017**, *9*, 143–164. [CrossRef]
11. Mufson, S. Shell's Brutus production platform spills oil into Gulf of Mexico. *Washington Post*, 13 May 2016.
12. Grant, N. Gulf Coast Oil Spill May Be Largest Since 2010 BP Disaster. *Bloomberg*, 16 October 2017.
13. DOI Secretary Zinke Announces Plan For Unleashing America's Offshore Oil and Gas Potential. Available online: https://www.doi.gov/pressreleases/secretary-zinke-announces-plan-unleashing-americas-offshore-oil-and-gas-potential (accessed on 2 February 2018).
14. Tabuchi, H. Trump Administration Drops Florida From Offshore Drilling Plan. *New York Times*, 9 January 2018.
15. Sissine, F. *Energy Independence and Security Act of 2007: A Summary of Major Provisions*; Congressional Research Service (Library of Congress): Washington, DC, USA, 2007.
16. Dixon, M.; Ritchie, B. GOP congressmen say Ryan to push permanent moratorium on eastern Gulf drilling. *Politico*, 8 January 2018.
17. Nelson, J.R.; Grubesic, T.H. Oil spill modeling: Risk, spatial vulnerability, and impact assessment. *Prog. Phys. Geogr.* **2017**, *42*, 112–127. [CrossRef]
18. Olita, A.; Cucco, A.; Simeone, S.; Ribotti, A.; Fazioli, L.; Sorgente, B.; Sorgente, R. Oil spill hazard and risk assessment for the shorelines of a Mediterranean coastal archipelago. *Ocean Coast. Manag.* **2012**, *57*, 44–52. [CrossRef]
19. Madrid, J.A.J.; García-Ladona, E.; Blanco-Meruelo, B. *Oil Spill Beaching Probability for the Mediterranean Sea*; The Handbook of Environmental Chemistry: Berlin, Germany, 2016; pp. 1–20.
20. Canu, D.M.; Solidoro, C.; Bandelj, V.; Quattrocchi, G.; Sorgente, R.; Olita, A.; Fazioli, L.; Cucco, A. Assessment of oil slick hazard and risk at vulnerable coastal sites. *Mar. Pollut. Bull.* **2015**, *94*, 84–95. [CrossRef] [PubMed]
21. Castanedo, S.; Juanes, J.A.; Medina, R.; Puente, A.; Fernandez, F.; Olabarrieta, M.; Pombo, C. Oil spill vulnerability assessment integrating physical, biological and socio-economical aspects: Application to the Cantabrian coast (Bay of Biscay, Spain). *J. Environ. Manag.* **2009**, *91*, 149–159. [CrossRef] [PubMed]
22. Lan, D.; Liang, B.; Bao, C.; Ma, M.; Xu, Y.; Yu, C. Marine oil spill risk mapping for accidental pollution and its application in a coastal city. *Mar. Pollut. Bull.* **2015**, *96*, 220–225. [CrossRef] [PubMed]
23. Lee, M.; Jung, J.-Y. Pollution risk assessment of oil spill accidents in Garorim Bay of Korea. *Mar. Pollut. Bull.* **2015**, *100*, 297–303. [CrossRef] [PubMed]

24. Nelson, J.; Grubesic, T.; Sim, L.; Rose, K.; Graham, J. Approach for assessing coastal vulnerability to oil spills for prevention and readiness using GIS and the Blowout and Spill Occurrence Model. *Ocean Coast. Manag.* **2015**, *112*, 1–11. [CrossRef]
25. French McCay, D.; Rowe, J.J.; Whittier, N.; Sankaranarayanan, S.; Schmidt Etkin, D. Estimation of potential impacts and natural resource damages of oil. *J. Hazard. Mater.* **2004**, *107*, 11–25. [CrossRef] [PubMed]
26. Cama, T. *Zinke Talks with More Governors about Offshore Drilling Plan*; The Hill: Washington, DC, USA, 2018; Available online: http://thehill.com/policy/energy-environment/368813-zinke-talks-with-more-governors-about-offshore-drilling-plan (accessed on 5 February 2018).
27. Klein, Y.L.; Osleeb, J.P.; Viola, M.R. Tourism-generated earnings in the coastal zone: A regional analysis. *J. Coast. Res.* **2004**, *20*, 1080–1088. [CrossRef]
28. FLGov Governor Scott Applauds Florida's Tourism Marketing. Available online: https://www.flgov.com/governor-scott-applauds-floridas-tourism-marketing-2/ (accessed on 25 February 2018).
29. Pforr, C. Crisis management in tourism: A review of the emergent literature. In *Crisis Management in the Tourism Industry: Beating the Odds*; Pforr, C., Hosie, P., Eds.; Ashgate: Burlington, NJ, USA, 2009; pp. 37–52.
30. Ritchie, B.W. Chaos, crises and disasters: A strategic approach to crisis management in the tourism industry. *Tour. Manag.* **2004**, *25*, 669–683. [CrossRef]
31. Boufadel, M.C.; Abdollahi-Nasab, A.; Geng, X.; Galt, J.; Torlapati, J. Simulation of the landfall of the deepwater horizon oil on the shorelines of the Gulf of Mexico. *Environ. Sci. Technol.* **2014**, *48*, 9496–9505. [CrossRef] [PubMed]
32. Camilli, R.; Reddy, C.M.; Yoerger, D.R.; Van Mooy, B.A.S.; Jakuba, M.V.; Kinsey, J.C.; McIntyre, C.P.; Sylva, S.P.; Maloney, J.V. Tracking hydrocarbon plume transport and biodegradation at Deepwater Horizon. *Science* **2010**, *330*, 201–204. [CrossRef] [PubMed]
33. Passow, U.; Ziervogel, K.; Asper, V.; Diercks, A. Marine snow formation in the aftermath of the Deepwater Horizon oil spill in the Gulf of Mexico. *Environ. Res. Lett.* **2012**, *7*, 35301. [CrossRef]
34. Montagna, P.A.; Baguley, J.G.; Cooksey, C.; Hartwell, I.; Hyde, L.J.; Hyland, J.L.; Kalke, R.D.; Kracker, L.M.; Reuscher, M.; Rhodes, A.C.E. Deep-sea benthic footprint of the Deepwater Horizon blowout. *PLoS ONE* **2013**, *8*, e70540. [CrossRef] [PubMed]
35. Felder, D.L.; Thoma, B.P.; Schmidt, W.E.; Sauvage, T.; Self-Krayesky, S.L.; Chistoserdov, A.; Bracken-Grissom, H.D.; Fredericq, S. Seaweeds and decapod crustaceans on Gulf deep banks after the Macondo Oil Spill. *Bioscience* **2014**, *64*, 808–819. [CrossRef]
36. Fisher, C.R.; Demopoulos, A.W.J.; Cordes, E.E.; Baums, I.B.; White, H.K.; Bourque, J.R. Coral communities as indicators of ecosystem-level impacts of the Deepwater Horizon spill. *Bioscience* **2014**, *64*, 796–807. [CrossRef]
37. Etnoyer, P.J.; Wickes, L.N.; Silva, M.; Dubick, J.D.; Balthis, L.; Salgado, E.; MacDonald, I.R. Decline in condition of gorgonian octocorals on mesophotic reefs in the northern Gulf of Mexico: Before and after the Deepwater Horizon oil spill. *Coral Reefs* **2016**, *35*, 77–90. [CrossRef]
38. Dubansky, B.; Whitehead, A.; Miller, J.T.; Rice, C.D.; Galvez, F. Multitissue molecular, genomic, and developmental effects of the Deepwater Horizon oil spill on resident Gulf killifish (*Fundulus grandis*). *Environ. Sci. Technol.* **2013**, *47*, 5074–5082. [CrossRef] [PubMed]
39. Tran, T.; Yazdanparast, A.; Suess, E.A. Effect of oil spill on birds: A graphical assay of the deepwater horizon oil spill's impact on birds. *Comput. Stat.* **2014**, *29*, 133–140. [CrossRef]
40. Haney, J.C.; Geiger, H.J.; Short, J.W. Bird mortality from the Deepwater Horizon oil spill. II. Carcass sampling and exposure probability in the coastal Gulf of Mexico. *Mar. Ecol. Prog. Ser.* **2014**, *513*, 239–252. [CrossRef]
41. Antonio, F.J.; Mendes, R.S.; Thomaz, S.M. Identifying and modeling patterns of tetrapod vertebrate mortality rates in the Gulf of Mexico oil spill. *Aquat. Toxicol.* **2011**, *105*, 177–179. [CrossRef] [PubMed]
42. Schwacke, L.H.; Smith, C.R.; Townsend, F.I.; Wells, R.S.; Hart, L.B.; Balmer, B.C.; Collier, T.K.; De Guise, S.; Fry, M.M.; Guillette, L.J., Jr. Health of common bottlenose dolphins (*Tursiops truncatus*) in Barataria Bay, Louisiana, following the Deepwater Horizon oil spill. *Environ. Sci. Technol.* **2013**, *48*, 93–103. [CrossRef] [PubMed]
43. Carmichael, R.H.; Graham, W.M.; Aven, A.; Worthy, G.; Howden, S. Were multiple stressors a "perfect storm" for northern Gulf of Mexico bottlenose dolphins (*Tursiops truncatus*) in 2011? *PLoS ONE* **2012**, *7*, e41155. [CrossRef] [PubMed]

44. McCrea-Strub, A.; Kleisner, K.; Sumaila, U.R.; Swartz, W.; Watson, R.; Zeller, D.; Pauly, D. Potential Impact of the Deepwater Horizon Oil Spill on Commercial Fisheries in the Gulf of Mexico. *Fisheries* **2011**, *36*, 332–336. [CrossRef]

45. Bishop, R.C.; Boyle, K.J.; Carson, R.T.; Chapman, D.; Hanemann, W.M.; Kanninen, B.; Kopp, R.J.; Krosnick, J.A.; List, J.; Meade, N.; et al. Putting a value on injuries to natural assets: The BP oil spill. *Science* **2017**, *356*, 253–254. [CrossRef] [PubMed]

46. Duran, R.; Beron-Vera, F.J.; Olascoaga, M.J. Extracting quasi-steady Lagrangian transport patterns from the ocean circulation: An application to the Gulf of Mexico. *Sci. Rep.* **2018**, *8*. [CrossRef] [PubMed]

47. Wang, S.D.; Shen, Y.M.; Zheng, Y.H. Two-dimensional numerical simulation for transport and fate of oil spills in seas. *Ocean Eng.* **2005**, *32*, 1556–1571. [CrossRef]

48. White, H.K.; Hsing, P.-Y.; Cho, W.; Shank, T.M.; Cordes, E.E.; Quattrini, A.M.; Nelson, R.K.; Camilli, R.; Demopoulos, A.W.J.; German, C.R. Impact of the Deepwater Horizon oil spill on a deep-water coral community in the Gulf of Mexico. *Proc. Natl. Acad. Sci. USA* **2012**, *109*, 20303–20308. [CrossRef] [PubMed]

49. Grubesic, T.H.; Wei, R.; Nelson, J. Optimizing oil spill cleanup efforts: A tactical approach and evaluation framework. *Mar. Pollut. Bull.* **2017**, *125*, 318–329. [CrossRef] [PubMed]

50. Jensen, J.R.; Ramsey, E.W., III; Holmes, J.M.; Michel, J.E.; Savitsky, B.; Davis, B.A. Environmental sensitivity index (ESI) mapping for oil spills using remote sensing and geographic information system technology. *Int. J. Geogr. Inf. Syst.* **1990**, *4*, 181–201. [CrossRef]

51. Romero, A.F.; Abessa, D.M.S.; Fontes, R.F.C.; Silva, G.H. Integrated assessment for establishing an oil environmental vulnerability map: Case study for the Santos Basin region, Brazil. *Mar. Pollut. Bull.* **2013**, *74*, 156–164. [CrossRef] [PubMed]

52. Carmona, A.S.L.; Gherardi, D.F.M.; Tessler, M.G.; Carmonaf, S.L. Environment Sensitivity Mapping and Vulnerability Modeling for Oil Spill Response along the São Paulo State Coastline. *J. Coast. Res.* **2006**, 1455–1458.

53. Fattal, P.; Maanan, M.; Tillier, I.; Rollo, N.; Robin, M.; Pottier, P. Coastal Vulnerability to Oil Spill Pollution: The Case of Noirmoutier Island (France). *J. Coast. Res.* **2010**, 879–887. [CrossRef]

54. Boer, S.; Azevedo, A.; Vaz, L.; Costa, R.; Fortunato, A.B.; Oliveira, A.; Tomás, L.M.; Dias, J.M.; Rodrigues, M. Development of an oil spill hazard scenarios database for risk assessment. *J. Coast. Res.* **2014**, *70*, 539–544. [CrossRef]

55. Guillen, G.; Rainey, G.; Morin, M. A simple rapid approach using coupled multivariate statistical methods, GIS and trajectory models to delineate areas of common oil spill risk. *J. Mar. Syst.* **2004**, *45*, 221–235. [CrossRef]

56. Fernández-Macho, J. Risk assessment for marine spills along European coastlines. *Mar. Pollut. Bull.* **2016**, *113*, 200–210. [CrossRef] [PubMed]

57. Azevedo, A.; Fortunato, A.B.; Epifânio, B.; den Boer, S.; Oliveira, E.R.; Alves, F.L.; de Jesus, G.; Gomes, J.L.; Oliveira, A. An oil risk management system based on high-resolution hazard and vulnerability calculations. *Ocean Coast. Manag.* **2017**, *136*, 1–18. [CrossRef]

58. BOEM Offshore Statistics by Water Depth. Available online: https://www.data.boem.gov/Leasing/OffshoreStatsbyWD/Default.aspx (accessed on 5 December 2017).

59. BOEM Official Protraction Diagrams (OPDs) and Leasing Maps (LMs) & Supplemental Official OCS Block Diagrams (SOBDs). Available online: https://www.boem.gov/Official-Protraction-Diagrams/ (accessed on 2 December 2017).

60. Sturges, W.; Leben, R. Frequency of ring separations from the Loop Current in the Gulf of Mexico: A revised estimate. *J. Phys. Oceanogr.* **2000**, *30*, 1814–1819. [CrossRef]

61. Socolofsky, S.A.; Adams, E.E.; Boufadel, M.C.; Aman, Z.M.; Johansen, Ø.; Konkel, W.J.; Lindo, D.; Madsen, M.N.; North, E.W.; Paris, C.B. Intercomparison of oil spill prediction models for accidental blowout scenarios with and without subsea chemical dispersant injection. *Mar. Pollut. Bull.* **2015**, *96*, 110–126. [CrossRef] [PubMed]

62. Sim, L.; Graham, J.; Rose, K.; Duran, R.; Nelson, J.; Umhoefer, J.; Vielma, J. *Developing a Comprehensive Deepwater Blowout and Spill Model*; 2015 NETLTRS-9-2015; EPAct Technical Report Series; U.S. Department of Energy; National Energy Technology Laboratory: Albany, OR, USA, 2015; p. 44.

63. NOAA Naval Oceanographic Office Regional Navy Coastal Ocean Model (NCOM). Available online: https://www.ncdc.noaa.gov/data-access/model-data/model-datasets/navoceano-ncom-reg (accessed on 2 January 2018).

64. Network Common Data Form (NetCDF). NetCDF: Introduction and Overview. Available online: http://www.unidata.ucar.edu/software/netcdf/docs/index.html (accessed on 20 December 2017).

65. Lehr, W.J.; Cekirge, H.M.; Fraga, R.J.; Belen, M.S. Empirical studies of the spreading of oil spills. *Oil Petrochem. Pollut.* **1984**, *2*, 7–11. [CrossRef]

66. Rasmussen, D. Oil spill modeling—A tool for cleanup operations. In *International Oil Spill Conference*; American Petroleum Institute: Washing, DC, USA, 1985; Volume 1985, pp. 243–249.

67. Mackay, D. Calculation of the evaporation rate of volatile liquids. In Proceedings of the National Conference on Control of Hazardous Material Spills, Louisville, KY, USA, 13–15 May 1980.

68. Tkalin, A.V. Evaporation of petroleum hydrocarbons from films on a smooth sea surface. *Oceanol. Acad. Sci. USSR* **1986**, *26*, 473–474.

69. Kankara, R.S.; Arockiaraj, S.; Prabhu, K. Environmental sensitivity mapping and risk assessment for oil spill along the Chennai Coast in India. *Mar. Pollut. Bull.* **2016**, *106*, 95–103. [CrossRef] [PubMed]

70. Banerjee, T.; Gollub, J.O. The Public View of the Coast: Toward Aesthetic Indicators for Coastal Planning and Management. In *Behavioral Basis of Design*; Suedfeld, P., Russell, J.A., Eds.; ERDA Inc. and Russell Dowden, Hutchinson and Ross Inc.: Stroudsburg, PA, USA, 1976; pp. 115–122.

71. Brody, S.D.; Grover, H.; Bernhardt, S.; Tang, Z.; Whitaker, B.; Spence, C. Identifying potential conflict associated with oil and gas exploration in Texas state coastal waters: A multicriteria spatial analysis. *Environ. Manag.* **2006**, *38*, 597–617. [CrossRef] [PubMed]

72. Costanza, R.; d'Arge, R.; De Groot, R.; Farber, S.; Grasso, M.; Hannon, B.; Limburg, K.; Naeem, S.; O'neill, R.V.; Paruelo, J. The value of the world's ecosystem services and natural capital. *Nature* **1997**, *387*, 253–260. [CrossRef]

Journal of
Marine Science and Engineering

MDPI

Article

Simulation of the 2003 Foss Barge - Point Wells Oil Spill: A Comparison between BLOSOM and GNOME Oil Spill Models

Rodrigo Duran [1,2,*], Lucy Romeo [1,3], Jonathan Whiting [4], Jason Vielma [1,5], Kelly Rose [1], Amoret Bunn [4] and Jennifer Bauer [1]

[1] National Energy Technology Laboratory, Albany, OR 97321, USA; Lucy.Romeo@netl.doe.gov (L.R.); jason.mv81@gmail.com (J.V.); Kelly.Rose@netl.doe.gov (K.R.); Jennifer.Bauer@netl.doe.gov (J.B.)
[2] Theiss Research, San Diego, CA 92037, USA
[3] AECOM, South Park, PA 15129, USA
[4] Pacific Northwest National Laboratory, Richland, WA 99354, USA; jonathan.whiting@pnnl.gov (J.W.); amoret.bunn@pnnl.gov (A.B.)
[5] Oak Ridge Institute for Science and Education, Oak Ridge, TN 37831, USA
* Correspondence: r.duran@theissresearch.org; Tel.: +1-574-387-2677

Received: 26 July 2018; Accepted: 29 August 2018; Published: 11 September 2018

check for updates

Abstract: The Department of Energy's (DOE's) National Energy Technology Laboratory's (NETL's) Blowout and Spill Occurrence Model (BLOSOM), and the National Oceanic and Atmospheric Administration's (NOAA's) General NOAA Operational Modeling Environment (GNOME) are compared. Increasingly complex simulations are used to assess similarities and differences between the two models' components. The simulations presented here are forced by ocean currents from a Finite Volume Community Ocean Model (FVCOM) implementation that has excellent skill in representing tidal motion, and with observed wind data that compensates for a coarse vertical ocean model resolution. The comprehensive comparison between GNOME and BLOSOM presented here, should aid modelers in interpreting their results. Beyond many similarities, aspects where both models are distinct are highlighted. Some suggestions for improvement are included, e.g., the inclusion of temporal interpolation of the forcing fields (BLOSOM) or the inclusion of a deflection angle option when parameterizing wind-driven processes (GNOME). Overall, GNOME and BLOSOM perform similarly, and are found to be complementary oil spill models. This paper also sheds light on what drove the historical Point Wells spill, and serves the additional purpose of being a learning resource for those interested in oil spill modeling. The increasingly complex approach used for the comparison is also used, in parallel, to illustrate the approach an oil spill modeler would typically follow when trying to hindcast or forecast an oil spill, including detailed technical information on basic aspects, like choosing a computational time step. We discuss our successful hindcast of the 2003 Point Wells oil spill that, to our knowledge, had remained unexplained. The oil spill models' solutions are compared to the historical Point Wells' oil trajectory, in time and space, as determined from overflight information. Our hindcast broadly replicates the correct locations at the correct times, using accurate tide and wind forcing. While the choice of wind coefficient we use is unconventional, a simplified analytic model supported by observations, suggests that it is justified under this study's circumstances. We highlight some of the key oceanographic findings as they may relate to other oil spills, and to the regional oceanography of the Salish Sea, including recommendations for future studies.

Keywords: oil spill model; ocean trajectory; GNOME; BLOSOM; Salish Sea; Point Wells; Foss Barge; hindcast; windage; model comparison

1. Introduction

1.1. Oil Spill Models

The Department of Energy's National Energy Technology Laboratory (NETL) created BLOSOM, a comprehensive modeling suite that follows the fate and transport of both subsurface oil blowouts and surface spills. The National Oceanic and Atmospheric Administration's (NOAA's) Office of Response and Restoration's (OR&R) Emergency Response Division built GNOME to predict the potential trajectory of offshore pollutants at the sea surface. BLOSOM and GNOME have been utilized by government and industry to simulate the fate of hydrocarbon after release events. This study compares how the models handle a sea-surface spill. A comparison comprising many of the well-known models, and focused on a blowout and the use of subsea dispersants, can be found in Socolofsky et al. [1].

Both models use ocean currents, typically from ocean models, and wind, from atmospheric models or weather stations, to force the movement of oil at the sea surface. BLOSOM and the latest version of GNOME are also able to simulate a fully three-dimensional spill or blowout over time. Dispersion due to processes not captured through integrating ocean currents and wind, is simulated with a stochastic component, typically a random walk.

Although evaporation can be significant over short time periods when the crude is light, simulated trajectories are not affected by evaporation or any other type of weathering. The primary focus of this study was placed on model trajectories rather than weathering.

1.2. Blowout and Spill Occurrence Model

With the Deepwater Horizon incident, the need for an open source model capable of simulating the fate and transport of oil, from source to sink, throughout the water column became apparent. BLOSOM was designed to fulfill this need.

BLOSOM is an integrated, 4-dimensional model that enables users to simulate the fate, transport, and degradation of both subsurface oil blowouts as well as surface spills. Originally designed to handle deepwater blowouts, such as Deepwater Horizon, BLOSOM is the first to be designed as an open source 4-D hydrocarbon fate and transport model. BLOSOM offers users a flexible modeling suite written in Java [2] and rewritten in C++; this comparison was completed using the C++ version. Built to aid in the prediction, prevention, and preparation with both subsurface blowouts and surface spills, the C++ version of BLOSOM is available both as an online tool, and as a desktop tool [3]. In addition, the full source code of BLOSOM is available to the public. BLOSOM has the capability of implementing multiphase hydrocarbon releases from the seafloor. Applying buoyant jet plume dynamics to the release, BLOSOM tracks the location and characteristics of both the plume and each oil parcel throughout the water column.

Since initial development and internal release in 2012, both the jet/plume component, as well as overall outputs of BLOSOM, have been internally and externally evaluated, as explained in a technical report [2]. The jet plume component of BLOSOM has been evaluated against field experiments, which were conducted in the North Sea [4,5]. In addition, BLOSOM participated in a model comparison study where outputs from multiple blowout models predicted the effect of subsea dispersant application on subsurface plumes, which was sponsored by the American Petroleum Institute (API) and co-sponsored by the BP/Gulf of Mexico Research Initiative [1]. BLOSOM results were also evaluated by comparing simulation outputs to historical spill overflight data. For example, results from a BLOSOM simulation, using hydrodynamic data for conditions during the Deepwater Horizon, were compared to spill extent data provided by NOAA's Experimental Marine Pollution Surveillance Report (EMPSR) [2].

An online version of BLOSOM is privately available, along with other NETL tools, through a common operating platform hosted by NETL. The common operating platform is set for public release in the future, but accessibility can currently be granted by requesting the use of BLOSOM through NETL.

As of 2018, BLOSOM integrates the ADIOS oil library [6], also known as NOAA's OilLibrary (https://github.com/NOAA-ORR-ERD/OilLibrary).

J. Mar. Sci. Eng. **2018**, *6*, 104

1.3. General NOAA Operational Modeling Environment

GNOME is the modeling tool developed and used by the NOAA Office of Response and Restoration's (OR&R) Emergency Response Division to predict the possible trajectory a pollutant might follow in, or on, a body of water. The first version of GNOME was released on 16 March 1999 as a replacement to the On-Scene Spill Model (OSSM) that had been used by the NOAA Emergency Response Division since 1979 [7]. For nearly 20 years, GNOME has been used by emergency responders on behalf of industry, government, and organizations to track oil spills, chemical spills, marine debris, and more. Designed to be a multipurpose trajectory model used by both experts and the public, GNOME has different modes that allow either greater control or a more simplified interface, respectively. Catering to ease of use, there is a suite of location files that provide generalized information about the tides, currents, and shorelines in the region it covers, allowing the user to quickly run example scenarios in a region [8]. During an oil spill, NOAA simulates the trajectory with the best available information and then corrects the trajectory from overhead flight observations at different intervals throughout the spill, guiding clean-up efforts and preventative measures.

At a fundamental level, GNOME uses winds, currents, and diffusion to move particles and generate a predicted trajectory, or "best guess". Uncertainty can be specified for each input, creating a "minimum regret" solution, representing other possibilities where the spill might go [9]. GNOME also predicts rudimentary weathering, where the spilled oil undergoes chemical and physical changes, though NOAA provides another tool called Automated Data Inquiry for Oil Spills (ADIOS) that has better evaporation and oil fate estimates from an extensive oil library. It should also be noted that NOAA is currently developing the next generation of GNOME, adding features such as a web interface for location files, three-dimensional plus time deep-water blowout capabilities, integration with ADIOS weathering, and enhanced output interaction with GIS. The latest version of GNOME is a full suite of modeling tools (https://response.restoration.noaa.gov/oil-and-chemical-spills/oil-spills/response-tools/gnome-suite-oil-spill-modeling.html). Before this suite, the rest of the tools used to be available separately from GNOME. A release of GNOME called PyGNOME is also available at https://github.com/NOAA-ORR-ERD/PyGnome. This version includes improvements in several algorithms including interpolation and integration. There is also a web configuration (WebGNOME) available at http://gnome.orr.noaa.gov/.

During the model comparison, GNOME version 1.3.9 was used (the most recent officially released version), and run in diagnostic mode to maximize control over inputs. Currents and boundary information were imported from a hydrodynamic model, rather than by loading a location file.

1.4. Foss Barge—Point Wells Oil Spill

The Foss Barge—Point Wells Spill was the basis of a comparison between two offshore spill trajectory models: BLOSOM and GNOME.

At 00:05 on 30 December 2003, heavy marine fuel oil #6 (IFO 380) spilled into the Puget Sound as it was pumped onto Foss tank barge 248-P2, from a Chevron/Texaco loading terminal at the Point Wells Asphalt facility near Richmond Beach in Shoreline, Washington. Based on gauge readings, 5712 gallons of fuel were accidentally released by overtopping; around 1075 gallons were recovered from the deck, and an estimated 4600–4800 gallons (about 110–114 bbl) spilled into the Puget Sound [10,11]. A timeline of oil spill movement is provided in Figure 1. A map showing the area of interest in the Salish Sea is provided in Figure 2, and the approximate spill path, including the points of interest mentioned in Figure 1 and in this section, are depicted in Figure 3.

Initially, oil drifted approximately 6 miles south along the eastern shore of Puget Sound, then began moving northwest towards the western shore. By 09:00, oil was observed within a mile from Port Madison, and was observed entering Port Madison 3 h later [10]. By the afternoon of 30 December, some of the oil slick beached between Point Jefferson and Indianola, and beaching increased by the morning of 31 December [10]. The oiled shoreline extended approximately 1.5 miles; an initial shoreline oiling survey estimated 3.5 acres were covered. An assessment by the United States Coast Guard

(USCG), characterized over 68% of these 3.5 acres as heavily oiled. The heavily oiled shoreline included the upper and middle intertidal zone of the Doe-Kag-Wats marsh. No evidence was found of oiling on the eastern Puget Sound shoreline [11].

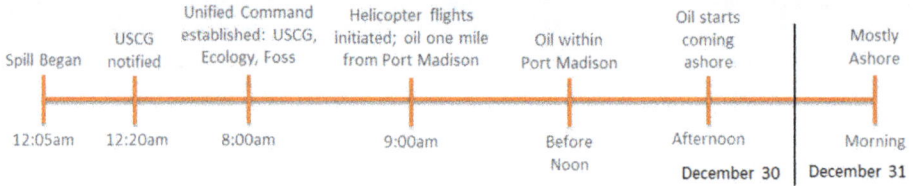

Figure 1. Approximate timeline of events, as recorded in ENTRIX 2005.

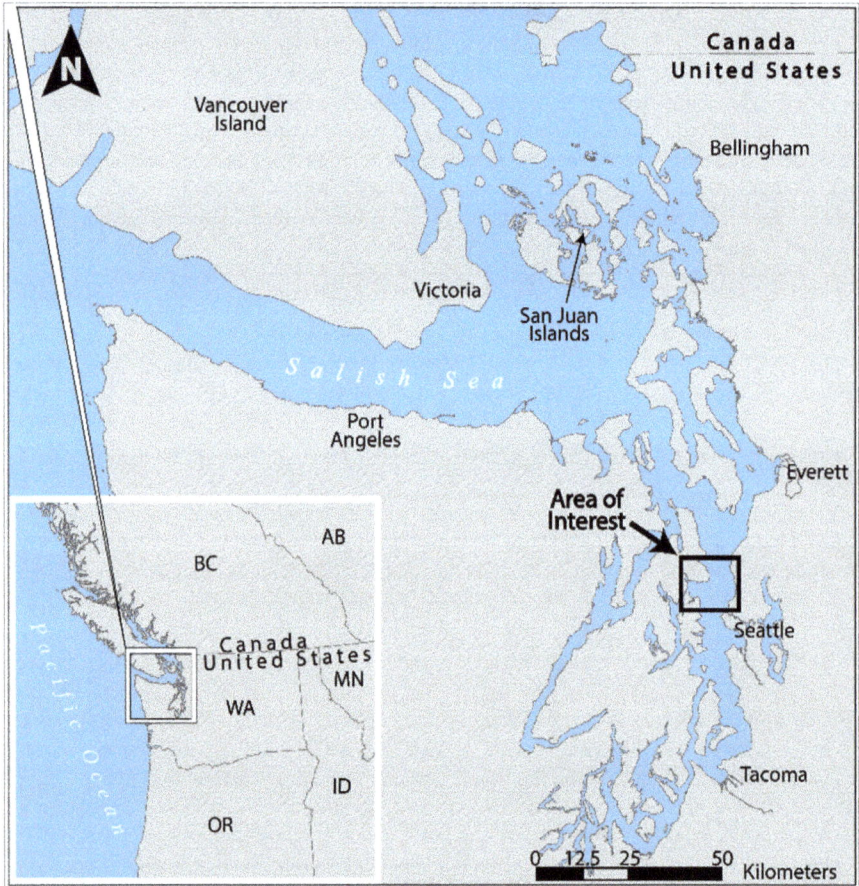

Figure 2. Maps showing the study area in the Salish Sea and surrounding areas.

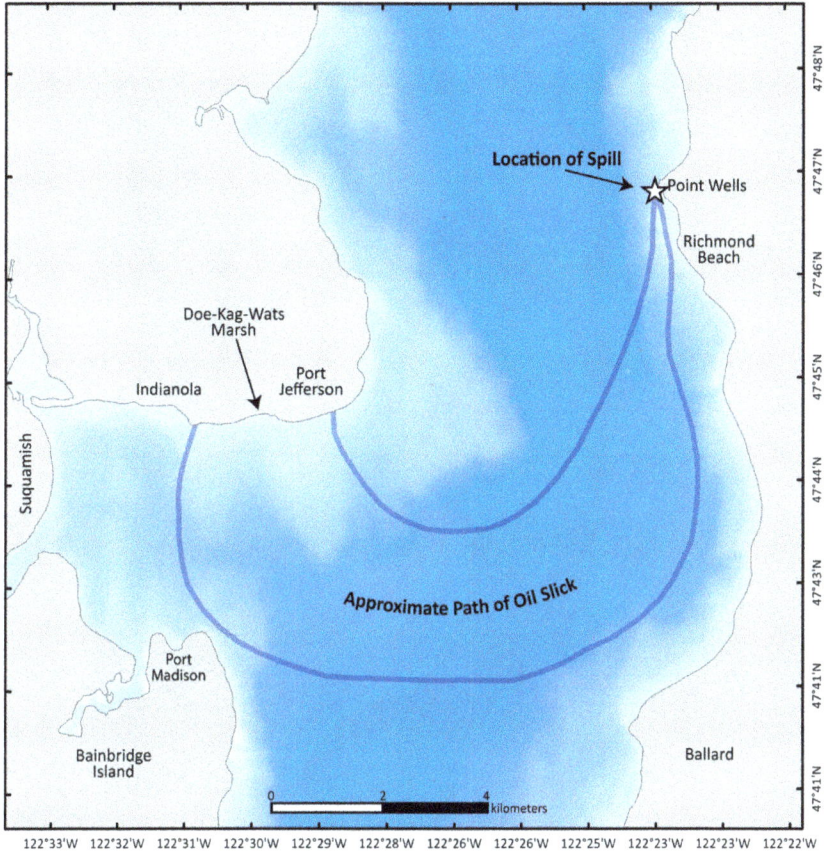

Figure 3. Project area map showing the approximate location and path of the oil surface slick, based on data recorded by ENTRIX, Inc., [10]; see area of interest in Figure 2.

According to the timeline presented in a report prepared by ENTRIX, Inc., [10] for the Natural Resource Damage Trustees, response to the spill began within an hour through the National Response Corporation (NRC) (Figure 1). By 02:00, the USCG and Washington Department of Ecology (WDOE) were in route, and by 06:00, some sensitive shoreline areas were boomed, and a Unified Command was created at the source location by the USCG, WDOE, and the responsible party, the Foss Maritime Company. By 09:00, overflight information was collected by the WDOE and Washington Department of Fish and Wildlife (WDFW). As shown in Figure 4, overflight information prepared by NOAA, and collected by NOAA and WDFW, illustrate some of the temporal variability through snapshots of the visible surface slicks, from 11:00 on 30 December 2003 to 10:55 on 31 December 2003. Overall response efforts included booming, cleanup, overflight data collection, and trajectory predictions. Reported agencies and corporations involved in these efforts include NRC, USCG, WDOE, WDFW, the Suquamish Tribe, Kitsap County, NOAA, United States Department of the Interior and Fish and Wildlife Service, Washington Department of Natural Resources, and the Foss Maritime Company.

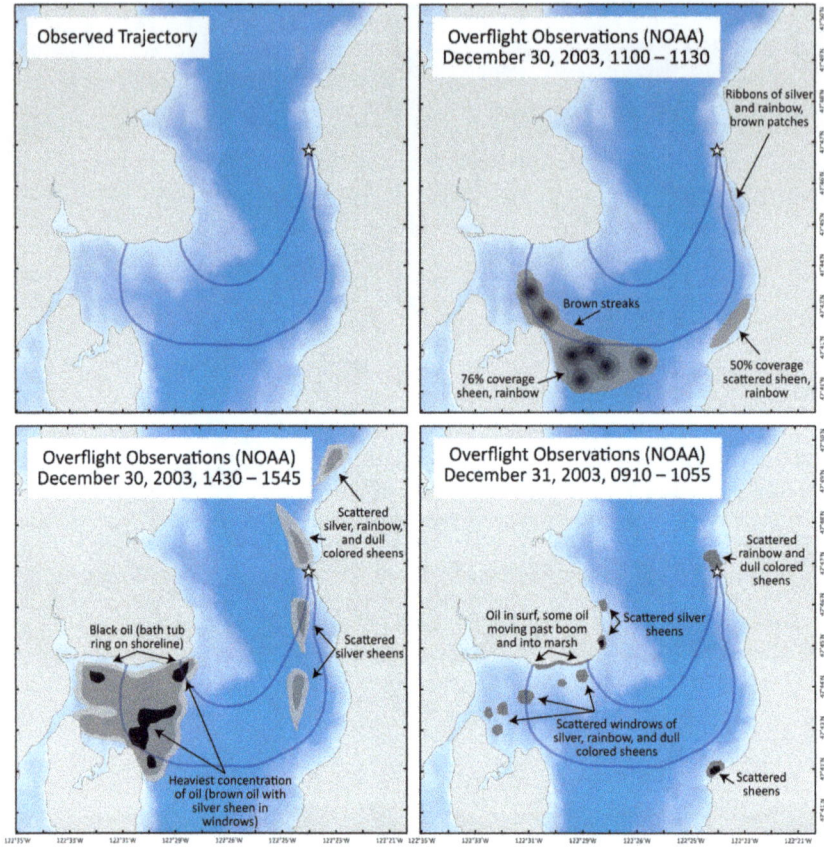

Figure 4. A schematic of the observed oil path (top left), and a time series of digitized maps from National Oceanic and Atmospheric Administration's (NOAA's) overflight observation records.

2. Data and Methods

2.1. Ambient Forcing for Oil Spill Models—The Hydrodynamic Model

One of the most challenging aspects of oil spill modeling is accurately simulating ocean currents. In 2011, the Pacific Northwest National Laboratory developed a hydrodynamic model of Puget Sound known as the Salish Sea Model. This model has been used as a tool for coastal estuarine research, nearshore restoration planning [12], water-quality management [13], and assessment of climate change effects [14,15]. The Salish Sea Model is one of the most detailed hydrodynamic models for the region, carefully calibrated for circulation and tidal exchange to primarily address water quality concerns, and thus, is a reasonable choice for oil spill modeling.

The Salish Sea Model was developed using the Finite Volume Community Ocean Model (FVCOM) framework [16]. The unstructured finite volume model grid extends to the west through the Strait of Juan de Fuca, to the north through Georgia Strait, thus covering all areas of the Puget Sound (Figure 5). Grid cells vary in size between 250 m within inlets and bays, to 3.5 km in the Juan de Fuca Strait. In the vertical, the model uses 10 terrain-following layers (11 sigma levels) with higher density near the surface, and the top layer is about 3.2% of the local depth. Bathymetry is obtained from the University of Washington's Digital Elevation Model (DEM) [17] and the Department of Fisheries and Oceans Canada. The model bathymetry is typically smoothed for water quality simulations to prevent pressure gradient errors in a sigma-stretched coordinate system [18]. However, for applications such as oil spill transport, the model was operated without smoothing to account for the effects of shallow depths along complex shoreline. Open boundary conditions at the Strait of Juan de Fuca and Georgia Strait are forced by 15 min XTide stations, a harmonic tide predictor based on NOAA's National Oceanic Service algorithms [19]. There are additional water inputs throughout the domain including: (a) 19 major rivers with flows determined by USGS gages, (b) estimated runoff from 45 watersheds, and (c) wastewater discharge from 99 industrial outfalls. The Weather Research and Forecasting (WRF) Model on a 12 km grid [20] was used for meteorological forcing. It should be noted that wind from WRF was only used for the hydrodynamic model, while local wind measurements were used to force the oil spill models. The wind with which the ocean model was forced was not used to drive trajectories, as the oil spill happened in a relatively small area (oil beached about 8 km away from where the spill originated) relative to the atmospheric model resolution. Also, wind observations are likely to be an accurate representation of the wind acting on the sea surface during the spill, as we show below.

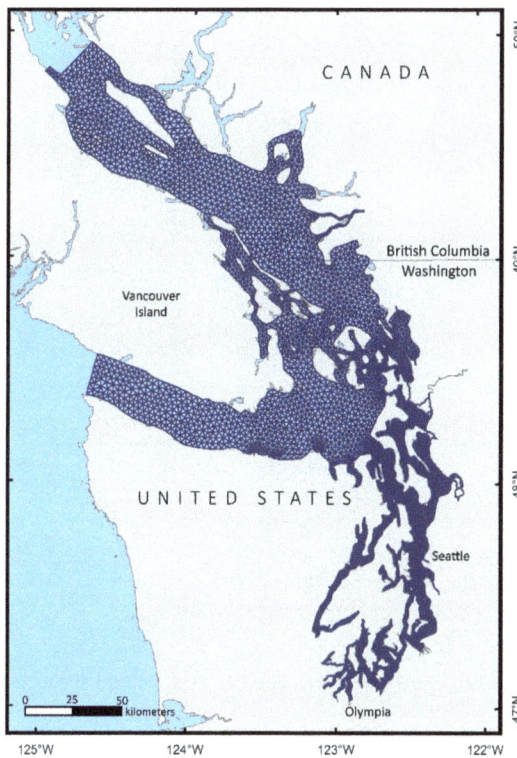

Figure 5. Extent of coverage for the Salish Sea Model.

The Salish Sea Model was run for the entire year of 2003, plus an additional 5 days in 2004, though only a few days near the end of December were used in the oil spill modeling. Figure 6 shows an evaluation of the model outputs between 29 December 2003 and 3 January 2004, compared to an XTide station located at Port Jefferson (47.745 N, 122.4767 W), near the oil trajectory. Table 1 shows the error statistics associated with the same five days; error statistics are deemed to be within an acceptable range.

Results from the model simulation were written at hourly intervals in NetCDF files that included information, such as sea-surface height, water velocity, salinity, and temperature at locations across the grid and at different depths. GNOME and BLOSOM each required a subset of the information; both used the sea-surface velocity, and while BLOSOM used an elevation raster to determine the coastline, GNOME used the ocean model boundary as the coastline.

GNOME has a readily available toolkit with scripts to convert outputs from a variety of hydrodynamic models into a NetCDF file directly compatible with GNOME. The script mapped variables, determined the grid boundary based on grid connectivity information, and removed excess information, such as water velocity at depths, since in this application, GNOME is only being used for surface trajectories. Meanwhile, BLOSOM was capable of directly reading Salish Sea Model outputs because BLOSOM developers included an option for the relevant variables directly within the model code.

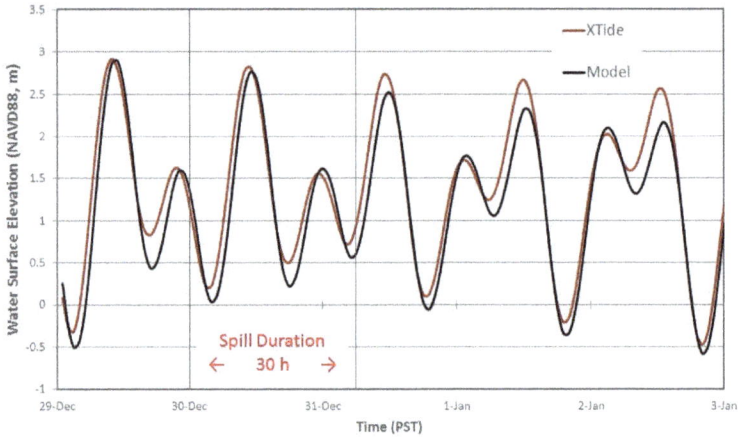

Figure 6. Comparison of sea-surface height between the model and a local XTide Station.

Table 1. Error statistics for sea-surface height during the evaluation period of 29 December 2003 to 3 January 2004.

Mean Absolute Error (MAE, m)	Root Mean Square Error (RMSE, m)	Relative Error (RE, %)	Correlation Coefficient (R)
0.2138	0.2711	6.3	0.977

2.2. Additional Ambient Forcing for Oil Spill Models

Oil spill models use ocean currents to force the trajectories of oil parcels. However, ocean models do not perfectly simulate all of the physical processes that may be responsible for transporting oil, and it is possible that missing processes may need to be included through parametrizations [21]. As mentioned above, we used an FVCOM ocean model implementation to simulate the Salish Sea that has typically been used for water quality, circulation, and habitat restoration purposes. While some work was performed to optimize the ocean model to drive surface trajectories in this study, additional optimization could include improving the vertical grid resolution, using higher-resolution wind forcing (if available), and, possibly, transitioning to smaller grid sizing at the area of the spill. It was deemed that deficiencies in the vertical resolution resulted in an underestimation of the wind forcing; consequently, wind data was used to parametrize the effects of both windage and turbulent transfer of momentum, as a linear combination. Observational and theoretical justifications for this approach are described and discussed below. Additional forcing in the form of a stochastic model was used to simulate diffusion.

2.3. Wind Data

NOAA has a wind station at West Point (NOAA NDBC-WPOW1) about 8 miles south of where the Point Wells oil spill originated (47.662 N, 122.436 W); wind data from this source for 30 and 31 December 2003 (Figure 7) was used. A concern with this data source was that the wind coming from the east would be underrepresented due to land elevating quickly next to the coast, thus potentially causing a shading effect. A different wind source on the opposite side of the channel (Kingston wind station at 47.7940 N, 122.4940 W) was therefore used to evaluate the wind from the NOAA station. At Kingston station, wind coming from north and south is underestimated due to the presence of land, however, we were able to confirm that the wind having an east-to-west component was consistent across the channel (Figure 8). The trajectory driven exclusively by 6% of the NOAA NDBC-WPOW1 wind strongly suggests the importance of wind forcing in replicating the oil's trajectory (Figure 9, compare to schematic trajectory in Figure 4). The locations for wind stations are also shown in Figure 9.

Additional wind data from the Loyal Heights weather station (47.690 N, 122.383 W; not shown) were tested but were not found to help explain the observed trajectory; Loyal Heights station is about 100 m above sea level, and is therefore unlikely to be representative of the wind forcing at the ocean surface (wind-stress force over the ocean is usually computed from wind at either 2 m or 10 m above the sea surface). We did not use this wind data in our comparison.

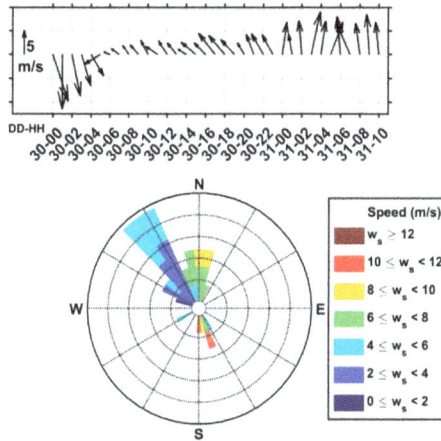

Figure 7. Hourly wind data from NOAA's WPOW1 station starting midnight 30 December 2003. Wind direction follows the oceanographic convention, i.e., the direction is towards where the wind blows.

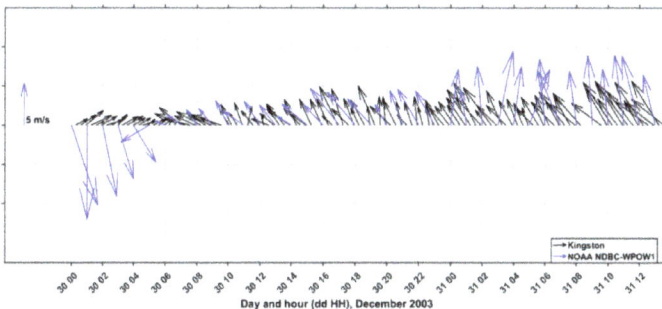

Figure 8. Comparison of wind data from NOAA (NDBD-WPOW1, black vectors) and Kingston (blue vectors) wind stations. The locations of these two stations can be seen in Figure 9.

2.4. Diffusion

To parameterize spreading of oil due to turbulent diffusion, GNOME uses a random walk with a constant diffusion coefficient. BLOSOM offers other options regarding both the diffusion scheme and the diffusion coefficient (e.g., using a diffusion coefficient read from the ocean model solution or computed through a Smagorisnky scheme, or options under development like a random flight scheme) as detailed in Duran [21]. To make the simulations directly comparable, a random walk with constant diffusion coefficient was also used for BLOSOM. Turbulent diffusion parameterized through a random walk is expected to cause oil parcels to spread around the trajectory computed without turbulent diffusion, and affect how oil beaches along the coastline. Different values for the diffusion coefficient were tested to identify when the simulated spread resembled the observed spread, as described below.

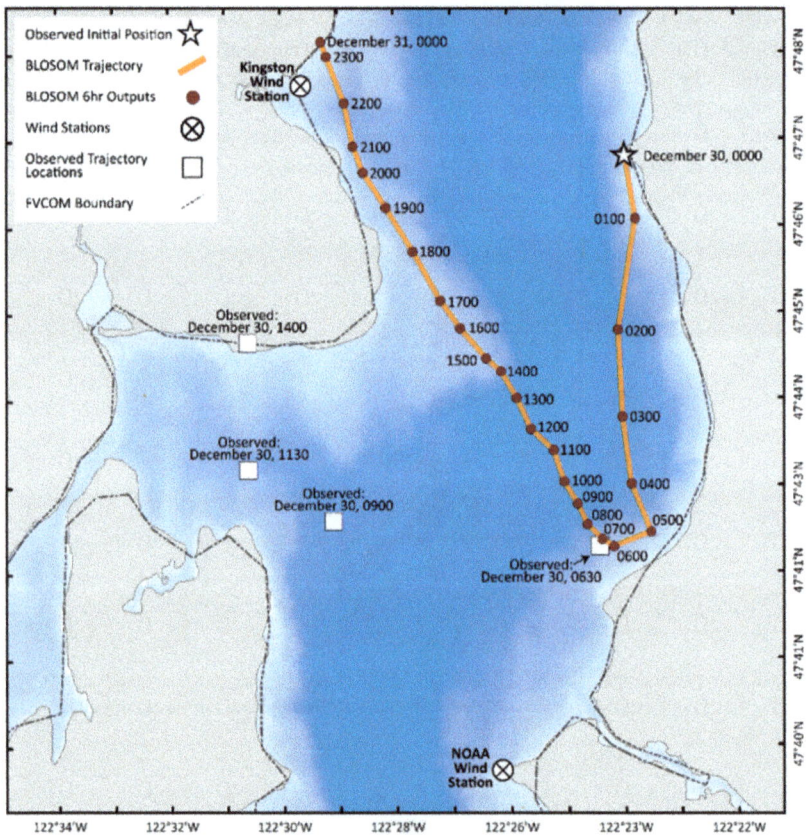

Figure 9. Trajectory (orange, red circles mark locations at hourly marks) initiated at the same time and location as the oil spill, resulting from forcing exclusively with 6% of the wind from the NOAA wind station. Also shown are the locations of NOAA (NDBC-WPOW1) and Kingston wind stations (white circles with black cross) and the approximate locations of oil at different times, as observed from overflights (white squares; see Figure 4).

3. Results

3.1. Exploring Differences in the Oil Spill Models through Simulations

We use diverse simulations to compare GNOME and BLOSOM. Gradually increasing in complexity, these simulations are designed to isolate and compare different aspects of the models, or the oil spill we wish to hindcast. Thus, each difference among the models is illustrated through a test. Using, as a starting point, the wind information described in the "Wind Data" subsection above, tests three through nine illustrate the Foss barge - Point Wells (from now on, Point Wells) hindcast.

3.2. Initial Conditions and Technical Details

Information on the Point Wells Spill detailed in the reports by ENTRIX, Inc., [10] and The Foss-Pt.Wells Natural Resource Trustees [11] was used to create a baseline for the comparison. Most tests use an instantaneous release of oil except where noted, and the duration of each simulation was set to 30 h, starting on 30 December 2003 at 00:05. Two spill locations were used: the one closer to shore accurately represents the location of the spill, another location further offshore was used for a sensitivity test.

Throughout our tests, we used a fixed computational time step of 6 min, and sensitivity tests (Appendix A.1) showed this value to be a good compromise between satisfying the numerical requirement known as Courant–Friedrichs–Lewy (CFL) condition (explained in Appendix A.1) and computational efficiency. Oil parcels' positions are plotted approximately every 15 min, to allow a clear visualization of the trajectories, and to make the trajectories from both models directly comparable. Baseline parameters used in the simulations are listed in Table 2.

Table 2. Baseline simulation parameters.

Actual initial location (x)	122.399274 W
Actual initial location (y)	47.780472 N
Offshore test location (x)	122.40899 W
Offshore test location (y)	47.782 N
Amount of oil released	4637 gallons
Release start date and time	30 December 2003; 00:05
Simulation duration	30 h (unless otherwise noted)
Release period	0 min (instantaneous, unless otherwise noted)
Currents	Salish sea hydrodynamic model (unless otherwise noted)
Wind	West Point NOAA station (unless otherwise noted)
Number of particles for simulation	1 (unless otherwise noted)
Diffusion	None (unless otherwise noted)
Computational time step	6 min
Trajectory points plotted every	15 min

3.2.1. Test 1: Coordinate System

The first couple of simulations are idealized test cases that allow for a direct comparison between the model's handling of advection due to ocean currents and wind, in the most simplified way possible.

BLOSOM uses, for its computations, an equidistant spatial reference system with units of meters; for this study, the Universal Transverse Mercator (UTM) Zone 10 North with the datum World Geodetic System 1984 was used. GNOME computes oil parcel trajectories using longitude and latitude directly, with an internal conversion factor that varies as a function of latitude, which makes their computations compatible with the metric system. Thus, BLOSOM assumes a tangent plane approximation, while GNOME computes advection on a sphere. These different approaches result in some divergence in the trajectories that can be assessed by computing a trajectory while keeping the ocean current velocity constant in space and time.

Our first test therefore consists of advecting a single oil parcel with a constant ocean sea-surface velocity set to $(-0.03, -0.1)$ m/s and no wind (Table 3). The constant velocity was chosen so that it would result in a trajectory comparable in length to the Point Wells oil spill trajectory, but without beaching. The difference in advection algorithms resulted in a trajectory separation of 76 m over 30 h (Figure 10).

Table 3. Parameters for test 1, ocean currents integration.

Parameters for Test 1, Ocean Currents Integration	
Currents	Spatially constant and steady currents = $(-0.03, -0.1)$ m/s
Wind source	None
Start location	Actual initial location
Wind advection coefficient	None

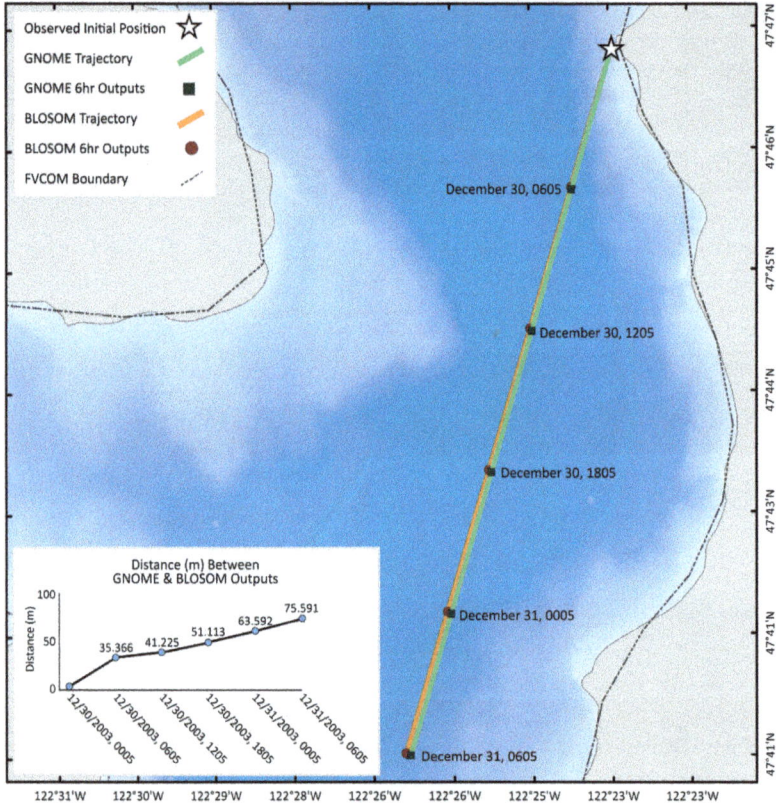

Figure 10. Comparison of advection algorithms using constant ocean currents; test 1.

3.2.2. Test 2: Wind Handling

The purpose of the second test (Table 4) was to test the advection due to wind. This test was kept comparable to the first one by requiring that the wind advective velocity were equal to the ocean advective velocity used in test 1. The wind was therefore set to $(-0.5, -1.667)$ m/s and the wind advection coefficient set to 6% of the wind velocity. To be directly comparable with GNOME, BLOSOM did not use deflection due to the effect of earth's rotation on the wind's transfer of momentum to the ocean's surface (this deflection is explained in test 3 below). Since GNOME handles wind inputs with limited precision (a characteristic motivated by the inherent uncertainty in atmospheric models), the resulting separation for this test was greater than in the previous test, reaching about 95 m over a 30 h simulation (Figure 11). The separation due to wind was about 26% greater than the difference due to ocean currents, over the same period.

Table 4. Parameters for test 2, wind integration.

Parameters for Test 2, Wind Integration	
Currents	None
Wind	Spatially constant and steady wind = $(-0.5, -1.666666666667)$ m/s
Start location	Actual initial location
Wind advection coefficient	6%
Deflection	None

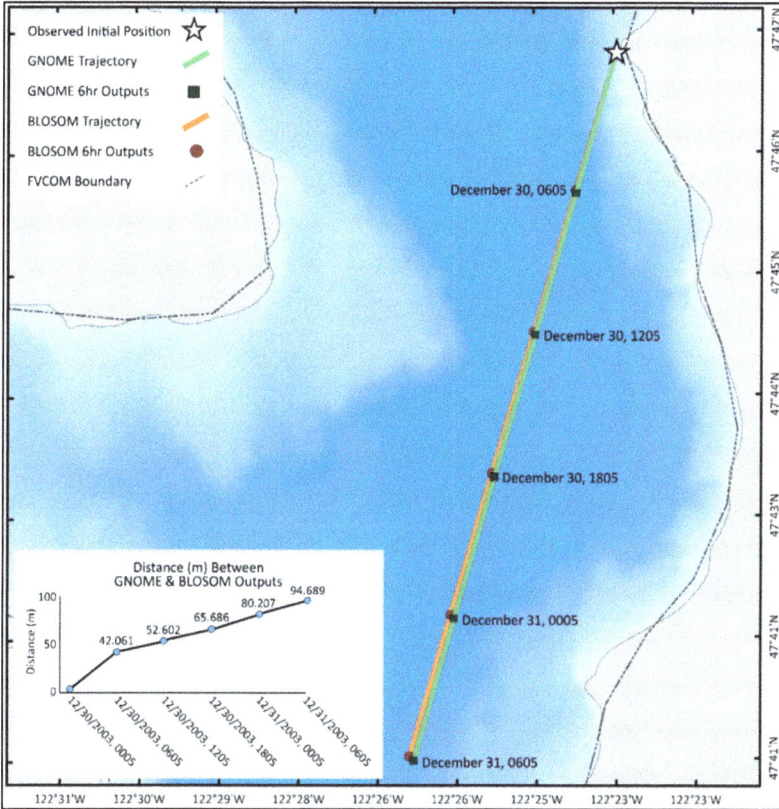

Figure 11. Comparison using constant wind, trajectories diverge with the separation between them is plotted in the bottom left inset; test 2.

3.2.3. Test 3: Wind Advection Scheme

Both models integrate the ocean currents using an Euler integration scheme, therefore, advection due to ocean currents should not result in any differences beyond those found in test 1. However, how BLOSOM and GNOME process wind data includes an additional distinction that could result in trajectory differences beyond those found in test 2. The turbulent transfer of momentum from wind into the ocean results in trajectories that are not aligned with the wind's direction, but are deflected to the right of the wind direction (in the northern hemisphere)—this is caused by Earth's rotation (Coriolis effect). BLOSOM simulates this deflection, while GNOME assumes that wind advection is in the same direction as wind. This aspect of GNOME's design is a consequence of the inherent uncertainty of numerically simulated wind, and how that may apply to emergency response scenarios.

This test aims to qualitatively understand the difference in wind advection schemes by computing, with each model, a trajectory for an oil parcel originating from the same location, advected by the ocean currents from the hydrodynamic model, and wind data from NOAA WPOW1 station (Table 5). The wind advection coefficient was set to 6%.

Table 5. Parameters for test 3, wind deflection part 1.

Parameters for Test 3, Wind Deflection Part 1	
Start location	Actual initial location
Wind advection coefficient	6%
Deflection for GNOME	Default (none)
Temporal interpolation for BLOSOM	Default (none)

BLOSOM's trajectory illustrates the deflection to the right of GNOME's trajectory, including a difference on the beaching location (Figure 12).

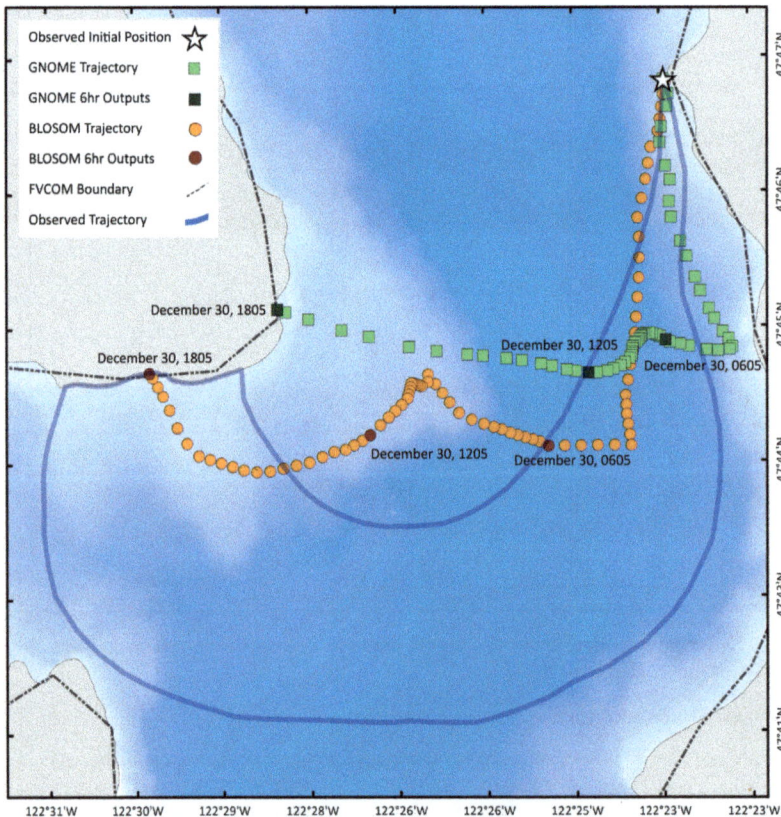

Figure 12. GNOME and BLOSOM trajectories diverge due to including, or not, the effect of earth's rotation on wind forcing; test 3, part 1.

To confirm if the totality of the trajectory difference was only due to the wind deflection, an additional test (Table 6) was then designed. Wind data was manipulated to produce a wind time series that GNOME could read, and that would include the same deflection that BLOSOM computes internally. We were able to confirm that most, but not all, of the difference found in Figure 12 was due to the deflection, yet a small difference remained (Figure 13); this difference is explored in test 4.

Table 6. Parameters for test 3, wind deflection part 2.

Parameters for Test 3, Wind Deflection Part 2	
Start location	Actual initial location
Wind advection coefficient	6%
Deflection for GNOME	Deflection added manually
Temporal interpolation for BLOSOM	Default (none)

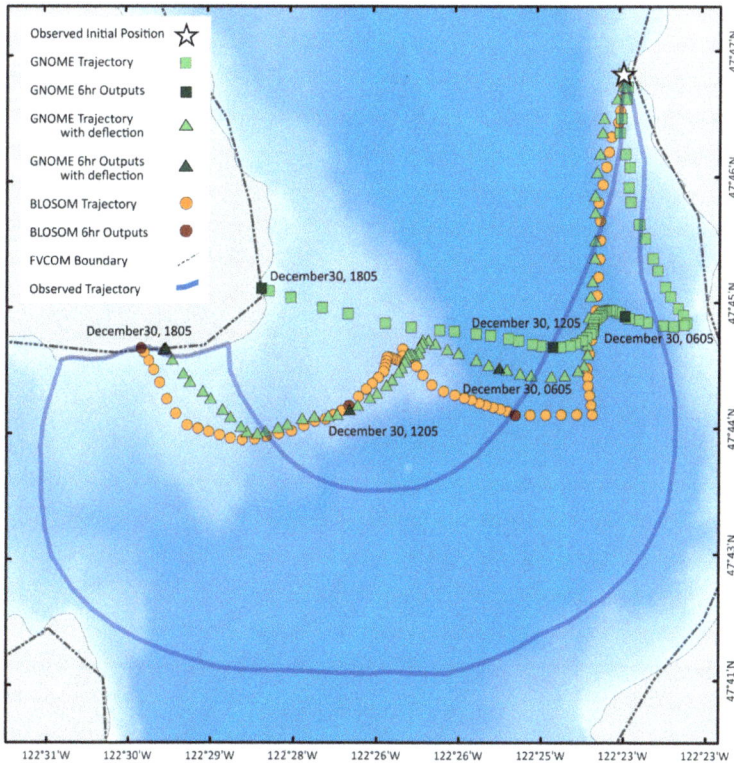

Figure 13. The same two plots shown in Figure 12 are shown along with an additional trajectory by GNOME, that now includes deflection due to earth's rotation; rotation was included directly to the wind data, forcing GNOME to replicate the deflection computed internally by BLOSOM. GNOME's trajectory with deflection agrees well with BLOSOM's trajectory, however, some difference remains; test 3 part 2.

3.2.4. Test 4: Temporal Interpolation

An additional distinction between BLOSOM and GNOME is that the latter includes temporal interpolation of the forcing fields by default. By adding temporal interpolation to BLOSOM as a beta feature, we were able to assess if the difference in trajectories that remained in the previous test could be explained. The parameters used for this test can be found in Table 7.

With both models using temporal interpolation, trajectories were now close enough (Figure 14) to where the remaining difference could be explained by the different approaches used to compute trajectories, as illustrated in the first two tests.

Table 7. Parameters for test 4, temporal interpolation.

Parameters for Test 4, Temporal Interpolation	
Start location	Actual initial location
Wind advection coefficient	6%
Deflection for GNOME	Deflection added manually
Temporal interpolation for BLOSOM	Temporal interpolation included

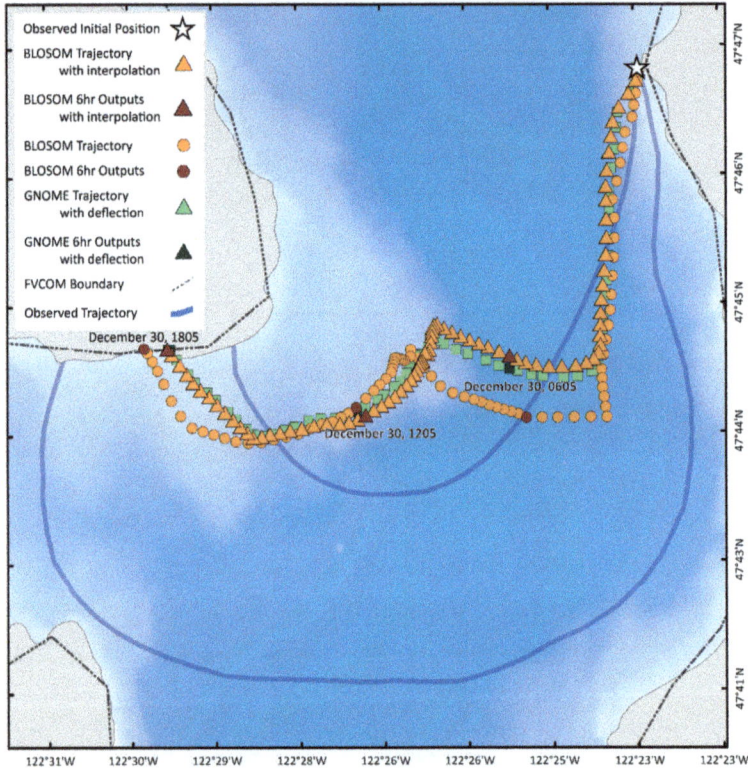

Figure 14. The trajectories for GNOME and BLOSOM, both with deflection included, as seen in Figure 13, are compared to the same BLOSOM trajectory, but now including temporal interpolation; test 4. The trajectory from GNOME with added deflection, and the trajectory from BLOSOM with added interpolation, now resemble each other closely.

3.2.5. Test 5: Differences in Beaching

Once the above differences were clarified, we turned out attention to other potential sources of discrepancy, including how the coastline and beaching are treated.

For this test, we explore how the models respond, with their default options, to ocean currents and a more typical wind coefficient value of 3%, thus further clarifying the role of wind as a driver for this oil spill. The parameters used for this test can be found in Table 8.

GNOME includes a refloating algorithm that empirically describes the adhesiveness of the oil to the shoreline; a "half-life" parameter can be set by the user, representing the number of hours over which half of the oil on a given shoreline can be removed by an offshore wind, diffusive transport, or from a sea level that is equal, or higher than when the oil originally beached [9]. Samaras et al. [22] present half-life values for different beach types. We use a half-life value of 24 h for GNOME simulations, which is representative of a sand or a gravel beach. While a marsh would typically have a much larger refloat half-life, the Doe-Kag-Wats marsh is mostly protected by a part-sand and part-gravel beach, with a small opening into the marsh. BLOSOM does not include a refloat option.

Table 8. Parameters for test 5, differences in beaching.

Parameters for Test 5, Differences in Beaching	
Start Location	Actual initial location
Wind Advection Coefficient	3%
Deflection for GNOME	None
Temporal interpolation for BLOSOM	None

In this simulation, BLOSOM's trajectory beaches in about an hour and a half, and about 200 m northeast of the initial location. GNOME also beached at a similar location, but in about half an hour. GNOME creates a rasterized shoreline map from loaded current data for the purposes of tracking the oil beaching. This generated shoreline has a finite resolution across the entire grid, so GNOME provides an option to restrict the model domain, thus creating finer resolution of the shoreline in the area of interest, however, the beaching still occurs slightly offshore from the FVCOM boundary due to this approximation (Figure 15a). Before the beaching, there was some divergence between the BLOSOM and GNOME trajectories (Figure 15b); this separation is somewhat higher than those detected in tests 1 and 2 (Figures 10 and 11), probably due to differences described in tests 3 and 4.

Due to the refloating algorithm, GNOME's oil parcel continued its trajectory about 18 h after beaching (at 18:53), subsequently moving north (Figure 15c), remaining within about 5 km of the spill's origin.

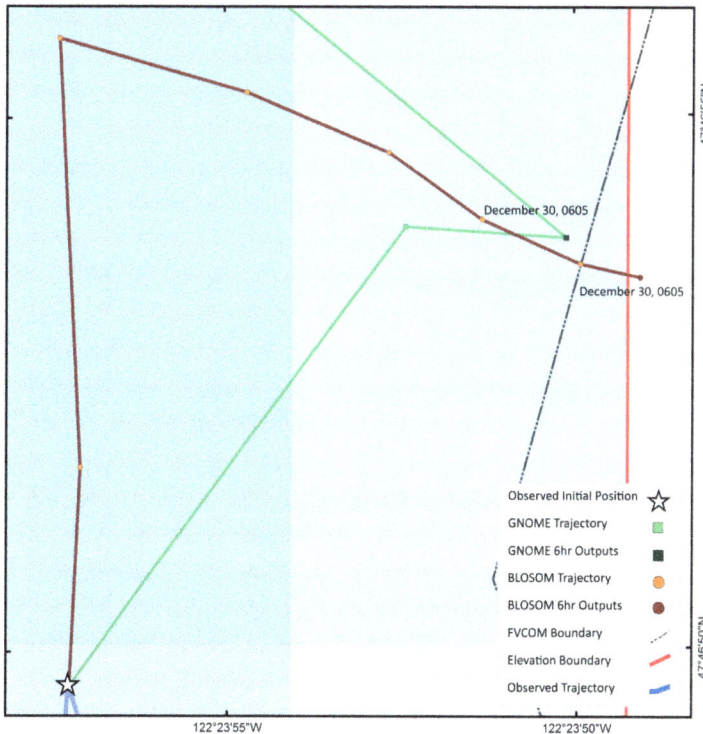

(a)

Figure 15. *Cont.*

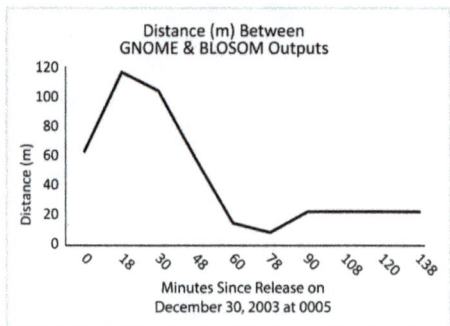

(b)

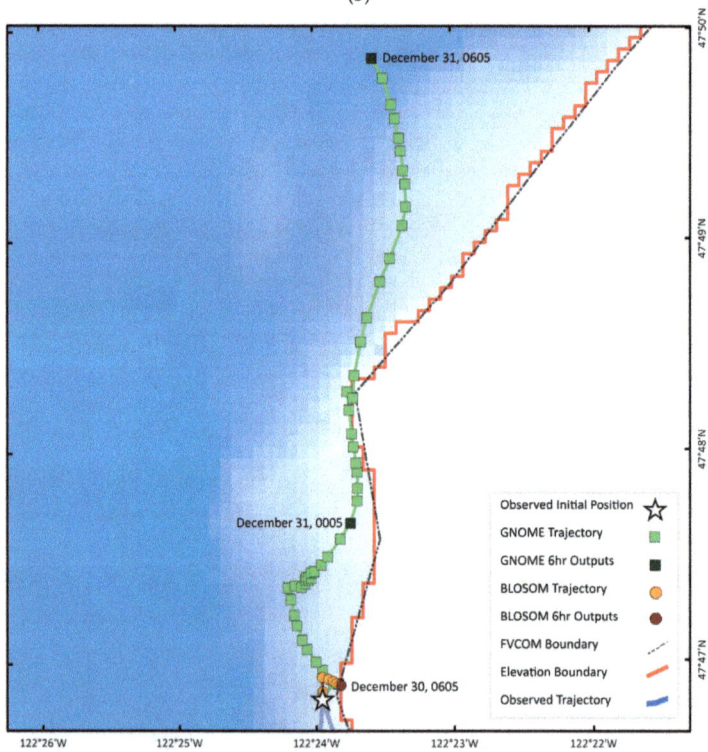

(c)

Figure 15. (a) Trajectories showing beaching differences, parameters for these simulations can be found in Table 8; test 5. (b) Distance between GNOME's and BLOSOM's trajectories as a function of time during the first two hours of the simulation for test 5, trajectories are plotted in (a). (c) GNOME's trajectory after refloating at 18:53 is shown; test 5.

3.2.6. Test 6: Sensitivity to Initial Position

As seen in the previous test, the simulated trajectories do not cross the channel with a 3% wind coefficient. An additional experiment exploring cross-channel transport was conducted, with no wind. Trajectories were initiated at both the actual initial location, and at an initial location that is about 730 m offshore, and 165 m north. This test was run on default values (i.e., BLOSOM does not include temporal interpolation); parameters used for these simulations are shown in Table 9.

Table 9. Parameters to test 6, sensitivity to initial location.

Parameters to Test 6, Sensitivity to Initial Location	
Wind	None
Start location	Nearshore vs. offshore location
Temporal interpolation for BLOSOM	None

When initiated offshore from the actual spill location, trajectories without wind forcing do cross the channel (Figure 16A). This is because an eddy just offshore from the coast, induced a relatively strong cross-channel component. Starting at about 18:00 on 30 December, the models begin diverging markedly. GNOME's trajectory does a south–north oscillation, while BLOSOM's trajectory goes on to beach at the correct location at 06:00 on 31 December, about 14 h later than the time when the actual beaching began. GNOME's trajectory has not beached by the end of the simulation.

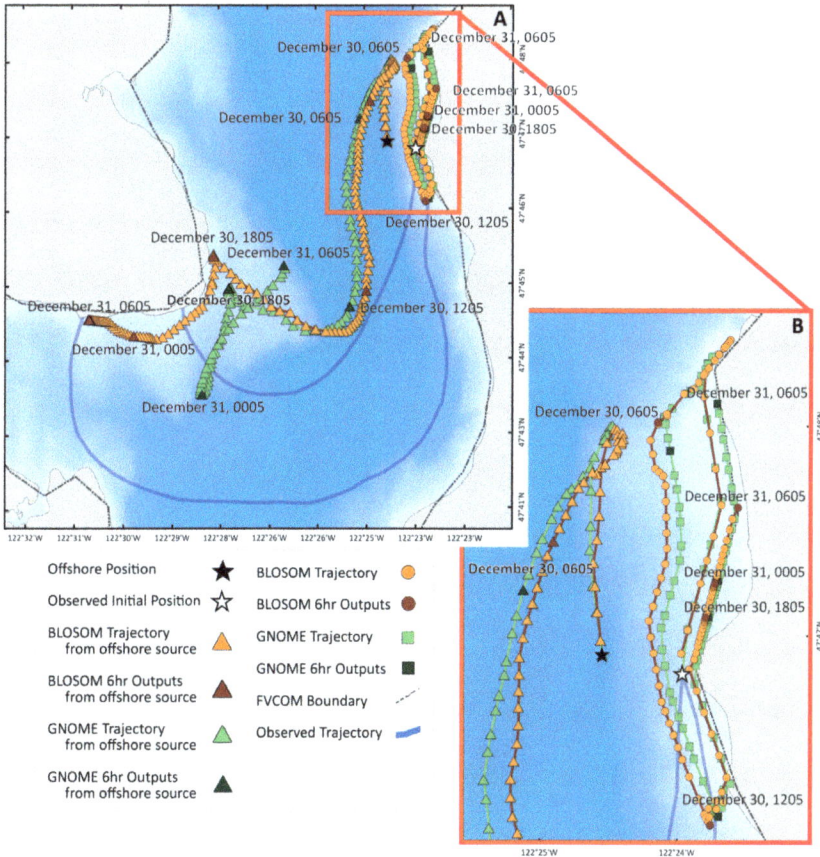

Figure 16. (**A**) Trajectories cross the channel without wind, as they are entrained by eddy-induced cross-channel transport when initiated offshore from the location of the oil spill; test 6. (**B**) Zoomed in view of the simulations that were initiated at the correct oil spill location.

By contrast, when using no wind but with both model's trajectories initiated at the correct location, oil moves northward up the coast, then southward slightly offshore, then again northward but closer to the shore, oscillating with the tides (Figure 16B). In this case, beaching for each model happens at locations about 700 m away.

3.2.7. Test 7: Number of Particles While Using Turbulent Diffusion

Two parameters were tested through some diffusion tests: the diffusion coefficient itself and the number of particles. Greater values of the diffusion coefficient cause greater spread, while the number of particles may also influence the amount of spreading. Using a fixed diffusion coefficient of 10,000 cm^2/s, we first test 30 oil particles (Figure 17) against 1000 particles (Figure 18); parameters for these simulations are listed in Table 10. We consider some of the implications of the number of particles that we use by presenting Oil Holding Capacity calculations in Appendix A.2.

This test suggests that thirty particles is a good approximation if we judge by the simulated spread at the beaching locations, which approximately matches the observed spread of beached oil, between Point Jefferson and Indianola (marked as observed trajectory in plots). BLOSOM's along-path spread is similar whether using 30 or 1000 particles. However, because GNOME's trajectory samples currents outside of the observed trajectory, there is spurious spreading, especially with 1000 particles. The greater the number of particles, the greater the possibility that trajectories sample a greater variety of ocean currents as they are diffused by the random walk displacements. GNOME's simulation beaches at additional locations when the number of particles is higher, also, some trajectories cross the channel further north.

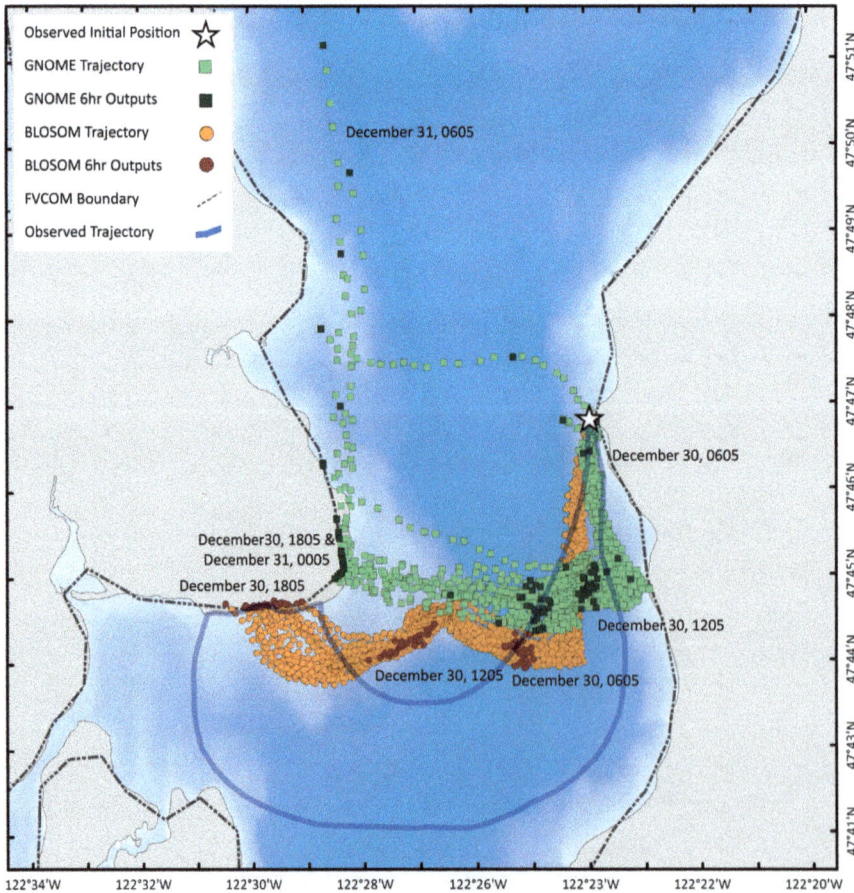

Figure 17. Simulation including diffusion with thirty particles; test 7.

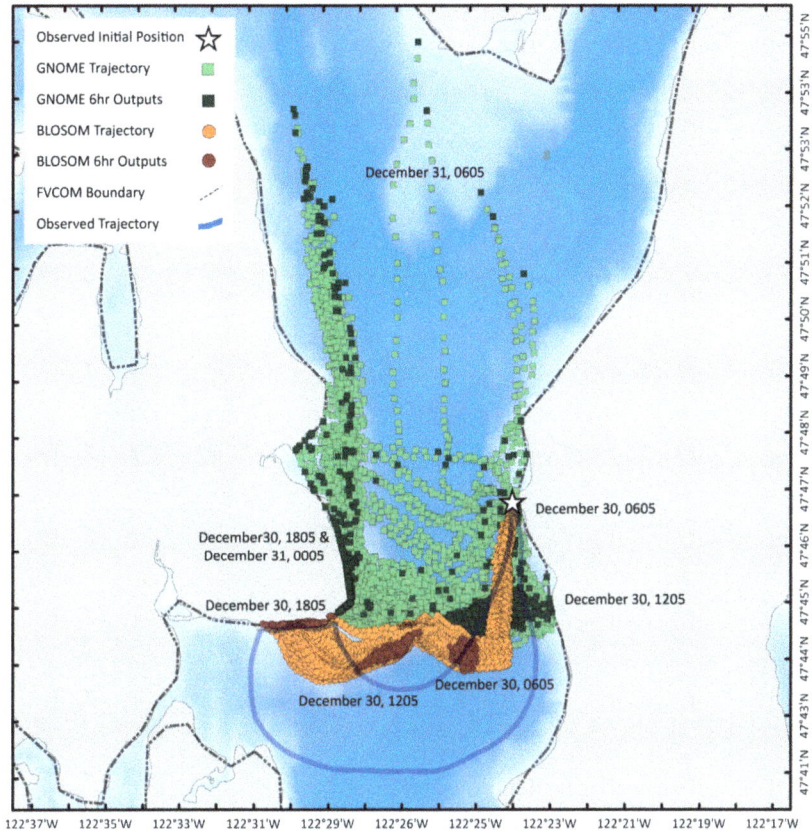

Figure 18. Simulation including diffusion with a thousand particles; test 7.

Table 10. Parameters for testing number of particles.

Parameters for Testing Number of Particles	
Start location	Actual initial location
Wind advection coefficient	6%
Deflection for GNOME	Default (none)
Temporal interpolation for BLOSOM	Default (none)
Number of particles	30 vs. 1000 particles
Diffusion coefficient	10,000 cm^2/s

As mentioned above, GNOME includes a rebeaching option, where particles that have beached can still move back into the ocean and continue their trajectory. This refloating ability explains the northward movement of particles from the GNOME simulation.

3.2.8. Test 8: Release Period

Once it was established that thirty particles would result in approximately the desired spread due to diffusion, we next tested if the period of release would make a difference. So far, all tests have used an instantaneous release, with oil released at 00:05. Here, we compare an instantaneous release to a release lasting 15 min, which is the estimated actual duration of the overflow (Table 11). The instantaneous release of the thirty-particle simulation (Figure 17) is very similar to a 15 min release (Figure 19) when using BLOSOM. GNOME's trajectories remained roughly the same as well, however, spreading increases with the 15 min release after crossing the channel (Figure 19); this difference can again be attributed to the random walk algorithm.

Table 11. Parameters for release period test.

Parameters for Release Period Test	
Start location	Actual initial location
Wind advection coefficient	6%
Deflection for GNOME	Default (none)
Temporal interpolation for BLOSOM	Default (none)
Number of particles	30 particles
Release period	0 min vs. 15 min
Diffusion	10,000 cm^2/s

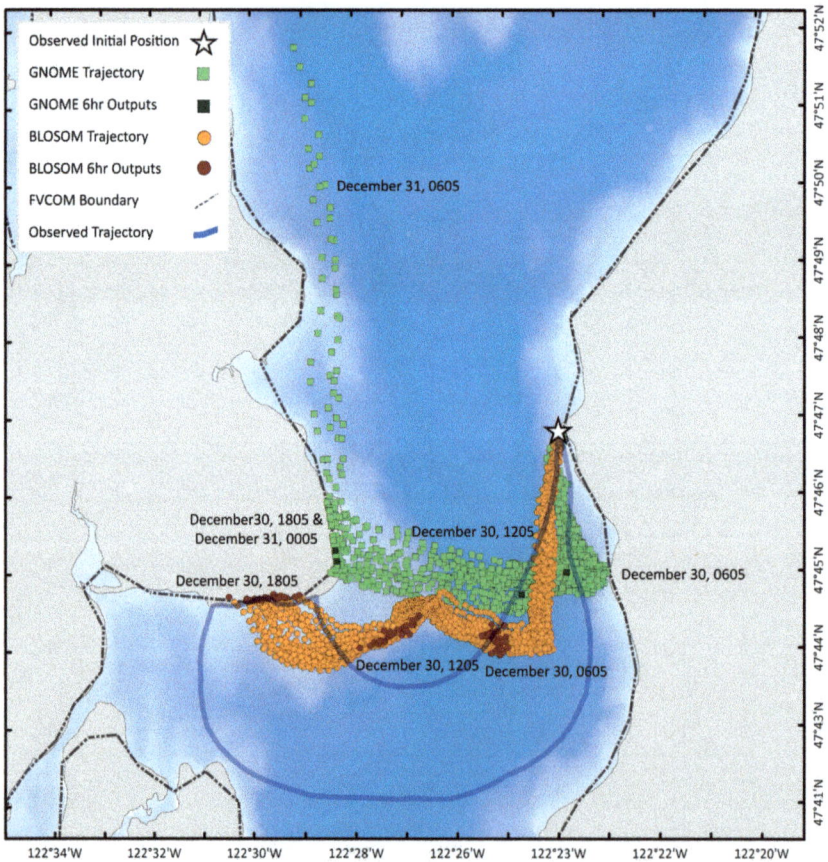

Figure 19. Thirty-particle simulation with diffusion, released over a 15 min period; test 8.

3.2.9. Test 9: Diffusion Coefficient

So far, our tests with diffusion have used a constant diffusion coefficient of 10,000 cm^2/s; here, we test simulations with a coefficient of 100,000 cm^2/s (Table 12). With the higher value, spreading along the BLOSOM trajectory seems to better match the width of the mid-channel observed trajectory, even if a few oil parcels diverge from the observed path (Figure 20). Likewise, the width of the oil in the

GNOME simulation seems closer to the observed path while crossing the channel, however beaching happens further north with the larger coefficient.

Table 12. Parameters for diffusion coefficient test.

Parameters for Diffusion Coefficient Test	
Start location	Actual initial location
Wind advection coefficient	6%
Deflection for GNOME	Default (none)
Temporal interpolation for BLOSOM	Default (none)
Number of particles	30 particles
Release period	15 min
Diffusion	10,000 vs. 100,000 cm^2/s

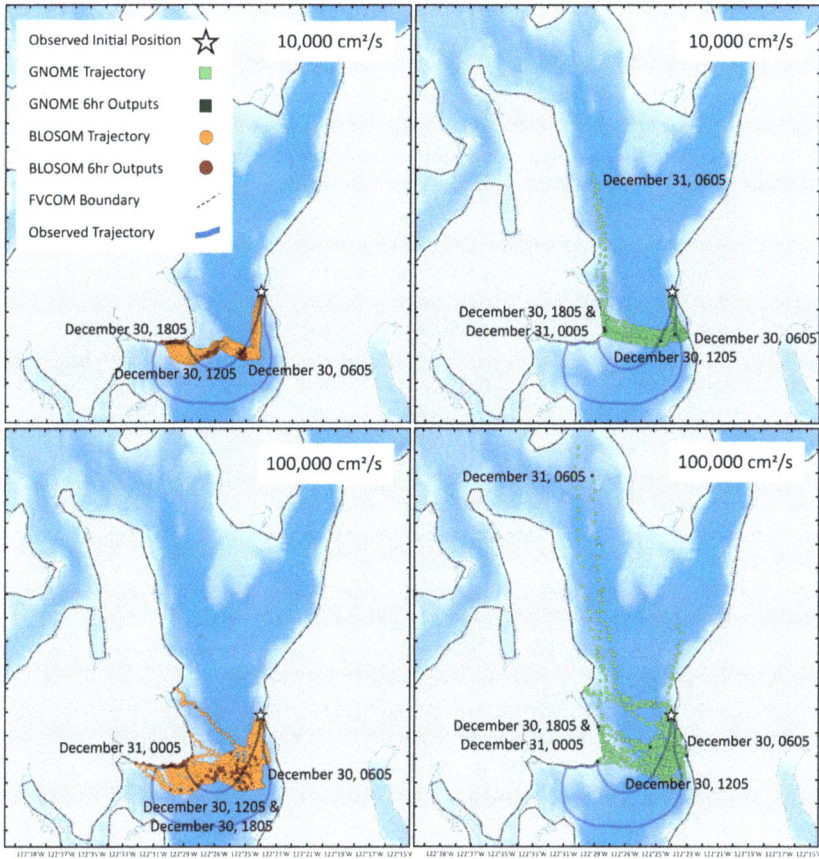

Figure 20. Comparison of diffusion coefficients; test 9.

4. Discussion

We first discuss the hindcast of the Point Wells oil spill, including some discussion of the regional oceanography as it may apply to other oil spills. We then discuss the differences between GNOME and BLOSOM.

4.1. Hindcasting the Historical Foss Point Well Spill

Further physical information will be needed to confirm our findings, however, we are able to make some compelling suggestions regarding what drove the Point Wells oil spill, insight that, to our knowledge, had so far remained elusive.

The trajectories are mainly forced with an accurate representation of the tides (which is a major driver of ocean currents in our region; Figure 6), as well as observed measurements of the wind velocity (Figures 7 and 8). A quick inspection of the wind speeds during the spill, immediately suggest that wind was a major driver for the Point Wells spill: Meier and Höglund [23] (pp. 101–129) show that the advection of oil at the very surface tends to be dominated by wind whenever wind speeds reach or exceed about 4 m/s. Indeed, the trajectory of a parcel initiated at the location and time of the spill, and forced with only wind, is quite suggestive of the actual path (Figure 9).

Under these circumstances, the advection of oil at the surface is often successfully modeled as a linear combination of ocean currents at the surface, and a velocity derived from wind, often called wind drift or windage. A wind drift velocity is typically calculated as a coefficient of 3%, multiplying the wind velocity. However, there are two points to be made regarding this general statement:

(1) Main components of ocean currents at the surface typically include a geostrophic component; in our case, this is likely well represented by the tidal motion from our FVCOM model, and a wind-driven component that is independent of the wind drift mentioned above. The wind-driven current is given by some representation of the turbulent transfer of momentum from wind to the ocean's surface. The solution to this problem is an ocean current that spirals while decaying exponentially with depth; it is, therefore, very sensitive to its vertical dependence.

(2) Determining a correct wind drift coefficient is not trivial, the difficulties arise from a wide variety of ambient and dynamical considerations, some of which are discussed in Duran [21]. However, that is not the end of the list. As an additional example: naturally occurring surfactant has been shown to increase the wind speed drift velocity by 25% [24]. If we consider a typical wind drift coefficient of 0.03, then, under the influence of surfactant, the wind drift speed, computed from the wind speed U, becomes 0.03U(1 + 0.25) = 0.0375U. This illustrates naturally occurring phenomena, that would be very difficult to detect without in situ measurements, and that would increase the effective wind drift coefficient from 3 to almost 4%. This is as far as windage is concerned, however, additional processes, such as Stokes drift, Coriolis–Stokes force, and Langmuir circulation, are often parameterized directly from the wind velocity as well, and using a similar parameterization (e.g., Weisberg et al. [25], and references therein).

In Appendix A.3, we show that the model we use has coarse vertical resolution, and therefore, underestimates the wind-driven component of the sea-surface velocity by about a factor of four. We show this using a simplified analytical model, which is in good agreement with very high-resolution observations of vertical shear in ocean surface currents [26]. To compensate, a wind-driven velocity is directly parameterized from the wind, in the same way that a wind-drift velocity would be parameterized (see e.g., the surface velocity in Figure A1). As mentioned above, the FVCOM model used in this study does an excellent job of simulating tides (Figure 5), however, for Lagrangian transport applications we can recommend a higher resolution near the surface to better resolve the turbulent transfer of momentum near the surface. This is because solutions for the turbulent transfer of momentum, from wind to the ocean, decay exponentially with depth, and are therefore sensitive to a coarse vertical resolution. Near the sea surface, the vertical resolution for ocean models used to simulate Lagrangian transport should be maximized, and is typically on the order of tens of centimeters (e.g., [27–29]), in marked contrast with the 4 to 6 m resolution of the uppermost layer of the model used for this study (except very near the coast, where the vertical resolution is higher). Indeed, even high vertical ocean model resolutions will need further parameterizations at times, since ocean models to not typically represent processes such as wind drift or Stokes drift. A natural follow up study towards calibrating this model for sea-surface trajectories could compare our results with an

FVCOM implementation that uses a higher vertical-resolution, while continuing to simulate tides with a high skill.

In our study, a relatively high wind coefficient of 6%, in conjunction with the tide-driven ocean currents, cause the simulated trajectory to match the observed trajectory in time and space (Figure 14). We do not find the high value for the wind coefficient surprising, because we are almost surely parameterizing at least two processes, the underestimated wind-driven current and windage. The former parameterization is remediating a coarse vertical resolution, while the latter is not typically included in ocean models, yet, is often a main driver when wind speed exceeds 4 m/s. It is also possible that additional wind-related processes (e.g., Stokes drift, Coriolis–Stokes force and/or Langmuir circulation) were at play. All of these processes can be parameterized with the same or a similar parameterization, a coefficient multiplying the wind velocity with a slight deflection to the right of the wind direction. Thus, the 6% coefficient likely represents at least two, perhaps more processes. Future studies should include detailed information, or at least a parameterization, of the processes related to surface waves, so as to gain insight to their role during this spill. We note that although Stokes drift is related to surface gravity waves, it can be effectively parameterized as a percentage of the local wind [30].

The inclusion of an angle of deflection for the parameterization of wind-driven processes was consequential in the hindcast of this oil spill. In this case, the angle computed internally by the default BLOSOM algorithm [21] did well. However, as discussed in Duran [21], the actual value of deflection does vary according to a number of conditions, making it difficult to simulate accurately. It is therefore recommended to have an option that allows the oil spill modeler to select an angle manually. Such an approach would allow the user to select an angle that matches observations, similar to the approach with the diffusion coefficient where the user selects a value to match the spread observed through overflights, satellite images or boat reports [9]. An example of an oil spill where wind drift was also very important, but where no deflection angle was appropriate, is given in Nakata et al. [31]. The value of 3.5% for the windage coefficient that they used, is consistent with our discussion, since their ocean upper layer was only 2 m deep, compared to the upper layer of the model we used, which is two to three times thicker (4–6 m).

When including deflection due to earth's rotation, tides, and a wind advection coefficient of 6% of the wind speed, the simulated trajectory does a good job of replicating the trajectory observed with overflights. The trajectory tends to hit the correct locations at the right times, although a bit too north during the first six hours of the simulation, and a few hours late after the first twelve hours (see BLOSOM's trajectory in Figure 12). Thus, BLOSOM's trajectory matches the observed path both in time and space, as can be seen by comparing to the timelines in Figures 1 and 9. The agreement strongly suggests that wind and tides were the major drivers during the 2003 Point Wells spill. This is because we are using observed winds (Figure 8), and because the tidal component of the ocean currents is almost perfectly simulated with the FVCOM model we use (Figure 6). Indeed, the trajectory forced only with 6% of the wind (Figure 9), is highly suggestive. Additionally, diffusion representing unresolved small-scale processes, helped match the observed spreading along that trajectory (Figures 19 and 20). It thus seems persuasive that winds, tides, and small-scale processes were the physical drivers during the 2003 Point Wells spill.

In the test where trajectories were initiated offshore from where the oil spill occurred and with no wind (Figure 16A), the cross-shore transport was due to eddies that spin off when the tidal currents interact with promontories (not shown). Thus, we have found two processes that may cause cross-shelf channel in our study region: cross-channel wind and eddies.

4.2. Integration Geometry

We now discuss the differences between GNOME and BLOSOM, as illustrated by the different tests we did, Table 13 includes a summary relating test numbers and the aspects being tested. The procedure an oil spill modeler would typically pursue for a hindcast or forecast is summarized in an infographic (Figure 21).

Table 13. Relation between test number, user-defined parameters, and model aspects being tested.

Test	Parameters Being Tested	Relevant Model Aspects Being Tested
1	N/A	Integration geometry
2	N/A	Integration geometry and wind handling
3	Wind advection coefficient	Effect of earth's rotation
4	N/A	Interpolation of wind forcing
5	Wind advection coefficient	Differences in beaching algorithms
6	Sensitivity to initial position and wind advection coefficient	N/A
7	Number of particles	Turbulent diffusion: number particles
8	Release period	N/A
9	Diffusion coefficient	Turbulent diffusion: diffusion coefficient

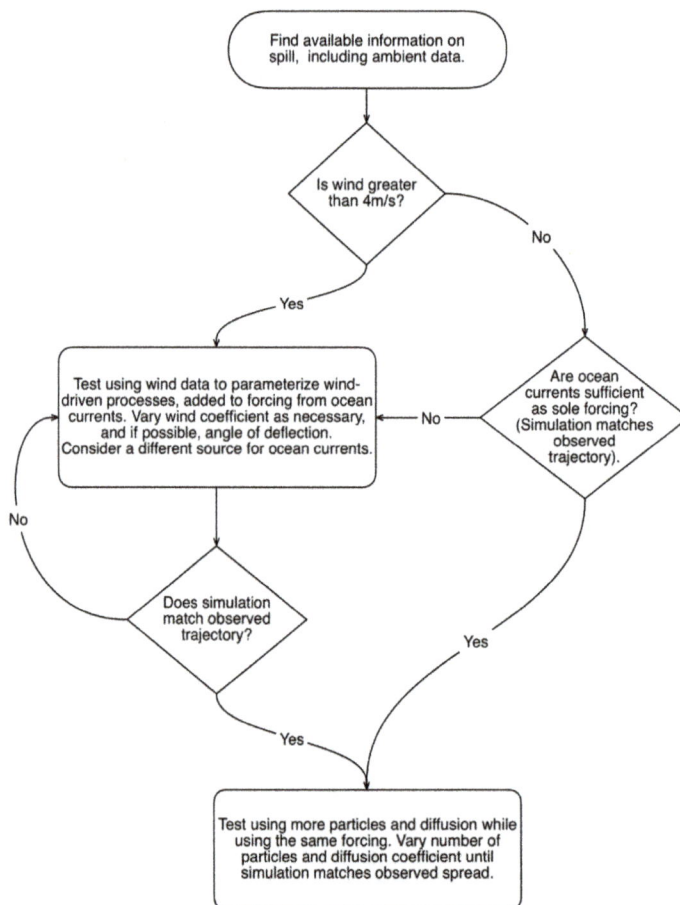

Figure 21. Diagram summarizing steps, as used in this study, for hindcasting an oil spill.

Differences were found between trajectories computed by GNOME and BLOSOM due to the geometry over which the trajectories are integrated. BLOSOM uses a tangent plane approximation while GNOME integrates over a sphere. The differences were small, at about 30 to 40 m in the first six hours, and 60 to 80 m during the first day. We note that integrating wind as the only forcing

J. Mar. Sci. Eng. **2018**, *6*, 104

mechanism (test 2), produced a somewhat higher difference than integrating only the ocean currents (test 1; compare Figures 10 and 11). The differences from tests one and two should be considered additive when the trajectory is forced by both ocean currents and wind forcing, as is often the case. As mentioned above, ocean currents are often unstable, consequently it possible that single trajectories may separate considerably over time due to these small initial differences. However, bulk calculations of trajectories (which is the typical approach during oil spill modeling) should give similar results.

The tangent plane approximation used by BLOSOM can be formally justified as an asymptotic expansion in terms of the length scale of interest, divided by the Earth radius, times the tangent of the latitude [32,33]. For the latitudes spanning the conterminous United States, about 25 to 49 to degrees North, the tangent plane approximation is appropriate for lengths of about one thousand kilometers or less. The Gulf of Mexico is approximately 1000 km in the zonal direction, and approximately 1000 km in the meridional direction, and therefore, BLOSOM is adequate for modeling a spill that encompasses the entire Gulf of Mexico. However, off Alaska's coasts, about 60 to 72 degrees North, the tangent plane approximation is adequate for distances up to about 500 km. Thus, due caution is needed when using BLOSOM for large spills in, say, Alaska, northern Europe or the Arctic.

4.3. Including the Effect of Earth's Rotation

The inclusion (BLOSOM) or not (GNOME) of the deflection caused by earth's rotation when computing the component of oceanic trajectories forced by wind (wind-forcing parameterization) is found to be consequential for this hindcast (test 3). The inclusion of earth's rotation, nudges the simulated trajectory to stay within the observed trajectory envelop for longer, and, importantly, it causes the oil parcel to beach at the correct location (Figure 12).

GNOME does not include the effect of earth's rotation because developers considered that, in the context of a quick response scenario, the uncertainty inherent to forecast winds would overcome the precision afforded by including this effect. However, because the wind used in our hindcasts comes from field measurements (Figures 7 and 8), the uncertainty due to forecasts is not an issue. In general, whenever the wind velocity is accurate, the inclusion of the deflection angle might result in an improvement to the simulation, this is because it approximates the physics of different wind-forced drivers that are missing in ocean models (e.g., [21]).

4.4. Interpolation of Wind Forcing

As a follow up to test 3, we used test 4 to understand the remaining difference between GNOME and BLOSOM trajectories, after including earth's rotation into both models (Figure 13). Thus, as in test 3, the same deflection that BLOSOM computed internally, was added into the wind data that GNOME used to force the trajectory, but additionally, in test 4, temporal interpolation of the forcing fields was added to BLOSOM. This was done to match GNOME, that includes temporal interpolation by default. This test revealed (Figure 14) that temporal interpolation explains an additional part of the difference between trajectories seen in Figure 13. We attribute the small difference remaining after the inclusion of temporal interpolation to the integration geometry as described in tests one and two.

Sensitivity to temporal interpolation when computing trajectories in the ocean is known, and has been discussed (e.g., [34]); it is especially important for coarse temporal resolutions. Future releases of BLOSOM will include temporal interpolation of the forcing fields.

Regarding spatial interpolation, the effect on currents from an ocean model with reasonable spatial resolution would likely be overcome by the use of a random walk scheme (turbulent diffusion) and would, therefore, only add to the computational cost without a benefit. Thus, both the GNOME and BLOSOM developer team currently do not prioritize using a spatial interpolation scheme beyond nearest neighbor. A study on the performance of integration and interpolation pairs for marine transport applications is ongoing, and will be reported elsewhere [35].

4.5. Differences in Beaching Algorithms

Test 5 was designed to understand differences in beaching algorithms. GNOME includes a refloat option based on an empirical parameter (half-life) that describes the adhesiveness of the oil to the shoreline. This half-life parameter is the number of hours in which half the oil on a given shoreline is expected to be removed with offshore wind or turbulent diffusion [9]. The spill described by Nakata et al. [31] is an example where the refloating of oil was important.

Based on results from test 4, and to make the trajectories directly comparable, wind deflection was introduced manually to the wind data for GNOME, thus, both models include wind deflection; likewise, BLOSOM uses temporal interpolation. We note that an additional difference between the models that affects beaching, is that GNOME uses the ocean model boundary as the coastline while BLOSOM uses an elevation raster provided by the user. In this test, the wind advection coefficient is reduced from 6% to a more typical value of 3%, therefore, this test also helps illustrate what a modeler would see, when using a typical value for the wind advection coefficient (Figure 15a). The trajectories diverge probably due to an initial perturbation caused by the additive effect of the different integration geometries detailed in tests one and two, that is amplified by the ocean currents (Figure 15b); however, both models end up beaching at similar locations (Figure 15a). A more in-depth discussion on coastline treatment for oil spill model beaching can be found in Samaras et al. [22]. It is important to note that neither GNOME nor BLOSOM are designed to simulate beaching with the level of detail that can be found in other pioneering studies. GNOME, in particular, is intended as a quick response tool, consequently, the beaching location is only approximated (Figure 15a). In Appendix A.2, we show that the number of particles we use for the diffusion tests (30 and 1000) are reasonable regarding the type, and dimension, of beach that was impacted. Calculations with 30 particles suggests that, if a detailed beaching study were desired, a greater number of particles might be needed. However, we also note that trajectories, especially GNOME's, may deteriorate considerably as the number of particles increases, as noted below (Section 4.7). Thus, there is a trade-off to be made. For our purposes (a study that is primarily focused on the oil's trajectory), 30 particles gives good results, while maintaining reasonable oil holding capacity (OHC) values (Appendix A.2).

After about eighteen hours, GNOME's trajectory initiates again after refloating, to then move north (Figure 15c). In a real-life situation, both GNOME's and BLOSOM's simulation would be rejected, since they do not replicate the observed path, suggesting that the mechanisms forcing these trajectories are inadequate.

During a response scenario, the refloating of oil can be modeled alternatively by re-initializing an oil spill simulation once a need has been detected (for example, through overflights). Thus, BLOSOM does not contemplate, at present, the inclusion of a refloating option in future releases.

4.6. Sensitivity to Initial Position

In test 6, we explore initializing the trajectories offshore from the actual incident locations, about 730 m offshore and 165 m north. These trajectories do cross the channel even without any wind, a marked difference from the trajectories using the correct initial position (Figure 16A).

To put this initial location offset into context, ocean models in the Gulf of Mexico are often 3 to 4 km resolution, with the higher-resolution models having resolutions of about 1 km. Thus, the correct and the offset initial locations, could have been within the same ocean model grid point of a high-resolution model of the Gulf of Mexico. However, the Salish Sea has much smaller spatial scales, and therefore requires a much finer ocean model. Thus, the distance between the actual and offset locations of the spill, spans about 4 ocean model grid cells. This is an example of the sensitivity to the initial location, a result of ocean currents varying considerably across the channel. As mentioned above, ocean currents are often unstable, and amplify exponentially any small, initial difference in the initial position [36,37]

BLOSOM and GNOME differ in their final positions when the trajectories are initiated at the offshore location. This illustrates how small differences in oceanic trajectories can cause a large difference in the end result. This is because ocean currents are chaotic at length scales that are often smaller than the distance traveled by most trajectories of interest, which implies that small differences at any time during the trajectory can be exponentially amplified, causing distinct differences even within a few hours (see e.g., [36] or [37]). This test does not include any wind forcing, thus, the small differences (that eventually resulted in larger differences) can be attributed to the distinct geometries used in the ocean current integration, as illustrated in test 1. Exponential separation of trajectories is generally not relevant in bulk trajectory computations, especially when using a random component to simulate small-scale processes.

The trajectories initiated offshore do not coincide with the observed trajectory envelope (Figure 16A). Since a trajectory can only be judged to be accurate if it visits the correct spatial locations at the correct times, BLOSOM's trajectory, in this test, is an example of getting the correct result (i.e., correct beaching location, although about 14 h late) for the wrong reasons (i.e., by initiating the trajectory offshore from the actual initial location).

In this test, we also run a simulation in which the trajectories are initialized at the correct location but, in contrast with test 5, without wind forcing (Figure 16B). In this case, the trajectories are not forced towards the coast as fast as in test 5, instead, they oscillate with the tides, approximately parallel to the coastline to then beach not far from where the spill originated. This confirms that no wind, or typical values of wind forcing, are inadequate to force this oil spill, given the ocean model used in this study, as discussed above.

4.7. Turbulent Diffusion: Number of Particles

Tests one through six suggest what physical processes drove the Point Wells oil spill trajectory, namely, ocean currents plus 6% of wind, as illustrated in test three. Having identified appropriate ambient forcing, the next step a modeler would typically pursue, would be to add turbulent diffusion in an attempt to match the spread of pollutants, ideally along an observed trajectory envelope. In this test, BLOSOM does not use temporal interpolation and GNOME does not include wind deflection; these features were only added, respectively, in previous tests to understand the model's differences. However, having illustrated such differences, and because these features, respectively, are not currently available in the oil spill models, the simulations that follow will use default configurations.

In BLOSOM and GNOME simulations, the spread due to diffusion is controlled mainly by the diffusion coefficient, which both models allow the user to select. However, the spread is also affected by the number of particles used. If there are not enough particles, then the random walk algorithm will not represent the correct solution to the diffusion equation.

In test 7, a fixed diffusion coefficient of 10,000 cm^2/s and 30 particles are selected. The turbulent spreading causes oil parcels to cover more of the area that was observed to be covered, providing an accurate estimate of the beaching, thus nudging the simulations towards greater realism (Figure 17). We note that the higher the number of particles, the greater the amounts of oil refloating in GNOME's simulation, causing spurious northward transport.

To assess the effect of the number of particles, this experiment is repeated while using 1000 particles (Figure 18). In this test, GNOME's trajectories diverge even further from the observed path, and the spread becomes less realistic, suggesting that 1000 particles is not an improvement. BLOSOM, however, does reach a desirable spread, suggesting that with 1000 particles, there is a modest improvement; note the width of BLOSOM's beaching approximately matching the observed width in Figure 18.

These results highlight the importance of simulating the diffusion-free trajectory as accurately as possible.

4.8. Release Period

All simulations, so far, used instantaneous releases at 00:05, therefore, the effect of releasing oil over a period of 15 min, an estimate of the duration of the actual spill, is assessed in test 8. No notable difference is found with this test, suggesting that an instantaneous release is adequate for simulation purposes.

4.9. Turbulent Diffusion: Diffusion Coefficient

In test 9, the diffusion coefficient is tested by using different values, while keeping the number of particles at 30. A coefficient of 100,000 cm^2/s causes the spread to be closer to the width of the observed path envelope (Figure 20), suggesting an improvement relative to a diffusion coefficient of 10,000 cm^2/s and 30 particles (Figure 17), but it is not necessarily an improvement relative to BLOSOM's simulation with a coefficient of 10,000 cm^2/s and 1000 particles (Figure 18). Since a smaller turbulent diffusion coefficient is preferable (if the same results can be obtained by increasing the number of particles), the conclusion is that a coefficient of about 10,000 cm^2/s and 1000 particles is the preferred setting for BLOSOM. Some fine-tuning is likely possible by testing numbers of particles between 10 and 100 thousand, but was not pursued.

Regarding GNOME's trajectories for both values of diffusion coefficients, the more significant factor continues to be that the trajectory does not match the observed path, due to not including the effect of earth's rotation when forcing with wind, as identified in test three. More trajectories diverge when using the 100,000 cm^2/s coefficient; these differences are inherent to the random walk algorithm and are therefore expected. Therefore, beyond the differences found in test 3 and test 7 (rebeaching of oil), this test does not find any further noteworthy differences between GNOME and BLOSOM.

This test finalizes the simulations that a modeler would need to understand what simulation parameters best replicate the Point Wells oil spill. Consequently, all relevant differences between GNOME and BLOSOM have been illustrated.

5. Conclusions

While not a rigorous proof, it is very compelling that the simulation, initiated at the correct location and time, forced with accurate wind observations and accurate simulated tides, replicates the observed trajectory. The simulated trajectory broadly reaches the correct locations at the correct times, through the duration of the spill, to finally beach at the correct location and time (beaching started around 14:30 and continued through the afternoon and night of 30 December, up to the morning of 31 December). Since it is highly unlikely that such a combination of events would happen by chance, we conclude that our hindcasts make a compelling case for what forced the Point Wells oil spill, despite the fact that it is unusual for oil spill modelers to use a wind coefficient as high as 6%. This conclusion is supported by observational studies, and a simplified analytic model. The analytic model was used to quantify the effect of the coarse ocean model vertical resolution on the surface velocity (Appendix A.3), and it agrees well with very high-resolution observations recently reported by [26].

Deflection for the full 6% of the wind speed coefficient was shown to be important; this is in agreement with multiple studies (e.g., references in [21]) that have found that deflection may be part of each of the physical processes that are parameterized from wind.

To the best of our knowledge, this paper is the first to build a compelling case for what drove the historical Point Wells spill. The drivers for particle motion were tidal currents (including their interaction with promontories), wind, and small-scale processes represented by diffusion.

A dedicated study will be needed to fully understand the physics driving the Point Wells spill. In this regard, we can recommend future studies to include the effect of waves, because Stokes drift is known to be an important driver at the sea surface [27,38]. Stokes drift can be parameterized directly from wind, despite it being due to surface waves [30]. Thus, the inclusion of waves should provide further insight into the nature of the wind forcing that we have found to be important. Likewise, ocean simulations with a higher vertical resolution, that continue to accurately replicate tides, should also be helpful. We note the strong wind (5–10 m/s) blowing consistently towards the impacted beach, during the period of time over which beaching was observed. There was also very strong wind (>10 m/s) during the first 6 h of the spill (Figure 7).

The Point Wells spill was difficult to forecast during response activities [39], presumably due to inadequate forcing. It is plausible that forcing was inadequate because of the strong dependence on wind. Understanding the drivers behind the Point Wells spill will allow oil spill modelers to be better prepared for future forecasts or hindcasts.

While initially developed for different purposes, GNOME and BLOSOM are similar models that are complementary in nature. For the particular hindcast used for this study, BLOSOM gives better results due to the inclusion of an internally computed deflection angle for the parameterization of wind-forced advection. However, we do not suggest that this implies that BLOSOM is a superior model. That this particular hindcast depends so heavily on a wind-driven parameterization, and that the wind we used comes from representative observations, may or may not be representative of other spills. Additionally, wind deflection could be included in a GNOME simulation by modifying the data, if so desired.

We are able to make two main recommendations for future releases: (1) that BLOSOM include temporal interpolation of the forcing fields, and (2) that GNOME include an option to allow for an angle of deflection when using wind-forced parameterizations.

Author Contributions: Conceptualization, R.D., L.R., K.R. and A.B.; Data curation, R.D., L.R., J.W. and J.V.; Formal analysis, R.D., L.R. and J.W.; Funding acquisition, K.R. and A.B.; Investigation, R.D., L.R. and J.W.; Methodology, R.D., L.R. and J.W.; Project administration, K.R., A.B. and J.B.; Resources, L.R. and J.W.; Software, R.D. and J.V.; Supervision, R.D., K.R. and A.B.; Validation, R.D., L.R. and J.W.; Visualization, R.D. and L.R.; Writing—original draft, R.D., L.R. and J.W.; Writing—review & editing, R.D., L.R., J.W., K.R. and J.B.

Funding: This study was funded by the Bureau of Safety and Environmental Enforcement.

Acknowledgments: This work was executed by NETL and PNNL under an inter-agency agreement, E14PG00045-M1, from the Bureau of Safety and Environmental Enforcement. Work conducted by AECOM and Theiss Research was performed in support of NETL's ongoing research under the RES contract DE-FE0004000. Special thanks to NOAA's Office of Response and Restoration for their collaboration. This Project was conducted by the Department of Energy, National Energy Technology Laboratory, an agency of the United States Government, Pacific Northwest National Laboratory, and with support via contract with AECOM. Neither the United States Government nor any agency thereof, nor any of their employees, nor AECOM, nor any of their employees, makes any warranty, expressed or implied, or assumes any legal liability or responsibility for the accuracy, completeness, or usefulness of any information, apparatus, product, or process disclosed, or represents that its use would not infringe privately owned rights. Reference herein to any specific commercial product, process, or service by trade name, trademark, manufacturer, or otherwise, does not necessarily constitute or imply its endorsement, recommendation, or favoring by the United States Government or any agency thereof. The views and opinions of authors expressed herein do not necessarily state or reflect those of the United States Government or any agency thereof.

Conflicts of Interest: The authors declare no conflict of interests. The funders had no role in the design of the study; in the collection, analyses, or interpretation of data; in the writing of the manuscript, and in the decision to publish the results.

Appendix A

Appendix A.1 Computational Time Step

GNOME and BLOSOM simulate and output the oil's location and characteristics given ambient conditions (e.g., ocean currents, wind, temperature etc.). How often the output is written and how often the output is visualized do not affect the models' solutions, although they affect how well the user is able to understand the solution; these time intervals depend on the user's needs and discretion.

The computational time step, however, is a quantity that the models use internally to compute the oil's trajectory as it evolves under ocean currents and wind forcing. Choosing a time step that is too big could result in spurious trajectories [34]; the criteria for "too big" is the well-known Courant–Friedrichs–Lewy (CFL) condition. Special care is needed when using ocean models for enclosed seas, because they often have a small grid size near the coast; this is especially true with unstructured grids (e.g., the FVCOM implementation that we use). For our purpose, the CFL condition C is a dimensionless ratio that compares the distance $|u|\Delta t$ traveled by an oil parcel moving under a velocity u over one computational time step Δt to the size of the spatial grid Δx (cf. [40] (pp. 171–175)). Thus, the CFL condition is defined as

$$C = \frac{|u|\,\Delta t}{\Delta x}. \tag{1}$$

We say that the CFL condition is satisfied whenever $C < 1$. If the CFL condition is not satisfied, then an oil parcel could overshoot through one or more grid cells without taking into consideration the velocity of those grid cells. Thus, the trajectory could artificially ignore valid information, resulting in a scrambled sampling of the true velocity field and, consequently, in spurious trajectories.

In numerical analysis' terminology, satisfying the CFL condition guarantees that the finite-difference scheme used to solve a differential equation has a numerical domain of dependence, on initial or boundary conditions, which includes the differential equation's domain of dependence, on initial or boundary conditions (e.g., [41] (pp. 98–100)).

From the CFL condition, we can derive an upper bound for acceptable computational time steps. This upper bound can be computed using the maximum velocity and a minimum grid size in the model

$$\Delta t < \frac{\Delta x_{min}}{|u_{max}|}. \tag{2}$$

The information needed for this upper bound is readily available from the ambient data used by any oil spill model.

It is in BLOSOM's development plan to compute this time step upper bound when initializing a simulation to inform the user of any possible violations of the CFL condition or, alternatively, of an inefficiently small time step. When familiarizing oneself with a new ocean model, it is convenient to run a few tests to determine the ideal time step; it should satisfy the CFL condition but also, it should not be smaller than necessary since this can easily result in run-times that are too long to be practical.

For these simulations, an estimate of the time step Δt was computed using a distance of 200 m and a velocity of 0.5 m/s (mean velocity + 1.5 standard deviations) in $\Delta t < \frac{\Delta x_{min}}{u_{max}} = \frac{200}{0.5} \approx 6.7$ min..

Thus, we obtain an a priori estimate of what the time step should be, this estimate can be confirmed running some simulations. The first test will be our benchmark, it uses a time step of one minute (such a small time step ensures convergence), and is shown in Figure A1.

Next a 6 min time step was used (Figure A2), the trajectories are identical to those with a time step of one minute, thus confirming our a priori estimate.

However, with a computational time step of 18 min, BLOSOM's trajectory started to deteriorate (Figure A3). Notice that GNOME and BLOSOM's trajectories differ due to advection algorithms (test 3 above) and, therefore, their sensitivity to the computational time step will be different. Any small differences can and will be amplified due to unstable currents.

With a 36 min computational time step, the trajectory is completely spurious (Figure A4), to the point that trajectories end well into land. This experiment was not repeated with GNOME, however, similarly bad results are expected as this spurious behavior is a property of the numerical method, and both models use the same method. The difference in the advection algorithm should be small compared to the error induced by such a large computational time step. Also, when the time step is very large, the diffusion does not cause much spreading, this is because the random increment is added every time step, consequently, there is insufficient random spread when the number of time steps is small.

In summary, with a large time step, the trajectory is sampling an inaccurate velocity field (by overshooting valid information); additionally, the random walk is not converging to the solution of the diffusion equation. These tests confirm that the time step estimate of less than 6.7 min was a good choice (because the trajectory using 6 min time step is very similar to the trajectory using a 1 min time step, confirming convergence).

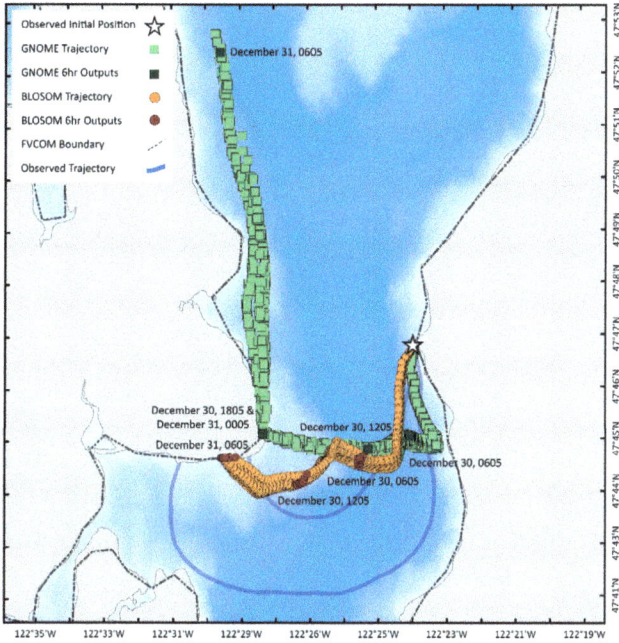

Figure A1. GNOME and BLOSOM simulations using a 1 min time step. This simulation includes 6% wind, ocean currents and some diffusion.

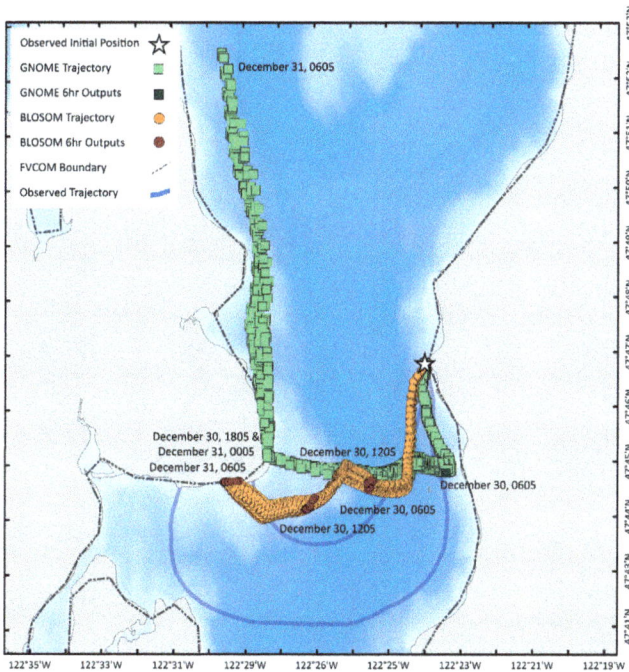

Figure A2. Same as plot A1 but both models using a 6 min time step.

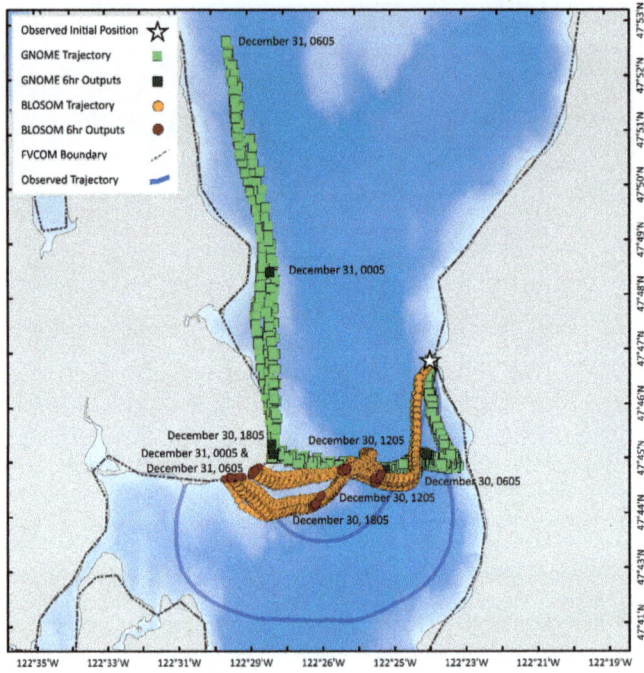

Figure A3. Same as Figure A1, with both models using a computational time step of 18 min.

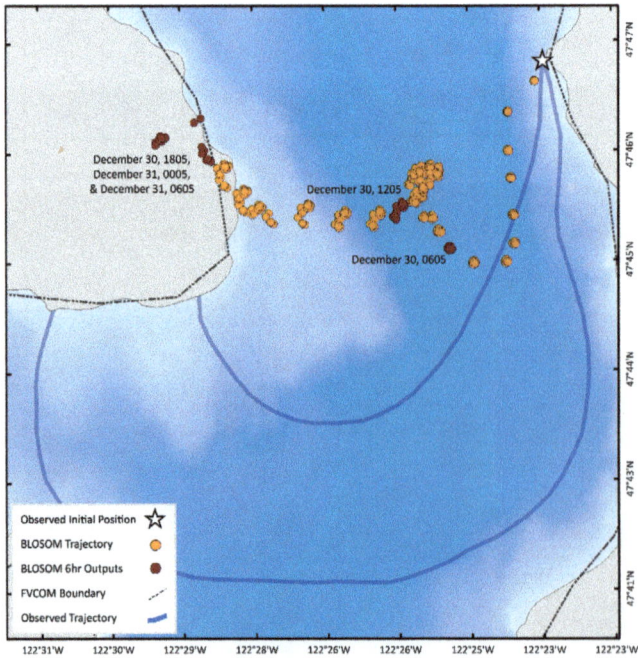

Figure A4. Same as Figure A1 but with a 36 min computational time step for BLOSOM.

Appendix A.2 Computations Related to the Number of Parcels in Our Simulations

In this section, we use data for the historical spill, such as impacted area and beach length, and volume of oil, to determine if our selections for number of parcels are reasonable.

Three and a half acres were impacted, or about 14,164 m^2. Roughly 2/3 of the total impacted area was heavily impacted. Half of the heavily impacted area was a gravel beach, and the other half a sand beach, roughly 4721 m^2 each half, for a total heavily impacted area of 9442 m^2. The maximum estimated amount of oil was 4800 gallons or about 18.2 m^3.

We can compute an upper bound for the oil holding capacity (OHC), by assigning all the oil at once, to the area that was heavily impacted: 18.2 m^3/9442 m^2 = 0.0019 m^3/m^2. This is just below the mean, and about an order of magnitude smaller than the maximum volume of oil per area reported for OHC in [22] (their Table 2). Thus, data from this spill is consistent with data collected and published elsewhere.

If we divide the total volume 18.2 m^3 by 30, the number of oil parcels used in some of our simulations, we get that each parcel would have a volume of about 0.061 m^3. For the 1000 parcel simulation, each parcel would represent about 0.0182 m^3 of oil. Using the upper bound for volume of oil per meter squared, computed above from Point Wells spill data, this means each parcel would impact, at least, an area of 0.061/0.0019 = 32.1 m^2. This is the minimum area that each parcel would need to impact. When using a thousand parcels, each parcel would need to impact at least about 9.6 m^2.

The length of the beach that was impacted was reported to be 2400 m (1.5 miles); considering the impacted area, this means the stretch of impacted beach was, in average, almost 6 m wide. Considering the minimum impacted area computed above for the 30 parcel scenario, this implies a reasonable stretch of beach of about 5 m, which is about 6 m wide. For the 1000 parcel scenario, each parcel would need to impact a stretch of beach of about 1.6 m by 6 m wide, which is also reasonable. Thus, we can conclude that our number of particles are not unreasonable, as far as the OHC is concerned.

In terms of the impacted area, we know a beach stretch of 2400 m was impacted, and if we divide this by 30 oil parcels, each parcel must stretch about 80 m long, which is also not an unreasonable quantity, and if we divide it by 1000 oil parcels, each parcel would need to stretch for about 2.4 m along the beach.

With the 30 parcel scenario, we arrived at a stretch of beach 5 m long when considering OHC, and a stretch of beach of about 80 m long when considering the length of the beach. This discrepancy suggests that, while neither of the values are unreasonable, it is likely, however, that if our goal was to model the beaching in more detail, a higher number of oil parcels would likely be desirable.

We emphasize that, with the level of detail that we are modeling, we are only looking for the trajectories to end at the correct beach. We do not suggest, or explore, anything regarding the distribution of oil on the beach. The computations in this appendix are merely to show that the number of particles we use are not unreasonable, given the data we know.

Appendix A.3 Parameterizing Missing Wind-Forced Physics

In this appendix, we explain the rationale behind using 6% of the wind speed.

The ocean model solution we used has a relatively coarse resolution in the vertical, using only ten terrain-following levels. The shallowest velocity in this FVCOM implementation is a velocity depth averaged over the upper 4–6 m (depending on the bathymetry, as the vertical coordinate is terrain-following). Such a vertically averaged Ekman velocity is considerably smaller than the Ekman velocity at the sea surface. To show this, we used a simplified model for the wind-driven velocity from [42] (pp. 22–24) using a constant wind forcing of (u, v) = (5, 5) meters/second, which is representative of the wind during the Point Wells spill. We use the solution of this model to depth-average the wind-driven ocean current velocity over the upper 6m, and we find that the magnitude is about one fourth of that at the magnitude at the surface (Figure A5). We therefore conclude that in our region of interest, the ocean model Ekman sea-surface circulation is likely underestimated by a factor of about 4. A typical wind coefficient for parameterizing the turbulent transfer of wind momentum is 3% of the wind speed, a value that can be derived approximately, from first principles (e.g., [43,44] (pp. 150–156)).

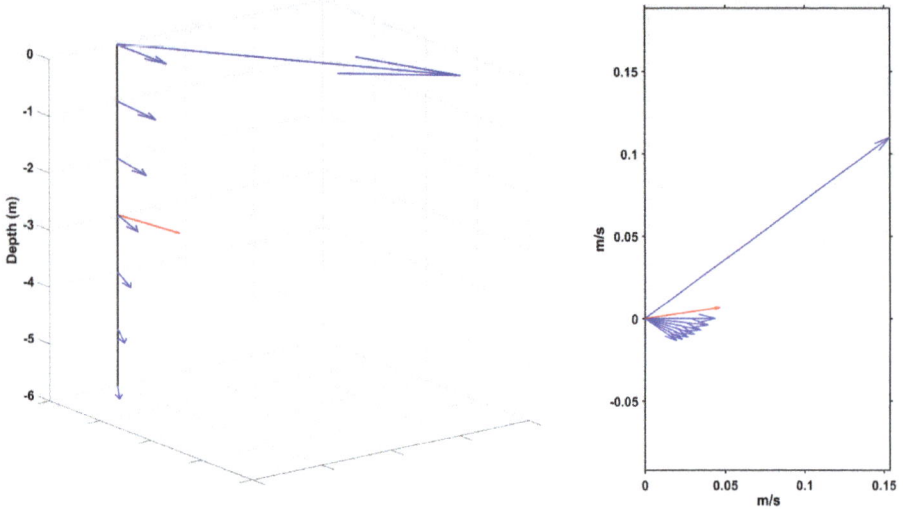

Figure A5. (Left Panel) The wind-driven velocity computed from a slip-velocity approach as specified in Csanady [42] (pp. 22–24) (blue arrows) as a function of depth (meters, vertical axis) when forced with a constant wind (u, v) = (5, 5) meters/second. Assuming a 6m deep surface cell, an approximation to the FVCOM's sea-surface velocity (in red) is computed by averaging the velocity vectors in blue. **(Right Panel)** Hodograph for the ocean current velocity solution (meters/second) that is plotted in the left panel. Results remain qualitatively similar when using other wind speeds that are likewise comparable to the wind speeds observed during the Point Wells 2003 spill.

However, wind does not only cause an Ekman circulation. An additional forcing from wind that moves oil parcels at the sea surface and is not usually included in ocean models is windage. Indeed, windage is known to dominate over ocean currents when wind speed exceeds 4–5 m/s [23] (pp. 101–129). Common values for windage range from 1–4% of the wind speed [9]. In our simulations, we found that adding 6% of the wind to the ocean currents reproduced the historical trajectory relatively well. While 6% is within bounds of what has been observed and suggested in the literature for parameterizations of wind-related processes (cf. references in [21]), it is rather large for most applications. We suggest that while windage was a necessary addition, part of the 6% we used was needed to make up for the coarse resolution of the FVCOM implementation, as explained above. Thus, the 6% percent we use is meant as a linear combination (e.g., [45] (pp. 109–110)) of at least two independent physical processes: windage and turbulent transfer of momentum from the wind to the surface of the ocean (i.e., Ekman drift). We note that there are other processes that could be at play, and would be parameterized in the same way, thus similarly (and additionally) contributing as a linear combination of forcing mechanisms—please refer to the discussion above. This would further support the use of a large coefficient.

The analytical model we used above, to show that the ocean model vertical resolution is too coarse, is supported by recent, very high-resolution observations focusing on the upper meters of the ocean. In their study, Laxague et al. [26] confirm that "the current magnitude averaged over the upper 1 cm of the ocean is shown to be nearly four times the average over the upper 10 m, even for mild forcing". This constitutes a reasonably good agreement with the simplified theoretical model we use, which includes a law-of-the-wall velocity for the first centimeter as described in Csanady [42] (pp. 22–24).

Averaging over the upper 10 m (as in the observations by Laxague et al. [26]), our simplified, theoretical model predicts that the velocity over the first centimeter would be 5 times greater than the average over the upper ten meters, which compares reasonably well to the almost 4-times greater velocity found by Laxague et al. [26] in the upper two centimeters.

Further research will be needed to isolate the physical mechanisms driving the Point Wells oil spill.

References

1. Socolofsky, S.A.; Adams, E.E.; Boufadel, M.C.; Aman, Z.M.; Johansen, Ø.; Konkel, W.J.; Lindo, D.; Madsen, M.N.; North, E.W.; Paris, C.B.; et al. Intercomparison of Oil Spill Prediction Models for Accidental Blowout Scenarios With and Without Subsea Chemical Dispersant Injection. *Mar. Pollut. Bull.* **2015**, *96*, 110–126. [CrossRef] [PubMed]

2. Sim, L.; Graham, J.; Rose, K.; Duran, R.; Nelson, J.; Umhoefer, J.; Vielma, J. *Developing a Comprehensive Deepwater Blowout and Spill Model*; Technical Report for NETL-TRS-9-2015 EPAct Technical Report Series; U.S. Department of Energy, National Energy Technology Laboratory: Albany, OR, USA, 2015.

3. Sim, L.; Vielma, J.; Duran, R.; Romeo, L.; Wingo, P.; Rose, K. BLOSOM—Release, 2017-09-26. *Energy Data eXchange* [Online]. 26 September 2017. U.S. Department of Energy National Energy Technology Laboratory. Available online: https://edx.netl.doe.gov/dataset/blosom-release (accessed on 14 August 2018).

4. Rye, H.; Brandvik, P.J.; Reed, M. Subsurface oil release field experiment-observations and modeling of subsurface plume behavior. In Proceedings of the 19th Arctic and Marine Oil Spill Program Technology Seminar, Ottawa, ON, Canada, 2–14 June 1996; pp. 1417–1435.

5. Rye, H.; Brandvik, P.J. Verification of Subsurface Oil Spill Models. In Proceedings of the International Oil Spill Conference, Fort Lauderdale, FL, USA, 7–10 April 1997; American Petroleum Institute: Washington, DC, USA, 1997; pp. 551–557.

6. Lehr, W.; Jones, R.; Evans, M.; Simecek-Beatty, D.; Overstreet, R. Revisions of the ADIOS oil spill model. *Environ. Model. Softw.* **2002**, *17*, 189–197. [CrossRef]

7. Beegle-Krause, C.J. GNOME: NOAA's next-generation spill trajectory model. In Proceedings of the OCEANS'99 MTS/IEEE. Riding the Crest into the 21st Century, Seattle, WA, USA, 13–16 September 1999; pp. 1262–1266. Available online: https://www.researchgate.net/profile/Cj_Beegle-Krause/publication/3821221_GNOME_NOAA%27s_next-generation_spill_trajectory_model/links/5aface04458515c00b6c3375/GNOME-NOAAs-next-generation-spill-trajectory-model.pdf (accessed on 14 August 2018).

8. *GNOME User's Manual*; National Oceanic and Atmospheric Administration, Office of Response and Restoration: U.S. 2002. Available online: https://response.restoration.noaa.gov/sites/default/files/GNOME_Manual.pdf (accessed on 14 August 2018).

9. Zelenke, B.; O'Connor, C.; Barker, C.H.; Beegle-Krause, C.J. *General NOAA Operational Modeling Environment (GNOME) Technical Documentation*; National Oceanic and Atmospheric Administration, Office of Response and Restoration: Seattle, WA, USA, October 2012. Available online: https://repository.library.noaa.gov/view/noaa/2621 (accessed on 14 August 2018).

10. *Data Collected to Support Response and NRDA Activities for the Foss 248-P2 Oil Spill of December 30, 2003*; Technical Report for ENTRIX, Inc. May 2005. Available online: http://www.cardno.com/en-us/Pages/Home.aspx (accessed on 1 January 2018).

11. *Draft Restoration Plan and Environmental Assessment for the Foss 248-P2 Oil Spill on 30 December 2003*; Technical Report for Foss-Pt; Wells Natural Resources Trustees: Lacey, WA, USA, 3 September 2009. Available online: https://www.google.com/url?sa=t&rct=j&q=&esrc=s&source=web&cd=1&cad=rja&uact=8&ved=0ahUKEwiKv8mCkbXOAhWFLmMKHdBsAo8QFggcMAA&url=http%3A%2F%2Fwww.cerc.usgs.gov%2Forda_docs%2FDocHandler.ashx%3Ftask%3Dget%26ID%3D554&usg=AFQjCNFwuNz2g-2HIj4L3nQJqObyZBA9mg&sig2=s6h3biHZNe6ilk2G6_R5Iw&bvm=bv.129389765,d.cGc (accessed on 1 January 2018).

12. Khangaonkar, T.; Yang, Z. A High Resolution Hydrodynamic Model of Puget Sound to Support Nearshore Restoration Feasibility Analysis and Design. *Ecol. Restor.* **2011**, *29*, 173–184. [CrossRef]

13. Khangaonkar, T.; Sackmann, B.; Long, W.; Mohamedali, T.; Roberts, M. Simulation of annual biogeochemical cycles of nutrient balance, phytoplankton bloom(s), and DO in Puget Sound using an unstructured grid model. *Ocean Dyn.* **2012**, *62*, 1353–1379. [CrossRef]

14. Roberts, M.; Mohamedali, T.; Sackmann, B.; Khangaonkar, T.; Long, W. *Puget Sound and the Straits Dissolved Oxygen Assessment: Impacts of Current and Future Nitrogen Sources and Climate Change through 2070*; Technical Report for Washington State Department of Ecology: Olympia, WA, USA, 2014. Available online: https://fortress.wa.gov/ecy/publications/documents/1403007.pdf (accessed on 14 August 2018).

15. Khangaonkar, T.; Long, W.; Sackmann, B.; Mohamedali, T.; Hamlet, A. Sensitivity of Circulation in the Skagit River Estuary to Sea Level Rise and Future Flows. *Northwest Sci.* **2016**, *90*, 94–118. [CrossRef]

16. Chen, C.; Liu, H.; Beardsley, R.C. An Unstructured Grid, Finite-Volume, Three-Dimensional, Primitive Equations Ocean Model: Application to Coastal Ocean and Estuaries. *J. Atmos. Ocean. Tech.* **2003**, *20*, 159–186. [CrossRef]

17. Finlayson, D.P. *Combined Bathymetry and Topography of the Puget Lowland*; University of Washington: Seattle, WA, USA, 2005. Available online: http://www.ocean.washington.edu/data/pugetsound/ (accessed on 18 August 2018).

18. Mellor, G.; Ezer, T.; Oey, L. The Pressure Gradient Conundrum of Sigma Coordinate Ocean Models. *J. Atmos. Ocean. Tech.* **1993**, *11*, 1126–1134. [CrossRef]

19. Flater, D. A Brief Introduction to XTide. *Linux J.* **1996**, *32*, 359–360.

20. Michalakes, J.; Dudhia, J.; Gill, D.; Klemp, J.; Skamarock, W. *Design of a Next-Generation Regional Weather Research and Forecast Model. Mesoscale and Microscale Meteorological Division*; Technical Report for National Center for Atmospheric Research: Boulder, CO, USA, 1998.

21. Duran, R. *Sub-Grid Parameterizations for Oceanic Oil-Spill Simulations*; Technical Report for NETL-TRS-9-2016 EPAct Technical Report Series; U.S. Department of Energy National Energy Technology Laboratory: Albany, OR, USA, 2016. Available online: https://edx.netl.doe.gov/ro/dataset/sub-grid-parameterizations-for-oceanic-oil-spill-simulations (accessed on 14 August 2018).

22. Samaras, A.G.; De Dominicis, M.; Archetti, R.; Lamberti, A.; Pinardi, N. Towards improving the representation of beaching in oil spill models: A case study. *Mar. Pollut. Bull.* **2014**, *88*, 91–101. [CrossRef] [PubMed]

23. Meier, H.M.; Höglund, A. Studying the Baltic Sea Circulation with Eulerian Tracers. In *Preventive Methods for Coastal Protection*; Springer: Heidelberg, Germany, 2013; pp. 101–129, ISBN 978-3-319-00440-2.

24. Soloviev, A.V.; Lukas, R. *The Near-Surface Layer of the Ocean: Structure, Dynamics, and Applications*, 2nd ed.; Springer: New York, NY, USA, 2014; p. 552, ISBN 978-94-007-7621-0.

25. Weisberg, R.H.; Lianyuan, Z.; Liu, Y. On the movement of Deepwater Horizon Oil to northern Gulf beaches. *Ocean Model.* **2017**, *111*, 81–97. [CrossRef]

26. Laxague, N.J.; Özgökmen, T.M.; Haus, B.K.; Novelli, G.; Shcherbina, A.; Sutherland, P.; Guigand, C.M.; Lund, B.; Mehta, S.; Alday, S.; et al. Observations of near-surface current shear help describe oceanic oil and plastic transport. *Geophys. Res. Lett.* **2017**, *45*, 245–249. [CrossRef]

27. Garraffo, Z.D.; Mariano, A.J.; Griffa, A.; Veneziani, C.; Chassignet, E.P. Lagrangian data in a high-resolution numerical simulation of the North Atlantic: I. Comparison with in situ drifter data. *J. Mar. Syst.* **2001**, *29*, 157–176. [CrossRef]

28. Carniel, S.; Warner, J.C.; Chiggiato, J.; Sclavo, M. Investigating the impact of surface wave breaking on modeling the trajectories of drifters in the northern Adriatic Sea during a wind-storm event. *Ocean Model.* **2009**, *30*, 225–239. [CrossRef]

29. Cucco, A.; Quattrocchi, G.; Satta, A.; Antognarelli, F.; De Biasio, F.; Cadau, E.; Zecchetto, S. Predictability of wind-induced sea surface transport in coastal areas. *J. Geophys. Res. Oceans* **2016**, *121*, 5847–5871. [CrossRef]

30. Clarke, A.J.; Van Gorder, S. The relationship of near-surface flow, Stokes drift, and the wind stress. *J. Geophys. Res. Oceans* **2018**, *123*. [CrossRef]

31. Nakata, K.; Sugioka, S.I.; Hosaka, T. Hindcast of a Japan Sea oil spill. *Spill Sci. Technol. Bull.* **1997**, *4*, 219–229. [CrossRef]

32. Rhines, P.B. The Dynamics of unsteady currents. *Mar. Model.* **1977**, *6*, 189–318.

33. Samelson, R. *The Theory of Large-Scale Ocean Circulation*; Cambridge University Press: Cambridge, UK, 2011; p. 39.

34. Keating, S.R.; Smith, K.S.; Kramer, P.R. Diagnosing Lateral Mixing in the Upper Ocean with Virtual Tracers: Spatial and Temporal Resolution Dependence. *J. Phys. Oceanogr.* **2011**, *41*, 1512–1534. [CrossRef]

35. Nordam, T.; Duran, R. Variable time-step integrators for marine transport applications. **2018**, unpublished work.

36. Samelson, R.M. Lagrangian motion, coherent structures, and lines of persistent material strain. *Annu. Rev. Mar. Sci.* **2013**, *5*, 137–163. [CrossRef] [PubMed]

37. Haller, G. Lagrangian Coherent Structures. *Annu. Rev. Fluid Mech.* **2015**, *47*, 147–162. [CrossRef]

38. Le Henáff, M.; Kourafalou, V.H.; Paris, C.B.; Helgers, J.; Aman, Z.M.; Hogan, P.J.; Srinivasan, A. Surface Evolution of the Deepwater Horizon Oil Spill Patch: Combined Effects of Circulation and Wind-Induced Drift. *Environ. Sci. Tech.* **2012**, *46*, 7267–7273. [CrossRef] [PubMed]

39. Barker, Chris. (NOAA, Seattle, WA, USA). Personal Communication, 2018.

40. Cushman-Roisin, B.; Beckers, J. *Introduction to Geophysical Fluid Dynamics: Physical and Numerical Aspects*; Academic Press: Waltham, MA, USA, 2011; pp. 171–175, ISBN 978-0-12-088759-0.

41. Durran, D.R. *Numerical Methods for Fluid Dynamics with Applications to Geophysics*, 2nd ed.; Springer-Verlag: New York, NY, USA, 2010; pp. 98–100, ISBN 978-1-4419-6411-3.

42. Csanady, G.T. *Circulation in the Coastal Ocean*; Springer: Dordrecht, The Netherlands, 1982; pp. 22–24, ISBN 978-90-277-1400-8.

43. Madsen, O.S. A Realistic Model of the Wind-Induced Ekman Boundary Layer. *J. Phys. Oceanorg.* **1977**, *7*, 248–255. [CrossRef]

44. Hearn, C.J. *The Dynamics of Coastal Models*; Cambridge University Press: Cambridge, UK, 2008; pp. 150–156, ISBN 9780511619588.

45. Fedorov, K.N.; Ginzburg, A.I. *The Near-Surface Layer of the Ocean*; CRC Press: Boca Raton, FL, USA, 1992; pp. 109–110, ISBN 978-1-4665-6453-4.

Journal of
Marine Science and Engineering

MDPI

Article

A Modeling Study on the Oil Spill of *M/V Marathassa* in Vancouver Harbour

Xiaomei Zhong [1,*], Haibo Niu [2], Yongsheng Wu [3], Charles Hannah [4], Shihan Li [2] and Thomas King [3]

1   Department of Environmental Engineering, Faculty of Engineering, Dalhousie University, Halifax, NS B3H 4R2, Canada
2   Department of Engineering, Faculty of Agriculture, Dalhousie University, Truro, NS B2N 5E3, Canada; haibo.niu@dal.ca (H.N.); sh987503@dal.ca (S.L.)
3   Ocean and Ecosystem Science Division, Fisheries and Ocean Canada, Bedford Institute of Oceanography, Dartmouth, NS B2Y 4A2, Canada; Yongsheng.Wu@dfo-mpo.gc.ca (Y.W.); Tom.King@dfo-mpo.gc.ca (T.K.)
4   Ocean Science Division, Fisheries and Ocean Canada, Institute of Ocean Science, Sidney, BC V8L 4B2, Canada; Charles.Hannah@dfo-mpo.gc.ca
*   Correspondence: xm976877@dal.ca; Tel.: +1-902-956-3821

Received: 24 August 2018; Accepted: 10 September 2018; Published: 17 September 2018

check for updates

Abstract: The *M/V Marathassa* oil spill occurred on 8 April 2015 in the English Bay. In the present study, the trajectory and the transport mechanism of the spilled oil have been studied by using the three-dimensional and particle-based Oil Spill Contingency and Response (OSCAR) model forced by the Finite-Volume Community Ocean Model (FVCOM). FVCOM provided the hydrodynamic variables used by the oil spill model of OSCAR. The results showed that the fraction of the oil on the water surface and on the shoreline, as well as the amount of oil recovered were affected by the time of the initial release, the overall duration of the discharge, wind and recovery actions. The hindcast study of the *M/V Marathassa* oil spill showed that the likely starting time for the discharge was between 14:00 and 15:00, on 8 April 2015. The release may have lasted for a relatively long time (assumed to be 22 h in this study). The results of modeling in this study were found reasonably acceptable allowing for further application in risk assessment studies in the English Bay and Vancouver Harbour. The trajectory of the spill was mainly controlled by the tidal currents, which were strongly sensitive to the local coastline and topography of First Narrows and that in the central harbour. The model results also suggested that a high-resolution model, which was able to resolve abrupt changes in the coastlines and topography, was necessary to simulate the oil spill in the harbour.

Keywords: oil spill model; FVCOM; OSCAR; *M/V Marathassa* oil spill; the English Bay; Vancouver Harbour

1. Introduction

Canada has the world's largest reserves of oil sands, which are deposits of bitumen in sand or porous rock [1]. The bitumen extracted from oil sands can be upgraded into various petroleum fuels (such as gasoline, diesel and aviation fuel) via proper hydro-treating processes. Due to the increasing bitumen and heavy oil production in Canada, the Trans Mountain Expansion Project (TMEP) was proposed to increase the capacity of bitumen and heavy oil transportation via pipeline from the province of Alberta, which has the majority of oil sands in Canada, to the west coastal province, British Columbia (BC). TMEP intends to triple the pipeline transportation capacity, which will consequently increase the oil tanker traffic by seven times on the BC coast, as well [2].

The biggest import and export port on the BC coast, Port of Metro Vancouver (PMV), consists of 34 anchorages (20 in the English Bay, 8 in Vancouver Harbour, 4 in the Indian Arm, 1 in Robert

Bank, and 1 in Sand Heads), as shown in Figure 1. In 2017, PMV handled about 142 million tonnes of cargo, which is 5% more than the previous year (2016) [3]. This busy and growing vessel traffic in PMV increases the potential risk of oil spill. The Canadian Coast Guard (CCG) receives about 600 pollution reports on the BC coast every year, nearly 40 of which occur in the PMV [4]. For instance, a small oil spill took place in the English Bay (one of the PMV anchorages) on 8 April 2015, which resulted in at least 2800 L of oil released from the cargo vessel, *M/V Marathassa* [4].

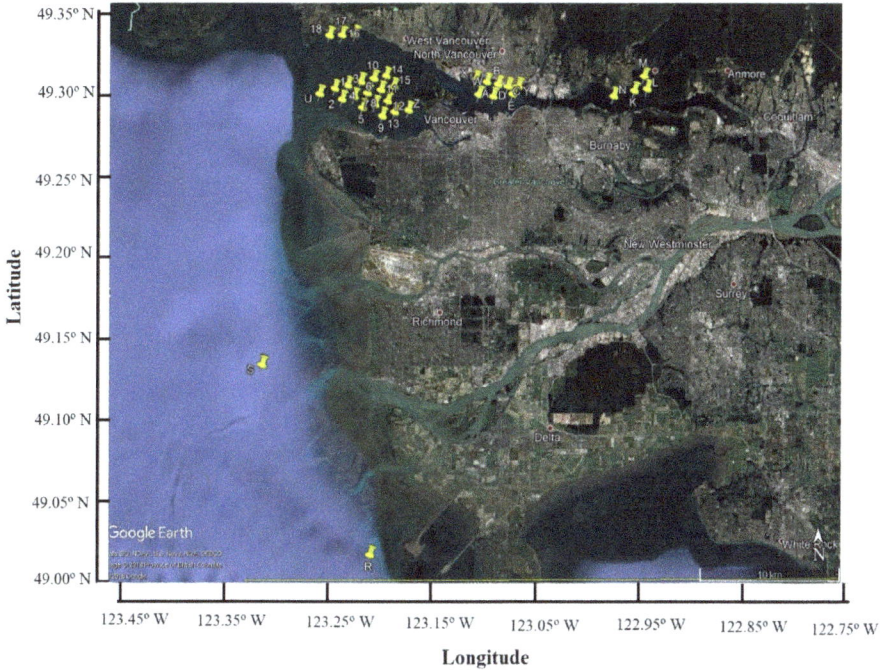

Figure 1. Anchorages' position at the Port of Metro Vancouver (PMV). Modified based on [5].

The oil spill was first reported by the public at 16:48 Pacific Time on 8 April 2015 [6]. It was suspected that IFO-380 (Intermediate Fuel Oil 380) was spilled from the *MV Marathassa* vessel, which anchored at the location of latitude: 49°17.5167′ N, longitude: 123°11.2333′ W (Anchorage #12) [7]. During this spill event, several aerial overflights, including the National Aerial Surveillance Program (NASP) flights provided by Transport Canada, were conducted to estimate the pollutant on the water surface and shoreline as shown in Table S1. At 12:20 on 9 April, it was estimated that about 2800 L of spilled oil remained on the water surface [4]. This estimate did not include any weathered and previously recovered fuel oil [4]. It was estimated that the Western Canada Marine Response Corporation (WCMRC) recovered 1400 L of spilled oil by using three vessels with skimmer equipment (Table S2) [4]. However, the type and efficiency of the skimmers were not clearly recorded. Later on, the Shoreline Cleanup Assessment Technique (SCAT) teams surveyed over 85 km of shoreline between 9 April and 23 April 2015 and determined that the most contaminated shoreline was the west side of Stanley Park, North Vancouver, and West Vancouver [7]. On 14 April 2015, the City of Vancouver provided the spilled oil distribution map shown in Figure 2, which clearly showed the observed spilled oil on the water surface and the contamination on the shoreline [8]. Unfortunately, the specific cause of this spill was not clear, and the exact volume of spilled oil was unknown.

To understand the fate/trajectory of spilled oil in the marine environment, oil spill models may be used. Examples of this type of model includes: the SPILLCALC by Tetra Tech [9], the GNOME (General

NOAA Operational Modeling Environment) model by NOAA (National Oceanic and Atmospheric Administration) [10], the OSCAR model by SINTEF (*Stiftelsen for INdustriell of TEknisk Forskning ved NTH*—Foundation for Industrial and Technical Research) [11], the OILMAP and SIMAP models by RPS-ASA (Applied Science Associates, Inc.) [12,13], the MOHID water model by MARETEC (Marine and Environmental Technology Research Center) [14] and the MIKE Oil Spill model by DHI (Dansk Hydraulisk Institut) [15]. For application to the TMEP, several organizations and consulting companies have simulated the potential risk of the oil spill in the Burrard Inlet, which geographically includes the English Bay, Vancouver Harbour, as well as in the Salish Sea (the mouth of Burrard Inlet opens onto the Salish Sea) by using various oil spill models. For example, the SPILLCALC model was used to simulate the possible trajectory of spilled diluted bitumen (dilbit) in 2013 [16]; the GNOME model was used to simulate the potential dilbit spill trajectory in 2015 [17]. However, these previously used models were limited by the following aspects: the stochastic model in SPILLCALC was 2D, which only tracked the surface transport of oil and did not provide the probability of water column contamination, and the study using the GNOME model simulated the trajectory of oil based on rough wind conditions and currents' information, but not the fate/weathering processes.

Oil spill modeling typically incorporates the modeling of hydrodynamic forcing. H3D is a 3D hydrodynamic model that has been used in several studies of the oil spill in the Salish Sea and Burrard Inlet [16,18,19]. However, the resolution of this H3D model was relatively low in the study area, with a 1 km × 1 km horizontal grid space. In order to get a more accurate hydrodynamic forcing for the Salish Sea, the NEMO (Nucleus for European Modeling of the Ocean) model has been applied. The horizontal grid space of the NEMO model was almost uniform from 440 m × 440 m to 500 m × 500 m in the Salish Sea [20]. Unfortunately, this model was unable to simulate currents in the English Bay and Vancouver Harbour. Therefore, a high-resolution hydrodynamic model was needed for the modeling of the oil spill in the English Bay and Vancouver Harbour.

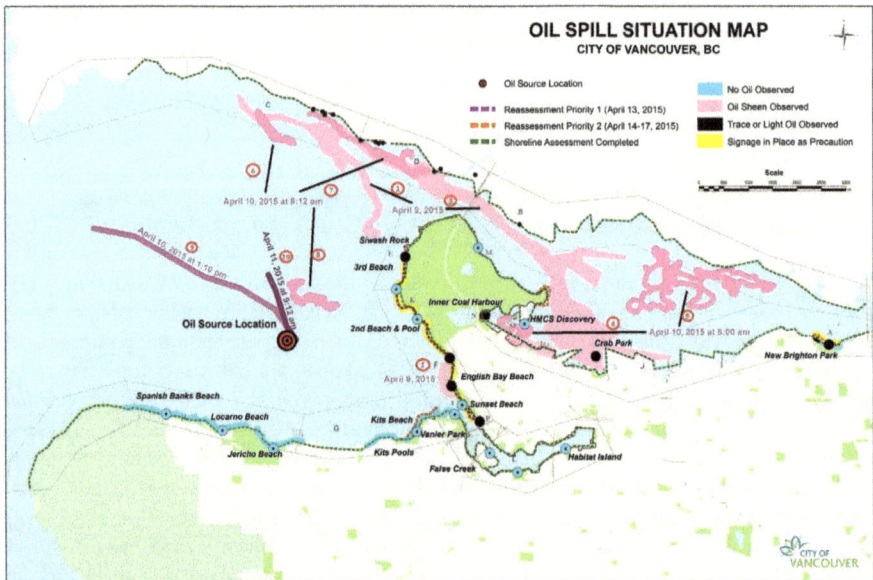

Figure 2. The *M/V Marathassa* oil spill situation map (map provided by the City of Vancouver, BC). This map shows the observed spilled oil trajectory on the water surface and the contamination on the shoreline in the English Bay and Vancouver Harbour from 8 April 2015 to 10 April 2015 [8]. Areas with oil sheen are numbered as 1–10, and contaminated shoreline areas are labeled as A–P.

This study aims to validate a three-dimensional (3D) and high-resolution hydrodynamic model (the Finite-Volume Community Ocean Model (FVCOM)) as the first step. Then, the validated FVCOM output was incorporated into a 3D oil spill model (the Oil Spill Contingency and Response model (OSCAR)) to model the oil spill in the English Bay. Forty numerical simulations were carried out to test this coupled oil spill model based on historical information from the *MV Marathassa* oil spill. Specifically, the mass balance and trajectory of *MV Marathassa* spilled oil were simulated by varying different factors, including the oil start of release time, discharge duration, wind forcing and recovery action.

2. Materials and Methods

2.1. Hydrodynamic Forcing: FVCOM

2.1.1. FVCOM Description

The hydrodynamic forcing used for this study was generated using the Finite-Volume Community Ocean Model (FVCOM). It is a 3D, finite-volume and unstructured grid ocean model, which was first developed by Chen et al. [21] and further upgraded by joint efforts from researchers at the University of Massachusetts, Dartmouth and Woods Hole Oceanography Institution [22–25]. FVCOM allows the use of different resolutions to fit complex coastline and topography by using the triangle mesh system. The model used in the present study was based on the model set up by Wu et al. [26]. The model was capable of achieving relatively high resolution in the region of interest (English Bay and Vancouver Harbour in this case), as shown in Figure 3. For instance, the horizontal grid spacing is about 10 m in Vancouver Harbour and about 2 m around the bridge bases in the Second Narrows. The vertical grid has twenty-one sigma levels that were stretched gradually, in order to gain higher resolution in the surface and bottom layers. More detailed information of the model can be found in Wu et al. [26].

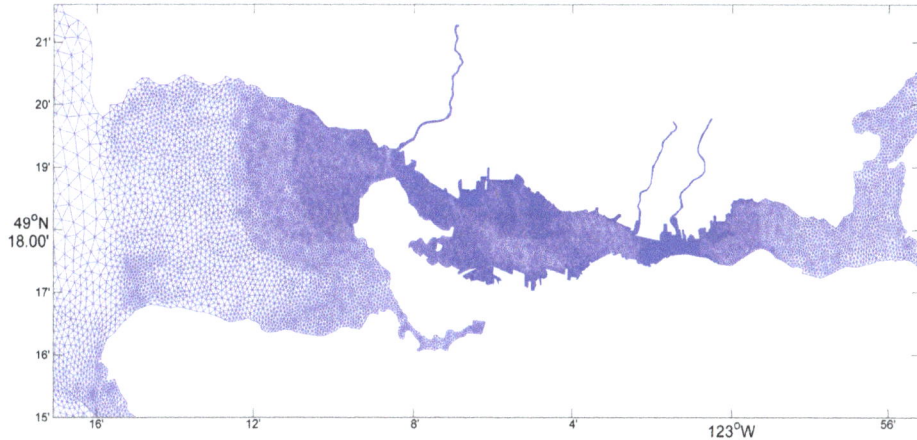

Figure 3. Finite-Volume Community Ocean Model's (FVCOM) horizontal grid in the English Bay and Vancouver Harbour. The horizontal grid space is 10 m in Vancouver Harbour and about 2 m around the bridge bases in the Second Narrows.

2.1.2. FVCOM Validation

The overall validation of the model has been done in Wu et al. [26] using tidal gauge water elevations and the ship-mounted Acoustic Doppler Current Profiler (ADCP) current data. Here, we further evaluate the model using surface drifter data, which were obtained from two Surface Current Tracker drifters (SCT). SCT is comprised mainly of wood for the structural support and cellulose sponge for floatation [27]. Four aluminum fins are mounted below the sponge to increase the surface area, and a zinc weight is installed at the very bottom of the unit to act as ballast [27]. There is also a thin aluminum disk installed above the cellulose sponge to facilitate labeling of the SCT with drifter ID and contact information [27]. SCT is a low-cost, low-impact, easily deployable drifter, tracks the surface currents and reports its location and timestamp [27]. Two SCT, named SCT1 and SCT2, were released in Vancouver Harbour (SCT1: 49°17.8812' N, 122°57.6414' W; SCT2: 49°17.8788' N, 122°57.6432' W) at 15:11 on 8 November 2015. The drifter's locations and velocities were recorded every 2–6 min. It is notable that a time step of five minutes was applied during simulations.

The modeled trajectory was compared with the observed drifters' trajectory. In addition, the prediction ability of FVCOM was statistically assessed by computing the following measures as shown in Equations (1)–(4) [28,29]: the Root-Mean-Square-Error (*RMSE*):

$$RMSE = \left\{ \frac{1}{N} \sum_{i=1}^{N} (X_{mod} - X_{obs})^2 \right\}^{\frac{1}{2}} \tag{1}$$

the relative average error (*E*):

$$E = 100\% \frac{\sum_{i=1}^{N} (X_{mod} - X_{obs})^2}{\sum_{i=1}^{N} (|X_{mod} - \overline{X}_{obs}|^2 + |X_{obs} - \overline{X}_{obs}|^2)} \tag{2}$$

the correlation coefficient:

$$R = \frac{\sum_{i=1}^{N} (X_{mod} - \overline{X}_{mod})(X_{obs} - \overline{X}_{obs})}{[\sum_{i=1}^{N} (X_{mod} - \overline{X}_{mod})^2 \sum_{i=1}^{N} (X_{obs} - \overline{X}_{obs})^2]^{\frac{1}{2}}} \tag{3}$$

and the quantitative agreement between model and observations:

$$Skill = 1 - \frac{\sum_{i=1}^{N} |X_{mod} - X_{obs}|^2}{\sum_{i=1}^{N} (|X_{mod} - \overline{X}_{obs}| + |X_{obs} - \overline{X}_{obs}|)^2} \tag{4}$$

where *X* is the variable being compared with a time mean $\overline{X}$. The subscripts "*mod*" and "*obs*" represent the model results and observations, respectively.

After validating FVCOM, it was run for 10 days (from 5 to 15 April 2015) with a time step of 1 s and saved every 1 h to generate the hydrodynamic forcing, which did not include the waves, because the study period was reported as "very calm", the surveillance photo showed no signs of breaking waves (white caps) and the non-breaking wave would also be very low due to low wind. The wave height used in the OSCAR model was computed from winds.

2.2. Oil Spill Model: OSCAR

The OSCAR model was used to simulate the mass balance and trajectories of the oil spill, based on the *MV Marathassa* oil spill's observation data in the English Bay. This is a 3D particle-based model, which is designed based on SINTEF's experimental field and laboratory data to support oil spill contingency and response decision making. The general structure of the OSCAR model is similar to most oil spill models as shown in Figure 4. The OSCAR model is capable of calculating the oil contamination on the sea surface and shorelines, in the water column and sediment, along with several oil weathering processes. Various oil weathering processes can be simulated by using the OSCAR model, including spreading, drifting, natural dispersion, chemical dispersion, evaporation, stranding, dissolution, adsorption, settling, emulsification and biodegradation of spilled oil. Overall, the OSCAR model has broad applications in oil spill modeling and has been validated in many related studies [30–33].

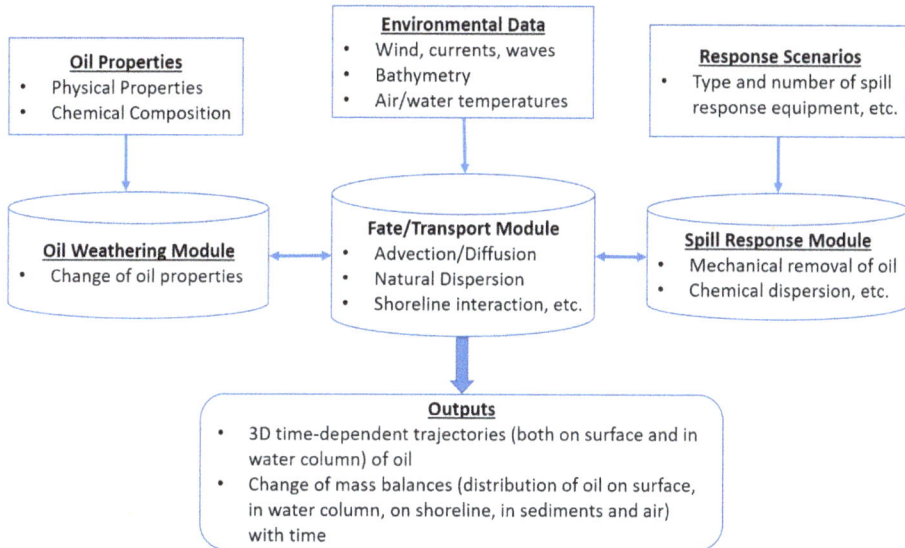

Figure 4. Schematic overview of a general oil spill model. Modified based on [30].

2.3. Wind Forcing: HRDPS Model

The present study used the wind forcing from the High-Resolution Deterministic Prediction System (HRDPS), which has been employed for weather prediction on the West Coast of Canada [34]. HRDPS is a set of the nested and Limited-Area Models (LAM) with forecast grids from the non-hydrostatic version of the Global Environmental Multiscale (GEM) model. This GEM has a 2.5-km horizontal grid spacing. The example of wind speed and direction at 16:00 on 8 April 2015 is shown in Figure 5. The dominant wind directions are south, southwest and southeast with speeds below 7 m/s near the release point from 5–12 April 2015, as shown in Figure 6.

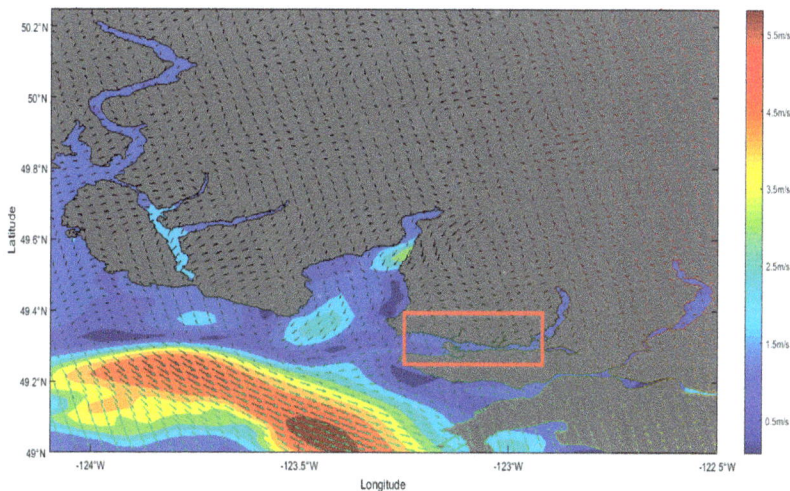

Figure 5. Wind speed and direction from the High-Resolution Deterministic Prediction System (HRDPS) (12 April 2015, 16:00). The area in the red square is the studied area.

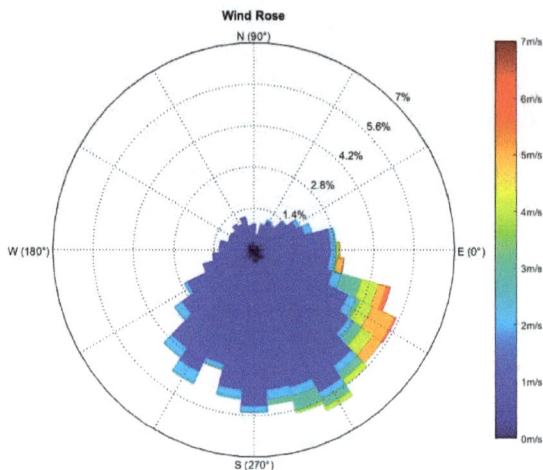

Figure 6. Speed, direction and frequencies of the wind near the oil released site (from 5–12 April 2015).

2.4. Hindcast Study of the M/V Marathassa Oil Spill

2.4.1. Identification of Spilled Oil

PMV collected the polluted samples and identified that the spilled oil had an API gravity degree of 13 and a density of 978–979 kg/m^3 (979 kg/m^3 at 15 °C). The oil chemical compositions were tested as well, and the results showed that the spilled oil contained about 96–99% of bunker fuel. Further testing on oil physiochemical properties illustrated that the spilled oil had comparable physical and chemical properties as IFO-380 [4].

IFO-380 is typically classified as a heavy fuel oil with an API gravity of 10–17.1 degrees (density of 950–1000 kg/m^3) [7,35]. It has a relatively high viscosity (maximum viscosity of 380 cSt [36–38]) and behaves as a semi-solid product at ambient temperature, which leads to a low rate of dispersion and evaporation [7,35]. The detailed chemical composition of IFO-380 was adapted from OSCAR's oil database as shown in Table S3.

2.4.2. Potentially Influential Factors

As reported, the oil probably began to spill between 11:00 and 16:48 on 8 April 2015, but the exact start time of the release is still unknown. Five possible starting times (12:00, 13:00, 14:00, 15:00 and 16:00) were explored in this study. Although the wind forcing can be obtained via the HRDPS model as illustrated in Section 2.3, the wind speed was reported as quite low (<2.6 m/s) during 8–11 April 2015 [7]. It is therefore interesting to study the spilled oil fate and trajectory without taking the influence of the wind into consideration. Because of the lack of information on the duration of oil release and the lack of documentation on the details of recovery actions, the duration of discharge and recovery actions was included in the model study as two additional factors. The discharge duration was assumed as 2 h (a case of a relative instantaneous release) and 22 h (a case of slow release over a long period of time). The case with or without recovery actions was studied to investigate the impact on the fate and trajectory of the spilled oil. It is notable that the assumptions of oil recovery actions were made based on the CCG's report, as shown in Table S4 [4] and the Western Canada Marine Response Corporation's (WCMARC) website. A summary of the above-mentioned factors that might influence the fate and trajectory of spilled oil is presented in Table 1, and detailed setup information for each simulation is shown in Table S5.

Table 1. The studied factors and their corresponding settings.

Factor	Setting				
Starting-releasing time	12:00	13:00	14:00	15:00	16:00
Wind forcing		With		Without	
Discharge duration		2 h		22 h	
Recovery action		Yes		No	

2.4.3. Deterministic Approach

The oil spill modeling can use both deterministic and stochastic approaches. A deterministic approach is used to simulate the fate and behaviour of oil from a single model run. This approach is helpful when studying a known historical oil spill event. A stochastic approach, on the other hand, is used to analyze the probability of oil contamination in the area of concern by overlaying a great number (tens to thousands) of individual deterministic simulations.

In this study, a deterministic approach was employed to study the mass balance and trajectories of the oil spill occurring on 8 April 2015. For each simulation, the oil was assumed to be released at Anchorage #12 (latitude: 49°17.5167′ N, longitude: 123°11.2333′ W) in the English Bay and then tracked for 3 days. A track duration of 3 days was used, because only a trace amount of spilled oil (5.9 L) remained on the water surface after 3 days, as reported by Transport Canada [4]. A time step of 20 min was selected to run the model. Since the hydrodynamic forcing was hourly, the use of a 20-min time step based on interpolation of current data helps to simulate a relatively smooth particle trajectory with less computation requirement compared with smaller steps (such as 1 min). The mass balances and trajectories for each individual simulation were saved every 1 h and represented by using 5000 particles. The chosen number of particles would affect the simulation to some extent. The use of 5000 is based on the preliminary test using 1000, 5000 and 10,000 particles. While the use of 1000 can produce a trajectory similar to that of 10,000, the use of a large number retains more details of the concentration field. Using 5000 can provide better details with less computational demand. This was also discussed in Reed and Hetland [39].

2.5. Statistical Analysis on Mass Balance

A full factorial design that incorporates the studied factors and their corresponding settings (Table 1) was generated by using Minitab software (version 18.1), resulting in $5 \times 2 \times 2 \times 2 = 40$ combinations in total. The mass balance (%) of oil calculated for the water surface, shoreline, water column, atmosphere, biodegradation and recovery was selected as the studied response. Analysis Of Variance (ANOVA) was carried out to evaluate the influence statistically of the studied factors on the mass balance. A *p*-value < 0.05 indicates that a certain factor has a significant influence on the mass balance. The normal distribution and constant variance on the error terms were assured during analysis, as well.

3. Results

3.1. FVCOM Validation

In order to validate the FVCOM, the simulated trajectory and velocities (U-velocity and V-velocity) from the model were compared with the observational data from the SCT drifters (SCT1 and SCT2). The simulated and observed trajectory are plotted in Figures 7 and 8 for SCT1 and SCT2, respectively. It can be noticed that both SCT1 and SCT2 moved from east to west (Central Harbour to Second Narrow to Vancouver Harbour), and the modeled trajectory was comparable to the observed trajectories for both SCT1 and SCT2.

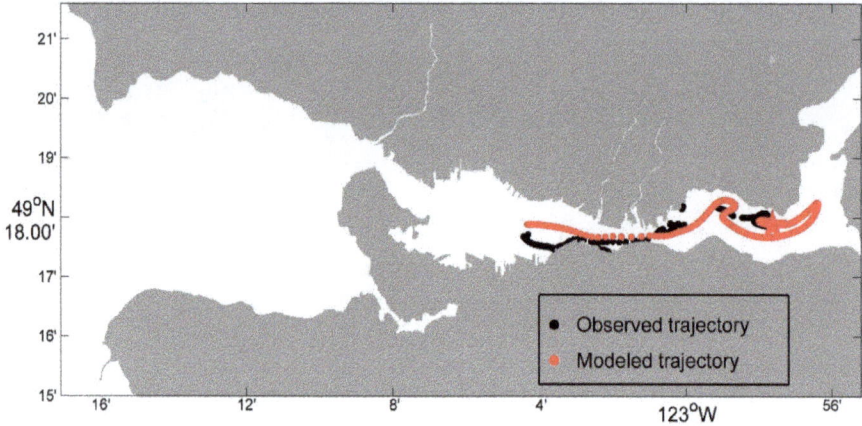

Figure 7. Trajectory comparison between Surface Current Tracker 1 (SCT1) drifter data and FVCOM data. The trajectory starts at the star.

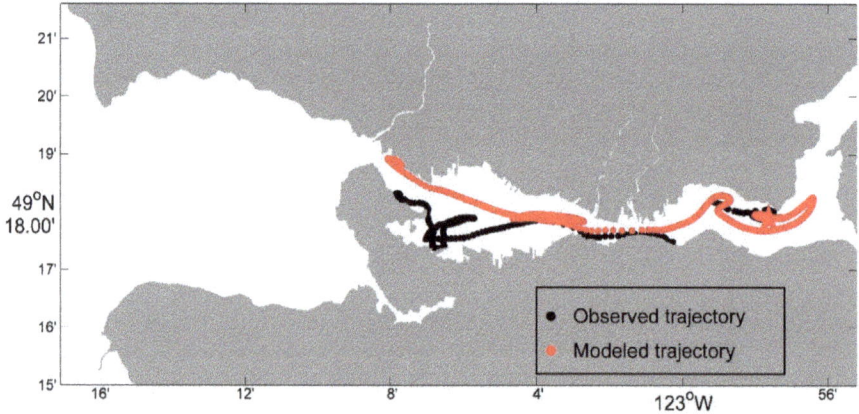

Figure 8. Trajectory comparison between SCT2 drifter data and FVCOM data. The trajectory starts at the star.

To quantify the prediction ability of FVCOM for velocities, statistical analysis, including the *RMSE*, relative average error (*E*), correlation coefficient (*R*) and skill, was carried out as presented in Table 2. Both *RMSE* and *E* for FVCOM were satisfactorily low (less than 0.16 m/s and 77%, respectively) indicating that only a slight difference existed between modeled and observed velocities. The correlation between the modeled and observed velocities was represented by *R* values, and their significance levels were indicated by *p*-values. As shown in Table 2, all *p*-values were lower than 0.05, which again demonstrated the satisfied correlation between modeled and observed velocities. The skill values were all greater than 0.51, which further verified the agreement between modeled and observed velocities.

The time series velocities from simulations and observations for SCT1 and SCT2 are plotted in Figures 9 and 10, respectively. In general, the simulated velocities matched well with that of observed data, even though some data were not recorded for unknown reasons. Overall, FVCOM was validated by using data from observed drifters, including trajectory and velocities in this study.

Table 2. Results of statistical analysis between model simulations and observations.

Statistical Measures	SCT1		SCT2	
	U-Velocity	V-Velocity	U-Velocity	V-Velocity
Root-Mean-Squared-Error (*RMSE*) (m/s)	0.149	0.056	0.158	0.052
Relative average error (*E*) (%)	48.65	76.38	73.42	68.44
Correlation coefficient (*R*)	0.618	0.256	0.310	0.383
p-value for *R*	0.000	0.001	0.003	0.000
Skill	0.719	0.514	0.577	0.551

3.2. Impacts of Studied Factors on Oil Mass Balance and Trajectory

After validating FVCOM for hydrodynamic forcing, it was incorporated into the OSCAR model to study the mass balance and trajectory of the oil spilled from the *M/V Marathassa*. Four potential factors mentioned in Section 2.4.2 might influence the spilled oil mass balance, including the release start time, oil discharge duration, wind forcing and recovery actions. The raw data on their influence on the mass balance of oil (e.g., water surface, shoreline, water column, sediments, atmosphere, biodegraded and recovered) are presented in Table S6. Since the mass balance for the water column, sediments, atmosphere and biodegraded were all less than 3% due to the very weak wind/waves, only the oil components at the water surface, on the shoreline and the oil recovered were statistically analyzed in this study. Analysis of Variance (ANOVA) was carried out, and the *p*-values for the influence of studied factors on the oil mass balance are presented in Table 3. The detailed mass balance distributions (after three days of tracking) are provided in Figure 11. In addition, the examples of trajectory comparison are shown in Figures S1–S4.

3.2.1. Influence of Release Start Time

From Table 3, it can be clearly seen that the oil start of release time had a significant impact on the mass balance of water surface, shoreline and recovered, as their *p*-values were less than 0.05. About 32.7% of spilled oil remained on the water surface and heavy contamination on the shoreline (63%) when the oil started spilling at 12:00, as shown in Figure 11a. In comparison with the 12:00 start of release time, the shoreline contamination was reduced (52.8%) along with an increased amount of spilled oil on the water surface (37.9%). If recovery was conducted, 7.46% of the oil was removed when it was released at 13:00. Interestingly, much more spilled oil (36.4%) can be recovered when the start of release time was 14:00, along with 32.9% contamination on the shoreline. The similar contamination on the shoreline was also observed if the oil spill started at 15:00 and/or 16:00, but larger amounts of spilled oil remained on the water surface (54.7% for and 70.5% for 16:00), rather than being recovered. In terms of oil trajectory, overall, the earlier the oil spill occurred, such as 12:00, the greater the contamination of the water surface and shoreline. However, this difference was not significant when the oil spill started at 14:00 and 15:00 (Figure S1).

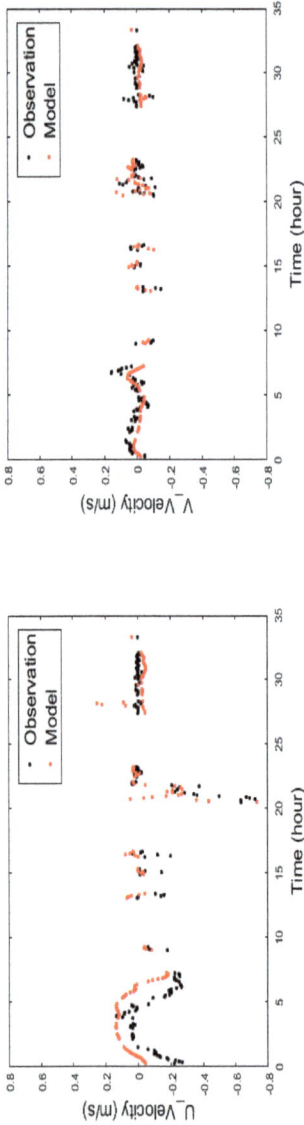

Figure 9. Time series velocities comparison between observed SCT1 drifter data and FVCOM data. U-velocity is on the **left**, and V-velocity is on the **right**.

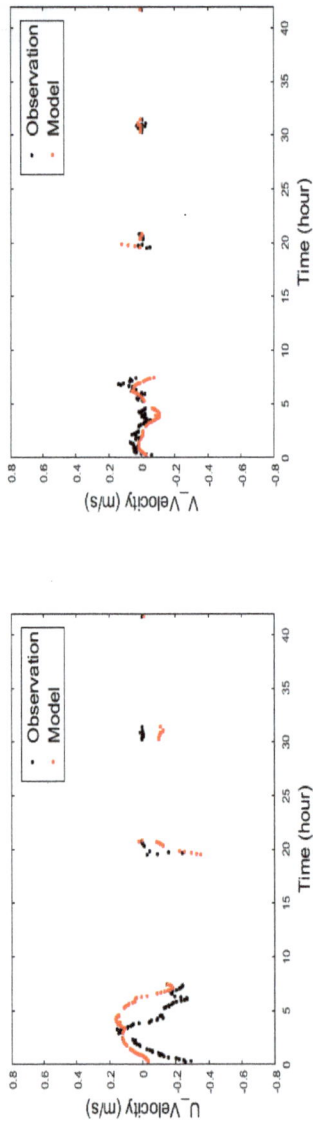

Figure 10. Time series velocities comparison between observed SCT2 drifter data and FVCOM data. U-velocity is on the **left**, and V-velocity is on the **right**.

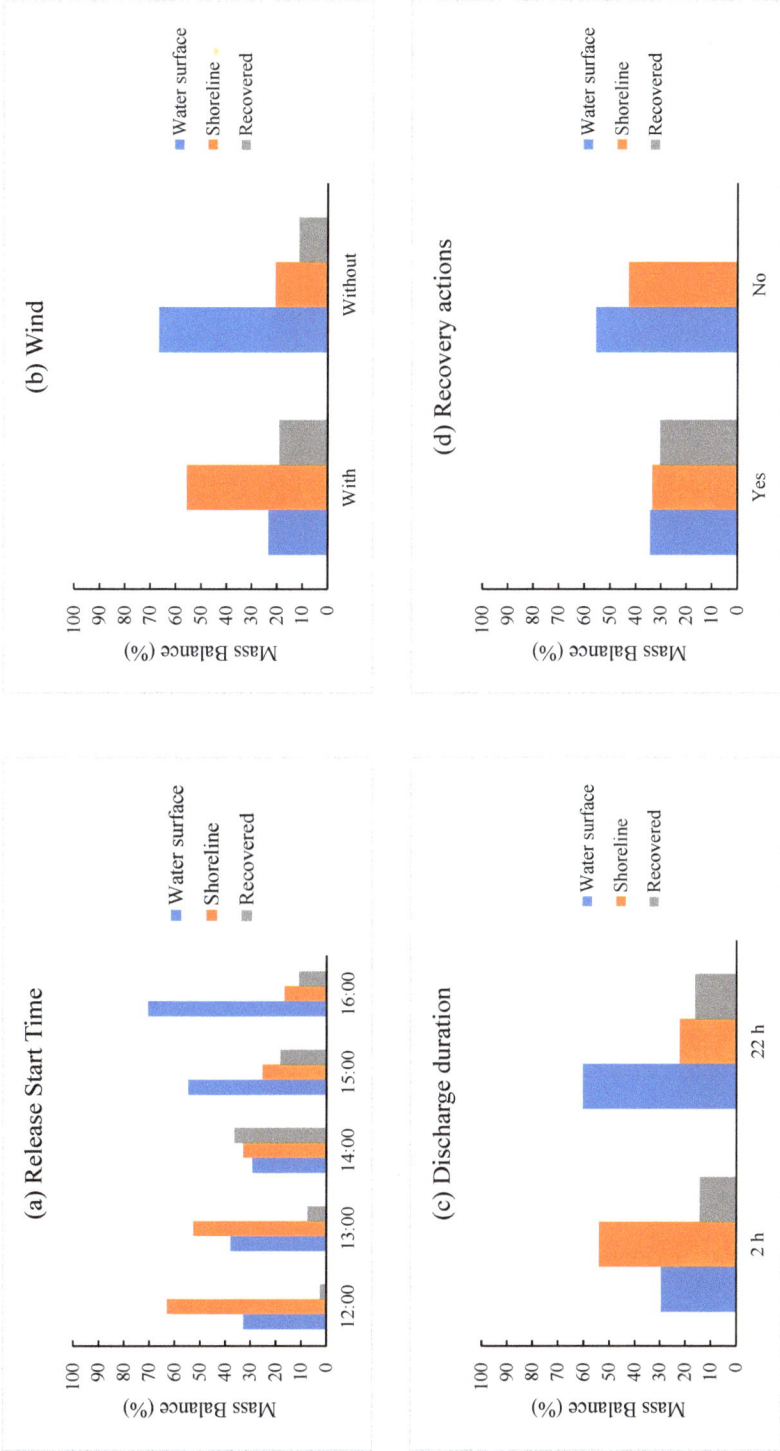

Figure 11. The influence of studied factors (a) release start time, (b) wind, (c) discharge duration and (d) recovery action on the mass balance of water surface, shoreline and recovered.

3.2.2. Influence of Wind

The effect of the wind was to increase the amount of oil on the shoreline and decrease the amount of oil on the water surface compared to the no-wind simulation (Table 3 and Figure 11b). Specifically, the fraction of oil remaining on the surface decreased from 67% to 23%, and the amount of oil on the shore increased from 21% to 56%. This can be seen in the trajectory results (Figure S2), which illustrate the heavy contamination on the shoreline of West Vancouver and the western side of Stanley Park. The amount of oil recovered stayed roughly constant at 15–20% and was not influenced by wind forcing in this study.

3.2.3. Influence of Discharge Duration

A short discharge duration (2 h) led to more serious shoreline contamination (54.1% vs. 22.2%) than that of long discharge duration (22 h) and resulted in less oil on the water surface (29.6% for 2 h vs. 54.1% for 22 h). Most of the contaminant was still concentrated on the water surface around the release location after 3 days tracking when a long discharge duration was taken into consideration (Figure S3). The discharge duration did not play a significant role in the amount of oil recovered.

3.2.4. Influence of Recovery Action

Whether the recovery action did not significantly influence the mass balance of the oiled shoreline (Table 3), as well as the oil trajectory (Figure S4), only about 34.4% of the spilled oil remained on the water surface, if the recovery action removed 30.2% of the oil, and 33.5% ended up on the shoreline, as shown in Figure 11d. When no recovery action was taken (0% of recovered oil), 55.6% of spilled oil remained on the water surface, and 42.8% contaminated the shoreline.

Table 3. The *p*-values for the influence of studied factors on oil mass balances. Significant influence (*p*-value < 0.05) is shown in bold.

Source	Water Surface	Shoreline	Recovered
Start-releasing time	0.000	0.001	0.008
Wind	0.000	0.000	0.196
Discharge duration	0.000	0.000	0.760
Recovery action	0.003	0.179	0.000

4. Discussion

4.1. FVCOM Validation

In general, the simulated trajectory and velocities from the FVCOM were comparable with that of SCT drifters in this study. However, it was relatively less capable of predicting SCT2 U-velocity, as shown in Table 2. This is likely due to the following three reasons: (1) The SCT drifter used in this study was a shallow water drifter that worked close to the water surface. This type of shallow drifter was therefore susceptible to the surrounding windage, which could potentially cause higher uncertainty on recorded data. This was supported by a similar statement that was proposed by Halverson et al. [40] to explain the inconsistency of radial and observed velocities. (2) Relatively more observed data of SCT2 velocity were missed, which resulted in a less thorough comparison of modeled and observed data. (3) The difference between the model and the drifters may also be due to the winds and waves, which are not included in FVCOM.

4.2. Hindcast of the MV Marathassa Oil Spill

4.2.1. Comparison of Oil Trajectory

The model simulations of oil trajectory were evaluated and compared with the observed oil distributions, as shown previously in Section 1 (Table S1 and Figure 2). The oil distribution map indicated the observed oil trajectory on the water surface and the contamination on the shoreline in the English Bay and Vancouver Harbour from 8 April 2015 to 10 April 2015 [8]. The contaminated water surface area was labeled as 1–10, and the contaminated shoreline area was labeled as A–P. The comparison of the results of modeled and observed for water surface and shoreline contamination are listed in Tables S7 and S8, respectively. Four scenarios achieved the highest matches with the observation data. The studied factors' setting in these four scenarios was: (1) oil started to release at 14:00, discharged continuously (22 h), with wind and without recovery actions (labeled as Scenario #4 in Table S5); (2) oil started to release at 14:00, discharged continuously (22 h), with wind and recovery actions (labeled as Scenario #8 in Table S5); (3) Scenario #4, which started to discharge at 15:00; (4) Scenario #8, which started to discharge at 15:00.

As described in Sections 3.2.1 and 3.2.4, there was almost no difference between the oil trajectories whether or not the recovery action was used, and the difference of trajectories was not significant when oil started to discharge at 14:00 and 15:00. Therefore, as shown in Tables 4 and 5, those four scenarios mentioned above have achieved 70% and 62.5% matches in the comparison of surface contamination and shoreline contamination, respectively.

The oil trajectory in Scenario #4 that started at 14:00 is plotted in Figure 12. It can be seen that oil was first transported east of the oil release point and then moved to the southwest in the next twelve hours under the forcing of hydrodynamics and wind. Spilled oil was forced and moved into the First Narrow and eventually entered into Vancouver Harbour, which resulted in heavy oil contamination on the water surface and the shoreline around Vancouver Harbour and the First Narrow. There was no oiled shoreline until 19:00 (9 April 2015) when the oil reached English Bay Beach, which conformed well to the observed information [4,7]. The majority of shoreline contamination was on the west side of Stanley Park, West Vancouver, and North Vancouver, which matched the observation data well [4,7].

Table 4. Examples of water surface contaminant comparison. The simulated results were compared with observation data.

Start-Releasing Time	Scenarios #	Labels of Surface Contaminant										Matches (%)
		1	2	3	4	5	6	7	8	9	10	
14:00	4	√	×	√	√	√	√	√	√	√	×	70
	8	√	×	√	√	×	√	√	√	√	×	70
15:00	4	×	√	√	√	√	×	√	√	√	×	70
	8	×	√	√	√	√	×	√	√	√	×	70

Scenario #4 represents oil discharged continuously (22 h), which then moves with the wind and without recovery actions; Scenario #8 represents oil discharged continuously (22 h), which then moves with wind and recovery actions. Detail factors' setting in each scenario is shown in Table S5. "×" means the simulated results do not match with the observed data; "√" indicates the simulated results match the observed data.

Table 5. Examples of shoreline contaminant comparison. The simulated results were compared with observation data.

Time to Start Spill	Scenarios #	Labels of Shoreline Contaminant															Matches (%)
		A	B	C	D	E	F	G	H	I	J	K	L	M	N	O	P	
14:00	4	×	√	√	√	√	×	√	√	√	√	×	√	×	√	×	×	62.5
	8	×	√	√	√	√	×	√	√	√	√	×	√	×	√	×	×	62.5
15:00	4	×	√	√	√	√	×	√	√	√	√	×	√	×	√	×	×	62.5
	8	×	√	√	√	√	×	√	√	√	√	×	√	×	√	×	×	62.5

Scenario #4 represents oil discharged continuously (22 h), which then moves with the wind and without recovery actions; Scenario #8 represents oil discharged continuously (22 h), which then moves with wind and recovery actions. Detail factors' setting in each scenario is shown in Table S5. "×" means the simulated results do not match with the observed data; "√" indicates the simulated results match the observed data.

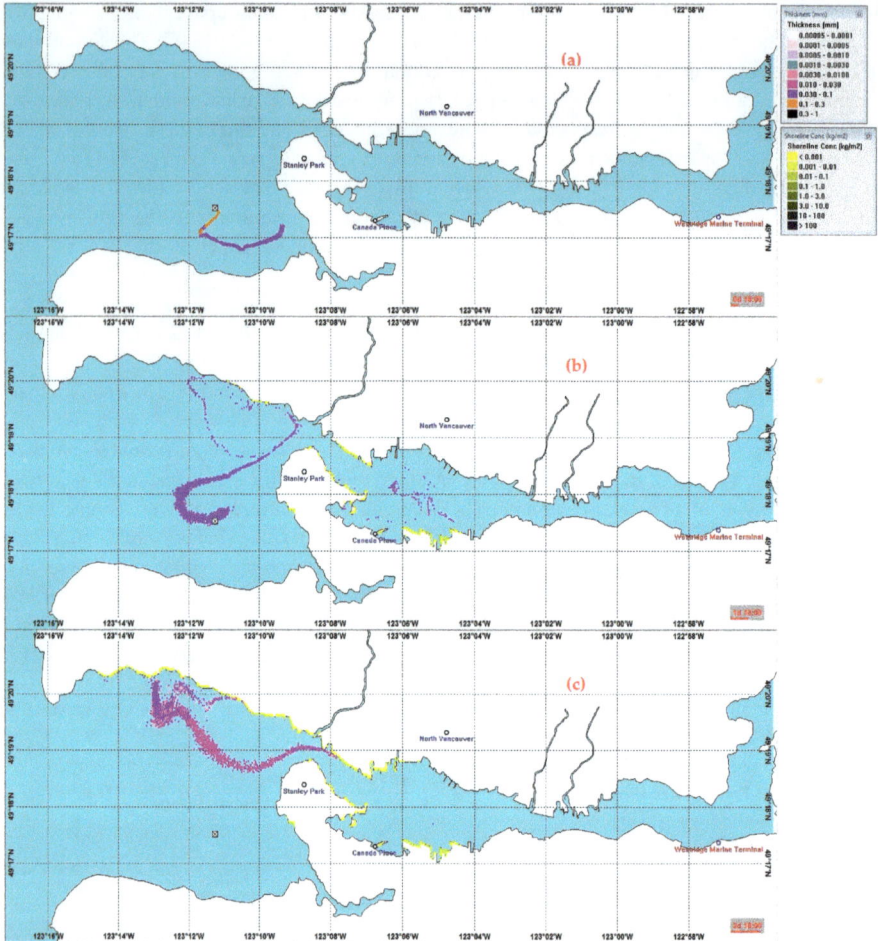

Figure 12. Example of oil trajectories for the oil spill that started to discharge at 14:00 on 8 April 2015. Spilled oil discharged continuously (22 h) and then tracked with wind and without recovery actions (labeled as Scenario #4 in Table S5). Figures from top to bottom are oil distribution at 8:00 on (**a**) 9 April, (**b**) 10 April and (**c**) 11 April 2015.

4.2.2. Comparison of Mass Balance

The oil mass balance in the simulations of the above-mentioned four scenarios was compared with that from the 2D, Automated Data Inquiry for Oil Spills (ADIOS, version 2.0) model (in CCG's report) [7]. In CCG's report, the ADIOS2 model was employed to study the mass balance of spilled IFO-380. Three metric tons (about 3067 L) of IFO-380 were assumed to be spilled at 4 Coordinated Universal Time (UTC) on 9 April 2015 (18:00 Pacific time on 8 April 2015) and then tracked for five days. A constant wind speed of 10 knots (about 5.14 m/s) was selected in CCG's modeling [7].

The modeling results from the ADIOS2 model (Figure 13) indicated that approximately 11% and 2% of the oil was expected to evaporate and disperse, respectively after three days post spill. The other 87% of spilled oil was expected to remain on the water surface. This proportion of evaporation and remaining oil was totally different from OSCAR's modeling. In Scenario #4, as shown in Figures 14 and 15, around 1.4% of spilled oil was predicted to evaporate after three days of tracking. About 24.9% and 43.1% of the spilled oil were expected to contaminate the shoreline in Scenario #4 that started

J. Mar. Sci. Eng. **2018**, *6*, 106

at 14:00 and 15:00, respectively. There are a number of reasons that could contribute to the different mass balance in the ADIOS2 and OSCAR models. The first main reason is that oil was trapped on shorelines as observed by oil spill responders. This process was included in the OSCAR model because it is a 3D fate/transport model that uses geographic and bathymetry data. By contrast, ADIOS2 is a weathering only model, and it has a limitation in accurately representing the significant onshore component. Another main difference is the evaporation, of which the rate is affected by wind, wave, currents and temperature [7]. The wind and currents conditions are very different in this study and CCG's modeling.

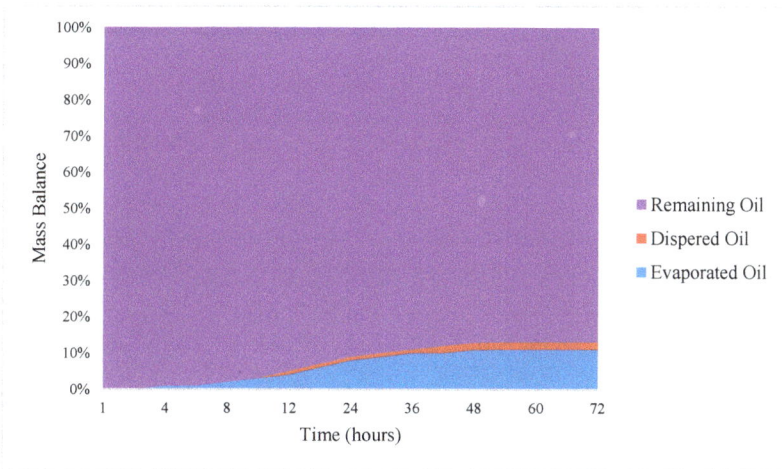

Figure 13. Automated Data Inquiry for Oil Spills (ADIOS, version 2.0) model's predictions of evaporated, dispersed and remaining (surface) Intermediate Fuel Oil 380 (IFO-380) oil after three days of simulation.

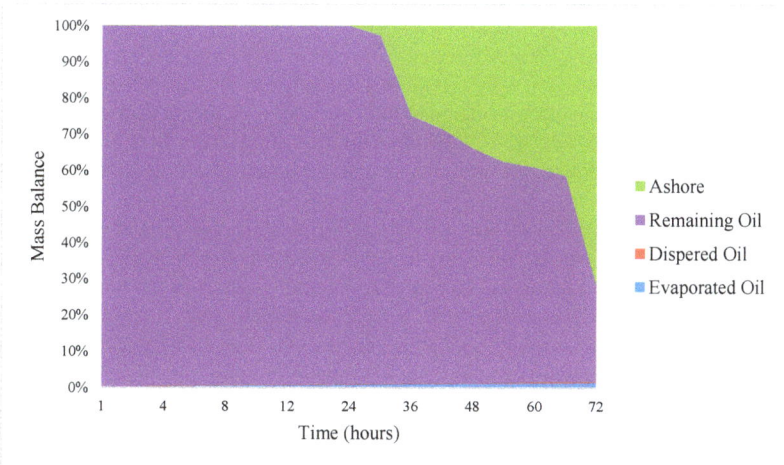

Figure 14. OSCAR model's predictions of evaporated, dispersed, remaining (surface contaminant) and ashore (shoreline contaminant) IFO-380 oil after three days of tracking in Scenario #4 and started at 14:00.

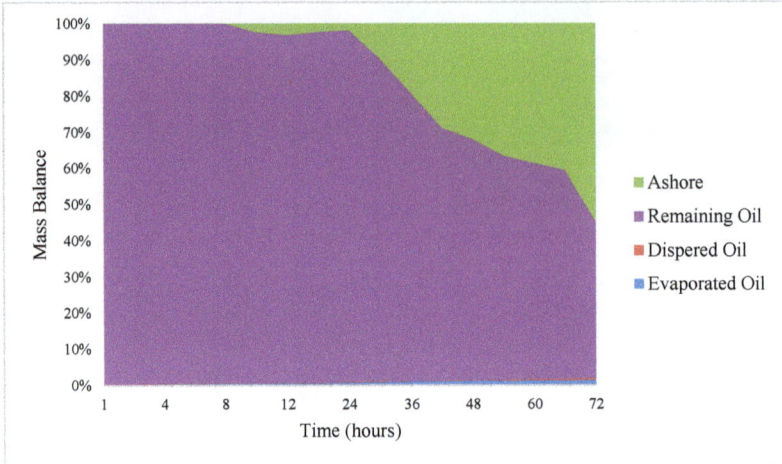

Figure 15. OSCAR model's predictions of evaporated, dispersed, remaining (surface contaminant) and ashore (shoreline contaminant) IFO-380 oil after three days of tracking in Scenario #4 and started at 15:00.

By comparison, a lesser proportion of the spilled oil remained on the water surface in Scenario #8 due to recovery actions. Nearly 61.8% (1730 L) and 65.5% (1834 L) of the spilled oil were recovered when oil was discharged at 14:00 and 15:00, respectively. The modeled recovered oil was more than the actual volume of spilled oil recovered, which was probably due to the lack of information reported by the response vessels.

Overall, among the total number of forty studied scenarios, the results from four scenarios agree well with observations. The results indicate that the *M/V Marathassa* oil spill was most likely started between 14:00 and 15:00 on 8 April 2015. This spill was most likely a continuous slow release for an unknown period (assumed to be 22 h in this study) instead of an instantaneous release.

5. Conclusions

The FVCOM implementation for English Bay and Vancouver Harbour was further validated in this study by comparing the simulated trajectory and velocities with that of observed data from SCT drifters (SCT1 and SCT2). This validated FVCOM was then used to generate the hydrodynamic forcing in English Bay and Vancouver Harbour, which was input in the state-of-art OSCAR model to simulate the *M/V Marathassa* oil spill.

The *M/V Marathassa* oil spill event was numerically simulated to assess the ability of the coupled oil spill model. Forty scenarios were performed using the OSCAR model to study the effects of various input parameters on the fate and transportation of spilled oil. The results were compared with the available data of the *M/V Marathassa* oil spill. The trajectories from four scenarios match well with the observed data. The assumed recovery actions were performed better in the scenario of oil discharged continuously (22 h) with winds at 14:00 than that in the other simulations. The combined results of trajectory and mass balance indicated that the *M/V Marathassa* oil spill probably started between 14:00 and 15:00 (8 April 2015) and kept discharging oil for a relatively long time (assumed to be 22 h in this study). The weathering processes and movement of spilled oil and contamination distribution in the surrounding waters and coastlines were affected by wind and currents.

In general, the oil spill model integrating the OSCAR and FVCOMs has effectively simulated the offshore and onshore distributions of the *M/V Marathassa* oil spill. To our best knowledge, this is the first study that modeled the oil spill in the English Bay and Vancouver Harbour by using the OSCAR model.

Supplementary Materials: The following are available online at http://www.mdpi.com/2077-1312/6/3/106/s1: Figure S1. Example of oil trajectories for the oil spill with different oil start-releasing time. Figures from top to bottom are oil start release at (a) 12:00, (b) 13:00, (c) 14:00, (d) 15:00 and (e) 16:00; Figure S2. Example of oil trajectories for spilled oil forced without wind (top) or with winds (bottom); Figure S3. Example of oil trajectories for oil discharge instantly (top) or continuously (bottom); Figure S4. Example of oil trajectories for oil spill without (top) or with (bottom) recovery actions; Table S1. Aerial overflight surveys for the *MV Marathassa* oil spill; Table S2. Western Canada Marine Response Corporation's (WCMRC) response to the spill; Table S2. The chemical composition of IFO 380 in the OSCAR model; Table S4. Assumptions for mechanical response strategies (recovery actions); Table S5. Factors' setting in each simulation; Table S6. The influence of studied factors on the mass balance of *MV Marathassa* spilled oil; Table S7. Water surface contaminant comparison. The simulated results were compared with observation data; Table S8. Shoreline contaminant comparison. The simulated results were compared with observation data.

Author Contributions: The paper was conceived of and written by X.Z., H.N. and Y.W. X.Z. carried out the model simulations and pre- and post-processed the results under the supervision of H.N and Y.W. C.H. provided drifter observation data. S.L. helped with the analysis of drifter data. T.K. provided guidance on oil weathering behaviors.

Funding: This research received funding from the Natural Sciences and Engineering Research Council of Canada (NSERC), the Marine Environmental Observation Prediction and Response Network (MEOPAR), the National Contaminants Advisory Group (NCAG) and the Fisheries and Oceans Canada (DFO).

Acknowledgments: The authors acknowledge the Natural Sciences and Engineering Research Council of Canada (NSERC), the Marine Environmental Observation Prediction and Response Network (MEOPAR), the National Contaminants Advisory Group (NCAG) and the Fisheries and Oceans Canada (DFO). SINTEF Marine Environmental Technology is thanked for providing the OSCAR oil spill model.

Conflicts of Interest: The authors declare no conflicts of interest.

References

1. Hein, F.J.; Leckie, D.; Larter, S.; Suter, J.R. Heavy oil and bitumen petroleum systems in Alberta and beyond: The future is nonconventional and the future is now. In *Heavy-Oil and Oil-Sand Petroleum Systems in Alberta and Beyond*; American Association of Petroleum Geologists: Tulsa, OK, USA, 2013; pp. 1–21.

2. Zmuda, K. Evaluation of the Regulatory Review Process for Pipeline Expansion in Canada: A Case Study of the Trans Mountain Expansion Project. Master's Thesis, Simon Fraser University, Burnaby, BC, Canada, April 2017.

3. Vancouver Fraser Port Authority Port of Vancouver Statistic Overview. Available online: https://www.portvancouver.com/wp-content/uploads/2018/03/2017-Stats-Overview-1.pdf (accessed on 17 July 2018).

4. Butler, J. *Independent Review of the M/V Marathassa Fuel Oil Spill Environmental Response Operation*; Martin's Marine Engineering Page: Nanaimo, BC, Canada, 2015.

5. Port of Metro Vancouver (PMV). Anchorage Positions, Port of Metro Vancouver. Available online: https://www.google.com/maps/d/viewer?mid=1TyRAenKnwvN-TQ3ZCtNR0h98t1Q&hl=en_US (accessed on 17 July 2018).

6. Gilbert, S.K. Concerns over Canadian Coast Guard Response to English Bay Oil Spill in Vancouver. Available online: http://www.oag-bvg.gc.ca/internet/English/pet_381_e_40802.html (accessed on 17 July 2018).

7. Wootton, B. *M/V Marathassa Fuel Spill Environmental Impact Assessment*; Hemmera Envirochem Inc.: Burnaby, BC, Canada, 2015.

8. Stormont, K. Stanley Park in the Wake of the English Bay Oil Spill. Available online: http://www.webcitation.org/70oOgbAj7 (accessed on 17 July 2018).

9. Tetra Tech EBA Inc. SPILLCALC Oil and Contaminant Spill Model. Available online: http://www.tetratech.com/en/projects/spillcalc-oil-and-contaminant-spill-model (accessed on 10 September 2018).

10. National Oceanic and Atmospheric Administration. GNOME User's Manual. Available online: https://response.restoration.noaa.gov/sites/default/files/GNOME_Manual.pdf (accessed on 10 September 2018).

11. SINTEF. OSCAR—Oil Spill Contingency and Response. Available online: http://www.sintef.no/en/software/oscar/ (accessed on 10 September 2018).

12. RPS-ASA (Applied Science Associates, Inc.). OILMAP Oil Spill Model and Response System. Available online: http://www.asascience.com/software/oilmap/ (accessed on 10 September 2018).

13. RPS-ASA (Applied Science Associates, Inc.). SIMAP Integrated Oil Spill Impact Model System. Available online: http://asascience.com/software/simap/ (accessed on 10 September 2018).

14. Fernandes, R. Risk Management of Coastal Pollution from Oil Spills Supported by Operational Numerical Modelling. Ph.D. Thesis, Universidade de Lisboa, Lisbon, Portugal, February 2018.

15. Danish Hydraulic Institute (DHI). DHI Oil Spill Model, Oil Spill Template, Scientific Description. Available online: http://manuals.mikepoweredbydhi.help/2017/General/DHI_OilSpill_Model.pdf (accessed on 9 September 2018).

16. Tetra Tech EBA Inc. *Modelling the Fate and Behaviour of Marine Oil Spills for the Trans Mountain Expansion Project Summary Report*; Tetra Tech: Vancouver, BC, Canada, 2013.

17. Genwest System Inc. *Oil Spill Trajectory Modeling Report in Burrard Inlet for the Trans Mountain Expansion Project*; Genwest System Inc.: Edmonds, WA, USA, 2015.

18. Trans Mountain Pipeline ULC. *Trans Mountain Pipeline ULC Trans Mountain Expansion Project Neb Heading Order Oh-001-2014 Reply Evidence*; Tetra Tech EBA Inc.: Vancouver, BC, Canada, 2015.

19. Stronach, J.A.; Hospital, A. The Implementation of Molecular Diffusion to Simulate the Fate and Behaviour of a Diluted Bitumen Oil Spill and its Application to Stochastic Modelling. In Proceedings of the 37th Arctic and Marine Oil Spill Program Technical Seminar on Environmental Contamination and Response, Canmore, AB, Canada, 3–5 June 2014; pp. 353–373.

20. Niu, H.; Li, S.; King, T.; Lee, K. Stochastic Modeling of Oil Spill in the Salish Sea. In Proceedings of the 26th International Ocean and Polar Engineering Conference, Rhodes, Greece, 26 June–2 July 2016; pp. 353–373.

21. Chen, C.; Liu, H.; Beardsley, R.C. An unstructured grid, finite-volume, three-dimensional, primitive equations ocean model: Application to coastal ocean and estuaries. *J. Atmos. Ocean. Technol.* **2003**, *20*, 159–186. [CrossRef]

22. Chen, C.; Beardsley, R.C.; Cowles, G.; Qi, J.; Lai, Z.; Gao, G.; Stuebe, D.; Xu, Q.; Xue, P.; Ge, J.; et al. *An Unstructured Grid, Finite-Volume Coastal Ocean Model: FVCOM User Manual*, 3rd ed.; Massachusetts Institute of Technology: Cambridge, MA, USA, 2011; pp. 303–315.

23. Chen, C.; Beardsley, R.C.; Cowles, G. An unstructured grid, finite-volume coastal ocean model (FVCOM) system. *Oceanography* **2006**, *19*, 78–89. [CrossRef]

24. Huang, H.; Chen, C.; Cowles, G.W.; Winant, C.D.; Beardsley, R.C.; Hedstrom, K.S.; Haidvogel, D.B. FVCOM validation experiments: Comparisons with ROMS for three idealized barotropic test problems. *J. Geophys. Res. Oceans* **2008**, *113*, C07042. [CrossRef]

25. MEDML (Marine Ecosystem Dynamics Modeling Laboratory). The Unstructured Grid Finite Volume Community Ocean Model (FVCOM). Available online: http://fvcom.smast.umassd.edu/fvcom/ (accessed on 14 September 2018).

26. Wu, Y.; Hannah, C.; O'Flaherty-Sproul, M.; MacAulay, P.; Shan, S. A modeling study on tides in the Port of Vancouver. *Anthropocene Coasts* under review.

27. Page, S.; Hannah, C.; Juhasz, T.; Spear, D.; Blanken, H. Surface Circulation Tracking drifter data for the Douglas Channel and the North Coast of British Columbia for April, 2014 to July, 2016. *Can. Data Rep. Hydrogr. Ocean Sci.* submitted for publication.

28. Spitz, Y.H.; Klinck, J.M. Estimate of bottom and surface stress during a spring-neap tide cycle by dynamical assimilation of tide gauge observations in the Chesapeake Bay. *J. Geophys. Res. Oceans* **1998**, *103*, 12761–12782. [CrossRef]

29. Warner, J.C.; Geyer, W.R.; Lerczak, J.A. Numerical modeling of an estuary: A comprehensive skill assessment. *J. Geophys. Res. Oceans* **2005**, *110*, C05001. [CrossRef]

30. Aamo, O.M.; Reed, M.; Downing, K. Oil spill contingency and response (oscar) model system: Sensitivity studies. *Int. Oil Spill Conf. Proc.* **1997**, *1997*, 429–438. [CrossRef]

31. Reed, M.; Aamo, O.M.; Downing, K. Calibration and testing of IKU's oil spill contingency and response (OSCAR) model system. *Int. Nucl. Inf. Syst.* **1996**, *28*, 689–726.

32. Reed, M.; Daling, P.S.; Brakstad, O.G.; Singsaas, I.; Faksness, L.-G.; Hetland, B.; Ekrol, N. OSCAR2000: A multi-component 3-dimensional oil spill contingency and response model. In Proceedings of the 23th Arctic and Marine Oil Spill Program (AMOP) Technical Seminar, Vancouver, BC, Canada, 14–16 January 2000; pp. 663–668.

33. Abascal, A.J.; Castanedo, S.; Medina, R.; Liste, M. Analysis of the reliability of a statistical oil spill response model. *Mar. Pollut. Bull.* **2010**, *60*, 2099–2110. [CrossRef] [PubMed]

34. HRDPS Data in GRIB2 Format. Available online: https://weather.gc.ca/grib/grib2_HRDPS_HR_e.html (accessed on 17 July 2018).

35. Srinivasan, R.; Lu, Q.; Sorial, G.A.; Venosa, A.D.; Mullin, J. Dispersant Effectiveness of Heavy Fuel Oils Using Baffled Flask Test. *Environ. Eng. Sci.* **2007**, *24*, 1307–1320. [CrossRef]

36. SL Ross Environmental Research Ltd. Spill Related Properties of IFO380 Fuel Oil. Available online: https://www.bsee.gov/sites/bsee.gov/files/osrr-oil-spill-response-research/506aa.pdf (accessed on 17 July 2018).

37. SL Ross Environmental Research Ltd. Spill Related Properties of IFO180 Fuel Oil. Available online: https://www.bsee.gov/sites/bsee.gov/files/osrr-oil-spill-response-research/506ab.pdf (accessed on 17 July 2018).

38. Meng, Q.; Wang, S.; Lee, C.-Y. A tailored branch-and-price approach for a joint tramp ship routing and bunkering problem. *Transp. Res. Part B Methodol.* **2015**, *72*, 1–19. [CrossRef]

39. Reed, M.; Hetland, B. DREAM: A Dose-Related Exposure Assessment Model Technical Description of Physical-Chemical Fates Components. In Proceedings of the SPE International Conference on Health, Safety and Environment in Oil and Gas Exploration and Production, Kuala Lumpur, Malaysia, 20–22 March 2002.

40. Halverson, M.; Gower, J.; Pawlowicz, R. *Comparison of Drifting Buoy Velocities to HF Radar Radial Velocities from the Ocean Networks Canada Strait of Georgia 25 MHz CODAR Array*; Institute of Ocean Sciences: Sydney, Australia, 2018.

Journal of
**Marine Science
and Engineering**

|MDPI|

Review

Main Development Problems of Vulnerability Mapping of Sea-Coastal Zones to Oil Spills

Anatoly Shavykin * and Andrey Karnatov

Engineering Ecology Laboratory, Murmansk Marine Biological Institute of the Kola Science Center of the Russian Academy of Sciences (MMBI KSC RAS), 183010 Murmansk, Russia; karnatov@mmbi.info
* Correspondence: shavykin@mmbi.info; Tel.: +7-921-160-2916

Received: 17 August 2018; Accepted: 3 October 2018; Published: 11 October 2018

check for
updates

Abstract: Vulnerability mapping of sea-coastal zones is an important element of oil spill response plans, environmental support for offshore projects, and the integrated management of the marine environment. The creation of such maps is a complex scientific problem. In their development, it is necessary to take into account differences in the nature of biotic and abiotic components existing in the cartographic area, dissimilarities in their relative vulnerability and significance, the seasonal variability of ecosystem components, and other factors. The purpose of this paper is to briefly review the main elements of international and Russian methods of mapping the vulnerability of sea-coastal zones to oil spills, and the development problems of such maps, including problems of using rank (ordinal) values, and to note possible solutions. Based on the analysis of key existing international and Russian approaches to vulnerability mapping, it was concluded that almost all methods of map calculations use rank (ordinal) values. However, arithmetic operations cannot be performed with them, as they lead to incorrect results. The paper shortly describes the main problems of mapping the vulnerability of sea-coastal zones to oil (the choice of the map scales and season limits for them, differences in the units of biota abundance, the calculation of relative vulnerability coefficients for the considered biotic components, the summation of the vulnerability of objects of different types, etc.). For some problems, possible solutions are outlined.

Keywords: oil spill response; sea-coastal zones; methods of vulnerability mapping to oil; the problems of vulnerability maps development; ordinal values; arithmetic operations with rank values

1. Introduction

Prospecting, extraction, and maritime transportation of oil require special attention to environmental safety, notably with respect to accidental oil spills in the shelf zones. In this regard, the problem of developing and using sensitivity/vulnerability maps of sea-coastal zones to oil is especially urgent. These maps should be used in the planning of oil spill response (OSR) activities, as well as in the course of these activities [1,2]. Such maps are able to minimize the potential damage from oil spills to natural and man-made environments. Specialists distinguish between two types of maps. Sensitivity maps represent the sensitivity of a shoreline to oil, which is ranked by the environmental sensitivity index (ESI). There is a corresponding well-developed procedure for the creation of such maps; these are widely used outside Russia [1–5]. Vulnerability maps represent the integrated vulnerability of sea areas to oil spills, describing all negative consequences, possible damage to biological resources, social and economic objects, and nature conservation territories, should a spill occur. This paper considers the water areas near the coast (sea-coastal zones), although a similar approach to vulnerability assessment is fully applicable to the water areas that are more remote from the shoreline.

The recommendations of international organizations [1,2] and some other methods [6–8], emphasize that in order to minimize oil spill damage, it is necessary to take into account both the vulnerability of sea-coastal zones and the sensitivity of the shoreline by the ESI index. We do not dwell on examples of sea oil spills because there are a lot of reviews focusing on this problem (for example, extensive bibliography [9] and bibliography of accident with oil platform in the Gulf of Mexico [10]). Detailed analysis of important incidents is given in [11–13] and many other publications. Spills statistics concerned with tanker accidents are represented in ITOPF material [14].

In this paper we consider only issues of the methodology to construct the maps of water area vulnerability. There is still no consensus on the procedure of vulnerability maps construction. These maps are compiled by different methods in different countries; they are supposed to have some general correct provisions and principles. The best situation is when the sensitivity or vulnerability maps of neighboring countries that have access to the sea are prepared by a single or similar method, and both are used in OSR, including joint operations. It is important for coordinating the actions of these countries' liquidators in large-scale oil accidents affecting the neighbor states.

It should also be noted that a general methodology for constructing the maps of a marine environment's vulnerability to oil can be used to make vulnerability maps and assess possible damage to the environment in any offshore project, for any form of anthropogenic impact. A unified methodology can be employed for constructing the maps of marine environment vulnerability to spills of oil and oil products (spills in accidents with tankers, oil platforms, underwater oil pipelines), reservoir water (accidents in shelf drilling) or suspended matter (in dredging and dumping of the ground), and acoustical action (working oil platforms, tubing, freight by large-capacity vessels). With an appropriate common algorithm, differences can only be accounted for in vulnerability coefficients of the considered groups/subgroups/biota species. This general approach is determined by the following: (1) the main base of the natural and man-made environment data for vulnerability mapping is one and the same for different types of exposure; (2) the coefficients of biota vulnerability themselves from different anthropogenic impacts may differ in one of the parameters, e.g., sensitivity and/or potential effect (the details are given below); (3) the algorithm for calculating such maps is virtually the same.

Except for OSR operations, such vulnerability maps are necessary for: (1) conducting preliminary environmental studies in the area of the possible impact of a project (in Russia they are called engineering and environmental surveys), which is required for environmental support and justification of the environmental safety of planned economic activities; (2) assessing the zone of their possible impact during normal operation and in emergency situations; (3) integrated management of marine environments, including the integrated management of coastal zones; (4) planning of environmental monitoring.

In and outside of Russia, there is experience in constructing vulnerability maps [6–8,15–18], but the calculations of water area integrated vulnerabilities are usually based on ordinal values (ranks) and the application of associated arithmetic operations. However, the latter is infeasible with rank values [19–22]. Otherwise, vulnerability maps constructed on this basis are incorrect. At the same time, refusal to use ordinal values leads to several methodic problems. Therefore, any research in this area—developing a correct methodology for constructing vulnerability maps of water areas—is significant and rather complicated, especially with regard to the total vulnerability of a large number of biological objects in one area. The Appendix A provides a brief justification and an example of the fact that using rank values in arithmetic operations, including the calculation of vulnerability maps, leads to incorrect results.

The purpose of this paper is to briefly review the main elements of international and Russian methods of mapping the vulnerability of sea-coastal zones to oil spills, and the development problems of such maps, including problems of using rank (ordinal) values, and to note possible solutions.

2. A Brief Overview of Methods for Constructing Vulnerability Maps

Vulnerability maps for oil are based on different approaches in different countries. The main methods are referred to below. In their consideration, it is more important to pay attention to the correctness of vulnerability maps, their clarity, and comprehensibility for the user, rather than the complexity or simplicity of the methodology. A rigorous approach is also significant.

2.1. The International Organizations—International Marine Organization (IMO), International Petroleum Industry Environmental Conservation Association (IPIECA), International Association of Oil & Gas Producers (IOGP)

These organizations prepared reports [1,2] on the mapping of environmentally vulnerable zones for oil spill response. In this report [2], the terms "sensitivity" and "vulnerability" are not explained, no distinction is made between them, and the term "sensitivity" is used, with very few exceptions. The term "sensitivity" always refers to the effects of marine environment pollution associated with accidental hydrocarbon spills (verbatim: 'Within this guide, 'sensitivity' always relates to the effects of accidental marine pollution involving hydrocarbons').

Vulnerability maps are a key stage in the preparation of OSR, and an essential tool for liquidators. The map scale is a very important element for the methodology and the end product. The nature and volume of necessary initial data, the volume of cartographic materials, and the possibility of liquidators to use such maps depend on scale. Let us consider the report of 2012 [2]. The maps are prepared by the working group during the preparation of OSR plans. There are three levels of spills—from Tier 1 (a small spill) to Tier 3 (a large-scale spill)—and three map scales corresponding to them: 1:10,000–1:25,000 (object-related), 1:25,000–1:100,000 (tactical), and 1:200,000–1:1,000,000 (strategic). At different stages of the OSR, they use one or more sets of maps, depending on the level of the spill.

Tactical maps are developed in the first instance and are fundamental for operations managers and field coordinators. They show: (1) the type of the shoreline by the ESI index; (2) sensitive ecosystems, habitats, biological species, and key natural resources; data on their concentration can be expressed in a simplified manner (presence/absence in ranks/points from 1 (no information) to 5 (high abundance of species)); (3) social and economic objects: ports, aquaculture, etc.; (4) logistic and other important resources for OSR; (5) potential sources of oil spills.

Strategic maps are prepared based on tactical maps; they are intended for the OSR headquarter management team. They represent: (1) the most sensitive types of the coastline, for example, only ESI 8 index as high and ESI 9–10 indices as very high; and the other types (low sensitivity) may not be shown; (2) the ranked sensitivity of ecosystems and native resources (a five-point scale from very low to very high); (3) the ranked sensitivity of socio-economic resources in a manner similar to the previous parameter. To determine the rank sensitivity of the site, there is a recommended matrix (Figure 1). It is recommended that an assessment of the site's sensitivity be undertaken in terms of the diversity of sensitive species (the abscissa axis) and the sensitivity of those species (the ordinate axis). A similar scheme can be used to rank socio-economic sites and nature conservation areas.

This is an approach based on rank assessment values. It doesn't involve performing arithmetic operations with the values of sensitivity and species diversity, although neither of the two matrix scales is a metric ratio scale (the rational zero and unit of measurement are not defined).

In our opinion, the most important point here is that this approach does not take into account the abundance (biomass or number per unit area) of individual groups/subgroups/species of biota. Abundance also largely defines the impact of oil on the site. Thus, the greater the biota abundance within a site, the more (with other conditions remaining the same) negative the consequences (possible damage) of an oil spill for it, and the more priority should be given to such a site in terms of oil spill protection. The presented matrix and the approach described in the report do not allow for this fact, since sensitivity and vulnerability are not distinguished. It was already noted that the report [2] uses only the concept of "sensitivity", which is not defined in this document. This concept can be interpreted as the presence of one or another reaction to the effect of a negative factor on one or another organism. To our mind, it is more optimal to talk of area vulnerability as of the total negative effects from oil spills, taking into account not only biota sensitivity, but also a number of other characteristics, including the abundance of biota in the area and its recoverability rate after a spill. For example, we may compare two zones. With a "medium" level of sensitivity of one or two mass species ("very low" diversity), the sensitivity of the first zone will be "medium" (Figure 1). Also with a "medium" sensitivity of species, but their "high" diversity and "low" total abundance, the sensitivity of the second zone will be "high". At the same time, the vulnerability (overall negative consequences) of the first zone may be greater than that of the second zone, just because of the greater abundance of species. In addition, the restorability rate of certain species is not taken into account at all. With other conditions remaining the same, the consequences (vulnerability of the zone) are more serious, based on the recovery rate of the species after a spill.

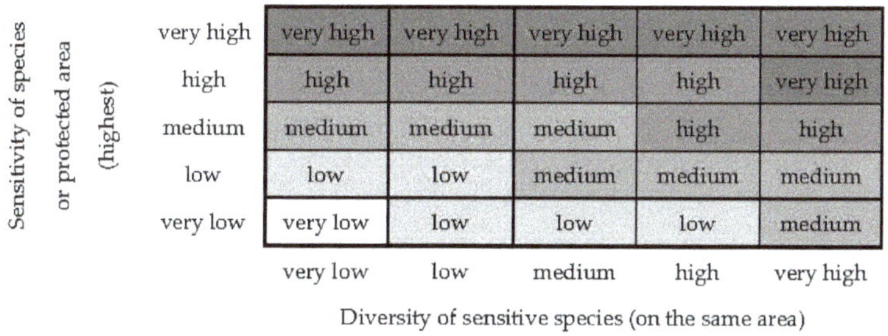

	very low	low	medium	high	very high
very high	very high	very high	very high	very high	very high
high	high	high	high	high	very high
medium	medium	medium	medium	high	high
low	low	low	medium	medium	medium
very low	very low	low	low	low	medium

Sensitivity of species or protected area (highest) — rows; Diversity of sensitive species (on the same area) — columns

Figure 1. The matrix for ranking the sensitivity of a site with a wide range of biological species and other objects [2]. This simple matrix can be used to establish a sensitivity ranking for an area where a diverse range of sensitive species is present, by comparing the sensitivity of the species/protected area with the diversity of species in that area [2] (reproduced with permission from IPIECA, 2018).

Operational maps are optional maps for liquidators and field coordinators; they are developed for the most sensitive areas and high-risk areas. These maps show detailed information about logistics and operational resources for OSR, data on the protection of specific vulnerable resources and areas, as well as information on the protection system for a specific object and details of planned OSR operations.

2.2. The Model of OSR Mapping in Norway (Modellen for Miljoprioriteringer—MOB)

The Norwegian Climate and Pollution Agency (Statens forurensningstilsyn) issued the document "Emergency Pollution Prevention—Model for prioritizing environmental resources for an acute oil spill in the coastal zone" [6]. This model is still in force. The authorities of coastal territories prepared maps according to this method. In case of an oil pollution threat to the sea areas adjacent to the territory, these maps must be used to coordinate the administration experts in oil spill response operations. The method is based on the classification of natural resources (biological, geographical, physical, chemical) and human activity objects according to four factors: natural occurrence, compensability, conservational value, and general oil vulnerability, which are estimated as ranks/points (Table 1).

Table 1. Estimation of the factor in the MOB model [6] (reproduced with permission from Miljødirektoratet, Norway, 2018).

Evaluation		Factor Value (V_X)			
		3	2	1	0
Natural occurrence?	I	-	Yes	No	-
Able to be compensated economically?	II	-	No	Yes	-
Conservational value	III	National/International	Regional	Local	Insignificant
General oil vulnerability	IV	High	Medium	Low	Insignificant

For all objects, vulnerability tables are shown on a point (or *rank*) scale: for biological objects (groups/subgroups/species of birds, marine mammals, fish)—in the integer range between 0 and 3; for the types of shores and objects of nature management—in the range between 0 and 2. After multiplying all the factors ($P = V_I \times V_{II} \times V_{III} \times V_{IV}$) for each object, they obtain one of the five priority protection categories: A ($P = 36$), B ($P = 18$ or 24), C ($P = 9$ or 12), D ($P = 4$ or 6 or 8), E ($P \leq 3$). These categories *Psx* (where *s* is the index of the calendar season, *x* is the reference number) are plotted on maps along with the boundaries of the corresponding objects' distribution or location. The maps are supplemented with a table containing a detailed description of each resource. The map scale is 1:100,000–1:200,000 (actually, they are tactical maps).

The maps take into account all of the most important biological, social, and economic resources, as well as the sensitivity of the coastline. All important information on resources is collected in a single format, which can be used both for OSR and for other environmental purposes. Two or more priority categories marked on the different areas does not change the protection priority of area where they may overlap. In addition, maps are developed for the whole year (although it is recommended that they be constructed for distinct seasons in exceptional cases). This, as well as the description of the resource priorities provided in the tables, may hinder the liquidators or coordinators in making a decision or undertaking action because they need to address both to the maps and to tables with resource descriptions in order to decide about the presence or absence of the corresponding resource in the area. Also, the use of rank values in calculating the priority category (*P*) of individual objects is not correct (see Appendix A).

2.3. The Method to Construct Environmental Vulnerability Maps in the Economic Zone of the Netherlands

The method was developed by the National Institute for Coastal and Marine Management (RIKZ, The Hague) [15]. Its experts prepared seasonal maps for the vulnerability of the Dutch Exclusive Economic Zone to various pollutants, including oil. The basis for their development is biota seasonal abundance in each map cell (5 × 5 km) and the vulnerability of biota to toxic agents. Ecosystem components should be represented by the major groups of biota (benthic and pelagic invertebrates, fish, birds, mammals), and related habitats. The method takes into account (1) biota species and their habitats for all vertical zones: near the seabed (zoo benthos), those moving freely in the water column (pelagic species), those found near the sea surface (birds) and the coast (phytobenthos); (2) the habit complex of these biota species (flight, swimming, diving, hunting) and the features of typical habitats (seabed, water column, sea surface, shore), in order to assess the potential impact of the pollutant on them.

The behavior of pollutants in the environment is divided, according to the Standard European Behavior Classification (SEBC code, Bonn Agreement), into five main types: gases, evaporators, floaters, dissolvers, sinkers. The mechanisms of their actions are based on the European criteria of persistence, bioaccumulation, and toxicity. *Vulnerability* is determined by three values: the potential effect (*E*) of the pollutant on the ecosystem component, its sensitivity (*S*) to the active factor, and the recoverability (*R*) of the component after exposure cessation. The values *E*, *S*, and *R* are calculated from a number of parameters; the final vulnerability (*V*) is calculated by the formula: $V = (E \times S)/R$. All the parameters for the calculation of these three values are estimated expertly on the basis of points/ranks, mainly in the integer range between 0 and 10. Seasonality is also considered. To construct biota distribution maps, initial abundance (spec/km², kg/m²) is normalized and reduced to non-dimension parameters. The scale of these maps is 1:100,000.

The negative point in this approach is that all the parameters used for calculations and mapping are assessed as ranks (if regarded as points, these are not "values" for the metric scale). Most likely, the maps provide strategic, small-scale reference points, especially since the objects of possible protection from oil spills and other pollutants within the area 25 km² are represented on the map as one element. The mapping model takes into account the biota and its habitats, as well as various socio-economic objects. However, there is no mention of how these two components are "added" when calculating the final area vulnerability. In a practical implementation, this method turned out to be difficult and is not used currently. Nevertheless, the very approach to the calculation of biological objects vulnerability proposed in this work seems important. In our view, it is advisable to use it in any methodology for constructing vulnerability maps.

2.4. The Method to Calculate Environmental Vulnerability to Oil Spills and Other Chemicals in the Baltic Sea (the BRISK Project)

In 2009–2012, the Baltic countries carried out the project "Sub-regional risk of spill of oil and hazardous substances in the Baltic Sea" (BRISK) [16]. It was a response to the concerns about the growth of accidents and environmental damage in the sea because of a significant increase in shipping, particularly oil tanker transportation. The project involved all the countries of the Baltic Sea, including Russia.

An integral part of the project is mapping the vulnerability of the Baltic Sea to oil and other hazardous chemical substances. Construction of vulnerability maps is a small part of the whole project; they serve as a foundation for calculating possible environmental damage in various scenarios of pollutant spills (*Risk of damage = Probability of oil spills × Vulnerability*). To reduce the risk in an area, possessing information on its vulnerability is essential.

The methodology for constructing vulnerability maps is very briefly described in the document [16]. It covers 17 different objects (natural and socio-economic resources), each with a distribution map. The maps (positions or areas without biota abundance quantities) are based on expert evaluation, taking into account the available information. The expert ecological seasonal vulnerability of each object is ranked as an integer from 0 to 4 (1—low, 4—very high). For each calendar season, all 17 maps are integrated into one map by summation of the distribution of corresponding objects, multiplied by their vulnerability coefficients. The vulnerability values on the integrated maps change in the range, approximately, between 0 and 40. This range is divided into 5 sub-ranges for better perception, from low to high vulnerability (or from dark green to red on the maps). The map scale is 1:500,000 (strategic maps).

The map resources were selected after project discussions. The vulnerability ranking is based on knowledge about the physical and biological characteristics of different ecosystems, organisms, or socio-economic resources, and their response to oil. Experts considered the behavior of oil, its potential impact on organisms and their habitats, and the recoverability of respective components after exposure; abiotic components were also taken into account. Given the incorrectness of the vulnerability calculation (because it is based on arithmetical operations with ranks), the assessments of pollutant spill damage risk are incorrect as well.

2.5. The Methodology to Construct Integrated Environmental Vulnerability Maps by CJSC "Ecoproject" and World Wildlife Fund (WWF) Russia

The leading author of the development in CJSC "Ecoproject" is Doctor of Biological Sciences V. Pogrebov [7]. The WWF methodology is the result of specialists' teamwork under the guidance of WWF Russia [8]. This methodology is completely based on the approach of CJSC "Ecoproject". The company has developed and improved it, but has not changed its principles. Starting from the 1990s, the works of V. Pogrebov and his team in creating vulnerability maps was practically unique in Russia. They greatly contributed to the development of this sphere in the country, and initiated broad discussions on the topic of working groups; their seminars were organized by WWF Russia.

According to the approach that gave a start to the Ecoproject method, the potential environmental vulnerability of a water area in a particular season is determined by the abundance of organism groups that inhabit that area and their varied vulnerability to oil [7]. The algorithm is as follows. Specialists determine the limits of seasons, define objects for evaluation (all environmental groups, from phytoplankton to birds), and make maps of abundance distribution for them (rank distribution based on points). The coefficients of biota vulnerability (in terms of sensitivity and recoverability) are evaluated expertly as integer points, in the range from 1 to 5; the potential impact of oil on the biota groups is not considered, but there are individual vulnerability coefficients for dispersed oil and oil films. In addition to biota, the method takes into account zones of special significance, e.g., water protection areas, vulnerable habitats, etc., but they do not constitute a separate group, and the vulnerability coefficients for them are not given. Initial maps are converted to geographic information system (GIS) data, represented in the form of layers on a regular grid, the cell size of which is based on the minimum size of the map contours.

Another step in the calculation of vulnerability maps is the spatial "summation" of all initial maps developed for the ranked ecological groups' abundance, taking into account their vulnerability coefficients as ordinal values. The obtained results are seasonal maps of vulnerability to oil (and/or other types of exposure). The integrated vulnerability is represented on maps by five color gradations—from green (low vulnerability) to red (high vulnerability), but the algorithm for dividing by subranges is not described. The maps have one specified scale. The maps developed according to procedure [7] are used in many Russian OSR plans and in offshore project materials. This process is based entirely on the use of ranks (non-metric points), which is not correct. Using cells instead of polygons for data representation distorts biota distribution and the positions of objects, and can hinder the orientation of liquidators during OSR.

WWF Russia's method [8] recommends the use of polygonal distribution of sensitive objects, but it also involves rank evaluation of species abundance. Two groups of objects are evaluated separately: biotic components, or important ecosystem components, and vulnerable socio-economic objects, or areas of priority protection. The map scale depends on a particular purpose and the level of oil spill: there are plans (1:10,000–1:25,000), large-scale (1:25,000–1:100,000), and small- and medium-scale (1:100,000–1:1,000,000) maps. For vulnerability factors in this method, it is proposed to use a table, which is more detailed than in [7], but also the range of the ranks is the same, from 1 to 5. The method calculations use polygonal shapefiles, but all sample vulnerability maps are presented in [8] with a division of the calculation area into individual cells, like in [7]. The coastline sensitivity maps should also support the ESI requirements.

Thus, given that these methods are based on the use of ranks, it is possible to speak of incorrectness of the maps developed with them.

2.6. The Methodology of Murmansk Marine Biological Institute (MMBI) for Constructing Vulnerability Maps

The methodology developed at MMBI [17,18] was initially based on the methods outlined in [7,8], but was fundamentally different from them in some aspects. The methods were similar in the following points: (1) all initial values for calculation (biota abundance distribution, biota vulnerability coefficients) were estimated in most cases as rank values (in the ranges 1–5 and/or 1–10); (2) the integrated vulnerability of a site was the total abundance of biota groups/subgroups multiplied by the corresponding vulnerability coefficients; (3) final vulnerability maps had the form of individual cells, although the calculation applied polygonal distribution.

However, this method differs from the Ecoproject /WWF Russia procedure in some significant points. Firstly, there were calculations of both relative and absolute integrated vulnerability [17]. The relative integrated vulnerability of the map area was presented as several (three-five) subranges of the total range of the area's integrated vulnerability in a given season. This range is different for each season. The absolute integrated vulnerability in a particular season is represented as 3–5 subranges of the total range of the area's integrated vulnerability for the whole year; the range of vulnerability is single for different seasons. Secondly, the selection of seasons was based on the limits of the periods of the year, i.e., the periods when the distribution densities of biota groups/subgroups are approximately constant [18]. With this approach, the number of seasons can differ from the number of conventional calendar periods. It should be noted that in the first version of the methodology to calculate the vulnerability maps for the eastern part of the Barents Sea [17], seasons were defined before the construction of maps for the biota groups' abundance distribution. Nevertheless, for large marine areas, it is only possible to use an approach that takes into account the seasonal variability of biota for various parts of the map area. Thirdly, it is the use of metric, dimensionless units of biota abundance (groups/subgroups distribution density), rather than rank values in subsequent specific calculations, that became possible due to normalizing the initial distributions to the average annual abundance of groups in the mapping area [18,23]. However, even partial use of rank values (relative vulnerability coefficients) also makes the methodology incorrect.

A comparison of the main elements of the vulnerability mapping method of sea-coastal zones to oil is represented in Table 2. It is possible to draw an overall conclusion from the comparison and short analysis of the considered methods: all these methods are not quite correct because they are based on the use of ordinal values (ranks) in different stages of calculations; however, this is not acceptable, and leads to incorrect maps. We recommend against the use of ranks in vulnerability maps calculation entirely.

We propose a new, more correct procedure of vulnerability mapping (briefly presented in [24,25], which is completely based on the use of only metric values. It also assumes a fundamentally different approach to assessing the specific vulnerability of biota groups with different interactions with water (see below Sections 3.4–3.6). This direction requires further research, since the "metric" methods still have unresolved questions and difficulties. Some of them are described below in Section 3.

Table 2. Main elements comparison of vulnerability mapping method of sea-coastal zones to oil.

Method Element	IPIECA, IMO, OGP [2]	Norway [6]	Netherlands [15]	BRISK Project [16]	WWF, Russia [8]	MMBI, Russia [17]
Map scale	Three different scales Operational: 1:10,000–1:25,000; tactical: 1:25,000–1:100,000; strategic: 1:200,000–1:1,000,000	One scale Less than 1:200,000, preferably 1:100,000 (separately for coast regions of each province)	One scale 1:100,000 (economic zone of the Netherlands is shown)	One scale Not indicated; probably 1:500,000 (assessment on drawing scale of maps)	Three different scales Plans: 1:10,000–1:25,000; large-scale: 1:25,000–1:100,000 small- and medium-scale: 1:100,000–1:1,000,000	One scale Maps of one scale without exact indication (mapped on 1: 1,000,000 basis)
Form of initial data presentation (polygons or cells)	Polygons	Not indicated	Cells of 5 × 5 km size	Polygons	Polygons	Polygons
Objects considered in vulnerability calculation (biota, biota habitats, social-economic objects, protected areas)	Biota, biota habitats, social-economic objects, protected areas Large list of significant ecological groups, subgroups, species of biota (except phyto- and zooplankton) and their habitats; abundance is not shown. Detailed list of social-economic objects, protected areas	Biota, biota habitats, social-economic objects, protected areas Specific species and groups of biota (shoals of fish, ichthyoplankton, benthos), habitats of sea birds and marine mammals, protected areas, recreation areas, objects of nature management	Biota, social-economic objects, protected areas Large number of main species of all biota groups (benthic and pelagic invertebrates, fish, birds, mammals) except unexpected species. Habitats of biota may be included	Biota habitats, social-economic objects, protected areas 17 objects (areas) coast habitats; areas of seagrass beds; habitats of fish, birds, marine mammals; protected areas; aquaculture objects	Biota, social-economic objects, protected areas Detailed list of significant ecological groups, subgroups, species of biota (including phyto- and zooplankton). Detailed list of priority protection areas (according to IMO, IPIECA, [26]): social-economic objects, protected areas	Biota Phyto-, zoo-, ichthyoplankton, benthos, fish (pelagic, bottom, migratory), birds, marine mammals
Mapping for various year periods (seasons): calendar seasons, months, other periods	Seasonal maps are not developed It's recommended to show the initial information about seasons and to indicate seasonal features on maps	Seasonal maps are not developed Period of object presence in the area shown on the map is given in a table; it may match or not to calendar seasons	Maps are developed for specific periods Number of periods for which maps are developed could be different	Maps are developed for calendar seasons: winter—XII-II; spring—III-V; summer—VI-VIII; autumn—IX-XI	Maps are developed for calendar seasons: winter—XII-II; spring—III-V; summer—VI-VIII; autumn—IX-XI	Maps are developed for 4 seasons: winter—I-III; spring—IV-VI; summer—VII-IX; autumn—X-XII
Units of biota abundance—ranks or natural measures (kg/m², sp/km²) and their using for vulnerability calculation	Ranks—not used in assessment (sensitivity assessment according to matrix on Figure 1) It's recommended to show on the map in the form: presence/absence or ranks from 1 to 5. Not used in sensitivity (vulnerability) mapping	Ranks—not used in calculations Biota abundance is not considered. Category of priority is calculated on the ranks basis	Natural measures—used in calculations after normalization Distribution of biota density is given in different units (kg/m², sp/km²), after normalization—non dimensional units	Absence—not used in calculations Distribution of biota density is not considered. Only habitats of biota are taken into account (see above)	Ranks—used in calculations Initial abundance distribution of significant ecological components—in natural measures (kg/m², sp/km²). They are ranked for calculations	Ranks—used in calculations Biota abundance on initial maps—ordinal values (ranks): 0, 1–3

Table 2. *Cont.*

Method Element	IPIECA, IMO, OGP [2]	Norway [6]	Netherlands [15]	BRISK Project [16]	WWF, Russia [8]	MMBI, Russia [17]
Vulnerability coefficients assessment of considered objects (biota, biota habitats, social-economic objects, protected areas)	Vulnerability coefficients are absence. Objects sensitivity is assessed as ordinal value according to matrix on Figure 1—five grades (from very low to very high)	Expert assessment. Resource sensitivity (ordinal value) is assessed according to tables	Calculated. Potential effect of pollutant, species sensitivity and its recoverability are considered for biota	Expert assessment. Assigned by ordinal values (0, 1–4). Expert table of ecological seasonal vulnerability of each resource is used	Calculated. Vulnerability of significant ecological components: multiplication of abundance (ranks) on vulnerability coefficients (ranks). Vulnerability of priority protection areas—priority protection (ranks). All ordinal values are expertly assessed on table basis. Range of ranks—0, 1–5	Calculated. Vulnerability coefficients are calculated based on lethal concentration of oil for considered ecological groups
Summation of vulnerability for objects of different nature (biotic and abiotic components)	Not carried out	Not carried out. Category of objects priority is not changed when their areas overlapping on the maps	Carried out	Carried out	Carried out	Not carried out
Total classification of separate areas (polygons or cells) with different sensitivity/vulnerability	Carried out. Sensitivity classification is carried out by matrix on Figure 1 with ordinal scales	Not carried out. Area importance is defined by value of priority category	Carried out. Values of total vulnerability are represented in grid cells on resulting maps using gradations of single color from light to dark	Carried out. Division of range into 5 classes (sizes of classes are not given)	Carried out. Size of subranges is expertly defined	Carried out. Method of equal intervals is used
Using of ranks (ordinal values) for calculations	Used for assessments of area sensitivity and biodiversity	Used for calculations of priority categories	Used for assessments of vulnerability coefficients	Used for assessments of seasonal vulnerability coefficients of each resource	Used for presentation of initial maps of significant ecological components and for vulnerability coefficients of significant ecological components and priority protection areas	Used for assessments of biota abundance

3. Main Problems in the Development of Vulnerability Maps

We believe that in order to construct sea-coastal vulnerability maps correctly, one should consider vulnerable objects and their specific (relative) vulnerability. For the correct assessment of coastal marine vulnerability to oil, it is necessary to have the following information: (1) the quantitative characteristics of seasonal spatial distribution of biota ecological groups/subgroups/species (their abundance) on all sites of the evaluated area; (2) the specific vulnerability of these biota groups/subgroups/species to oil; (3) the position of the considered abiotic (socio-economic, nature-conservative) objects in the evaluated area, which are not directly related to the biota abundance; (4) the degree of their significance for human beings. In fact, the latter parameter is the evaluation of coefficients of abiotic objects' significance, an analogue of biota specific vulnerability. All values must be appropriate for the metric ratio scale.

The development of vulnerability mapping procedures has different problems. These are the main ones, taking into consideration the absence of ranks and points.

3.1. Selection of Vulnerability Mapping Scale

It is recommended developing maps on different scales only in one method [2]. Our proposals for map scales are as follows (all scale values are tentative and require detailed discussion). Strategic small-scale maps (1:500,000 and less) give a general representation of the most vulnerable areas. For such maps (covering quite large areas of the sea), it is difficult to correctly identify the seasons, for which the maps are designed (see below). But they are important to OSR senior managers for general strategic planning, especially for large-scale spills. Tactical maps (1:100,000–1:250,000) should probably be prepared for all sea-coastal areas; they are the most numerous vulnerability maps for marine and coastal zones. Object-related maps (1:10,000–1:50,000) must be developed for the most significant marine and sea-coastal areas.

Some questions remain unanswered: what is shown on each of the vulnerability maps of different scales, what the strategic, tactical, and object-related maps have in common, and what their fundamental differences are. The work [2] gives comprehensive recommendations in this respect, and the content of information on maps with different scales is fundamentally dissimilar. We believe that vulnerability maps should be on different scales for cartographic areas. The difference between them should be specified by a generalization in transfer from large-scale to small-scale maps. In any case, this suggestion and recommendations on scales stated in report [2] should be discussed in detail.

3.2. The List of Objects for Evaluation

This is a debatable issue, especially if the development of maps involves specialists working in different fields. It is important to avoid inclusion into this list all or almost all the biota groups inhabiting the cartographic area; only the dominant, essential, and Red Book species should be considered. It is also necessary to take into account socio-economic objects and nature conservation areas. At this stage, the issue should be generally solved using methods which are in line with the recommendations given in the report of international organizations [2].

3.3. Adjustment of Limits for Seasons

It is important to decide about periods—will the map be created for the entire year, for each month, or for certain seasons (climatic, calendar, etc.)? The authors of [17] show that for large areas, e.g., for the whole eastern (Russian) part of the Barents Sea, there can be several proposed variants of such periods, based on different criteria. We believe that the most important criterion for the selection of seasons is the stability of the distribution density of vulnerable objects, primarily biological ones. While preparing the initial data, it is reasonable to start from the periods of the year within which the initial distribution (abundance) of the evaluated biota and the position of abiotic objects remain relatively constant. The seasons for mapping should be chosen with the help of these data on stability [18]. If you construct a series of vulnerability maps of a long-stretching coastal water area for a single season and deal with great variability in many parameters (there is an example of the Barents Sea coast of the Kola Peninsula), you will inevitably face discrepancies between the maps of neighboring sites of the cartographic area. The reason is differences in the time limits of the seasons for the western and eastern parts of the coastal regions. This issue also still requires discussion and solution.

3.4. The Units of Biota Abundance

Usually units of biota abundance are different—spec/km^2, kg/m^2, g/m^3, etc. Integrated map construction requires summarizing the vulnerabilities of individual biota groups (abundance of groups/subgroups/species multiplied by corresponding specific vulnerability). This cannot be done if biota distribution values have different units of measurement. That is why biota abundance must be in the same units as those that have been proposed in [15] for map calculations using cells instead of polygons. Another possible option is a transition to dimensionless units via normalizing the abundance of groups/subgroups/species to the average annual abundance of the corresponding group in the mapping area [23–25]. The following is a proposed procedure.

Determine the list of vulnerable components of the ecosystem: important biotic components (IBC), especially significant social-economic objects (ESO), and protected areas (PA). The required information on all these objects, such as results of expeditions, published works, and expert estimates, is assumed to have been collected for the area mapped.

Demarcate seasonal boundaries for this area. When boundaries of the seasons do not coincide with different biotic components (and probably for occurring ESOs), the final number and boundaries of seasons for maps of integral vulnerability are determined by boundaries of corresponding seasons of all biotic groups/subgroups/species and abiotic components that are taken into account.

Make seasonal maps of the distribution of ecosystem components for each adopted scale: IBC density (B^{sg}, s is the season index, g is the biota group index), locations of ESOs and PAs (C^e and D^f, where e and f are the indices of the corresponding abiotic objects).

Maps of distributing B^{sg} are constructed in the units that are accepted for biotic groups (benthos—g/m^2; birds—item/km^2).

Normalize seasonal maps (Equation (1)) of distribution of each gth biotic component (B^{sg}) to the annual average abundance of the corresponding group P^{gy} within the mapped area (the superscript y indicates that a year is the period under consideration):

$$B^{sg(y)} = B^{sg} / P^{gy}. \tag{1}$$

This procedure enables us to use identical units of measurement for densities of biota [23]: all biotic components that are taken into account are represented in the units of the share of the annual average abundance of the corresponding group in the mapped region per unit area: (kg/m^2)/kg = 1/m^2; (item/km^2)/item = 1/km^2 → 1/m^2.

Construct ESO maps, polygons of C^e (C^e = 1 for ESOs, the remaining water area is 0) and PA maps, polygons of D^f (D^f = 1 for PAs; the remaining water area is 0).

The normalization of the distribution density of biotic groups/subgroups/species to the annual average abundance (P^{gy}) of the corresponding groups makes it possible to represent all biotic components under study in identical units of measurement (shares of annual average abundance of the group within the mapped region per unit area).

Such an approach, like in the previous case, can lead to discrepancies between the maps of neighboring areas, because the maps for each of them are normalized, taking into account the abundance of biota groups/subgroups/species specific for each area. This issue is not finally resolved on this stage.

3.5. Coefficients of Biota (Relative) Vulnerability

Expert evaluation of vulnerability coefficients or the parameters (sensitivity, recoverability, the potential exposure to oil) required for calculating the coefficients of biota specific vulnerability to oil can easily be made in integer ranks or points [6–8,15–18]. The refusal to use rank values leads to a new problem. The indicated parameters are quite difficult to quantify, for both dominant environmental groups and well-studied individual species. Vulnerability coefficients (V) may be calculated as described (in simplified view) in [15]: $V = (E \times S)/R$. Considerable research has been devoted to sensitivity assessments of major biological groups/subgroups/species to oil (evaluation of the values LC_{50} and/or LL_{50}). The works were carried out both at the end of the last century (e.g., [27,28] and many others) and in recent years, when new methodological approaches were used (e.g., [29–32] and others). On this basis, it is possible to make quite a realistic estimation of the necessary metric values of S for most biota groups/subgroups/species. The values of S for the remaining objects will be chosen by expert evaluation, but also on the metric ratio scale. The coefficients of potential exposure (E, percentage) and recovery (R, years) are probably easier to manage, because the information about them is extensive and easily available, and the methodological problems are few. Priority protection coefficients for abiotic components should also be presented as metric values on the basis of their social and economic importance.

The choice of a certain scale of units for biota sensitivity (S) is an additional problem. It is possible to distinguish at least 4 habitats: pelagic (fish, plankton), bottom (fish, benthos), littoral (zoobenthos, macrophytobenthos, birds), and sea surface (birds), although this segmentation is quite nominal. Birds, mostly contacting with the water surface, are exposed to oil film (including exposure on littoral areas), but not to oil dissolved or dispersed in water. Fish and plankton are exposed to oil dispersed only in the water column. Thus, the sensitivities of fish/plankton and birds have different scales and cannot be fully compared. A possible solution to this problem is also briefly presented in [24,25].

Vulnerability coefficients for IBSs (V_b^g) should be calculated, and coefficients of priority protection for ESOs (V_c^e) and PAs (V_d^f) expertly evaluated. All values are given in the metric scale (the use of points and ranks is excluded).

For IBCs, the coefficients of V_b^g are estimated by three parameters (Equation (2)):

$$V_b^g = R^g \times E^g / S^g, \tag{2}$$

where R^g is expressed in years, E^g is in percent, and S^g is in units of values of oil concentrations in the water, or the thickness of an oil film on the water, that are maximum permissible for biota of the components that are taken into account; subscript b denotes the ratio of these parameters to the biota.

Initially, LC_{50} (the lethal concentration of oil in the water) or LL_{50} (the lethal load of oil) is taken as the sensitivity of the biota (parameter S_n^g) inhabiting the water column [27–32]; S_n^g is normalized to the maximum permissible concentration of oil in the water, MPC: $S^g = S_n^g / MPC$.

The thickness of an oil film that causes 50% death of biota (conventionally $S_n^g = LT_{50}$) is taken as the sensitivity of biota that mostly contacts the water surface, i.e., an oil film. Then LT_{50} is normalized to the maximum permissible thickness (MPT) of an oil film that does not produce a considerable effect on these biotic groups: $S^g = S_n^g / MPT$.

The values of coefficients of priority protection are expertly selected with respect to the ecological, economical, and/or other importance of objects for humans or the ecosystem of the region. In this case, the ratios between coefficients $V_c^{e_i}/V_c^{e_j}$ and $V_d^{f_i}/V_d^{f_j}$ (e_i, e_j, and f_i, f_j are the indices of ESO and PA objects) should reflect the ratio of importance between the corresponding objects that is the closest to reality, rather than the "ratios" of rank (ordinal) values.

The normalization of the values of S_n^g to the maximum permissible concentrations, or the maximum permissible thickness, removes the dependence of this parameter (sensitivity) from that it is related to the water column or to its surface; S^g is expressed in identical units of measurement for all biotic groups/subgroups/species.

This approach also requires more detailed separate consideration, and more precise determination of LC_{50} (LL_{50}) and LT_{50} values, including the issues of littoral and benthic communities.

3.6. Summation of Vulnerability for Objects of Different Nature

In constructing vulnerability maps, it is necessary to take into account not only the vulnerability of biota, but also the vulnerability (significance) of abiotic components. Their summation is inevitable, regardless of possible overlapping of biota distribution areas and socio-economic zones/nature conservation areas. We propose the following as a solution to this problem.

Maps of vulnerability for IBCs and maps of priority protection for ESOs and PAs (in fact, maps of ESO and PA vulnerability) based on the data obtained for each season and each scale adopted are constructed as follows.

For IBCs: $Y_b^s = \sum_g B^{sg(y)} \times V_b^g$ and normalize the values obtained for each season:

—to max Y_b^s per season for maps of relative vulnerability $Y_b^{[s]s} = Y_b^s / (max Y_b^s \ per \ season)$;

—to max Y_b^s per year for maps of absolute vulnerability $Y_b^{[y]s} = Y_b^s / (max Y_b^s \ per \ year)$.

For ESOs: $Y_c^s = \sum_e C^{es} \times V_c^e$ and normalize the values obtained for each season:

—to max Y_c^s per season for maps of relative vulnerability $Y_c^{[s]s} = Y_c^s / (max Y_c^s \ per \ season)$;

—to max Y_c^s per year for maps of absolute vulnerability $Y_c^{[y]s} = Y_c^s / (max Y_c^s \ per \ year)$.

For PAs, perform the same procedure as for ESOs.

Make seasonal maps of integral vulnerability of the region:

—for maps of relative vulnerability $Y_\Sigma^{[s]s} = K_b \times Y_b^{[s]s} + K_c \times Y_c^{[s]s} + K_d \times Y_d^{[s]s}$;

—for maps of absolute vulnerability $Y_\Sigma^{[y]s} = K_b \times Y_b^{[y]s} + K_c \times Y_s^{[y]s} + K_d \times Y_d^{[y]s}$,

where $K_{b,c,d}$ are coefficients (estimated expertly) that determine the contribution of IBCs, ESOs, and PAs to the integral vulnerability.

The range of values of vulnerability $Y_\Sigma^{[s]s}$ is divided into three subranges for each season (they are given ranks (1, 2, 3): each season has its own range of values of min ÷ max $Y_\Sigma^{[s]s}$. Here, ranks can be used, since this is the final stage of mapping and no further mathematical operations are performed with the data, except for comparisons of the obtained values shown in the maps.

The range of values of vulnerability $Y_\Sigma^{[y]s}$ is also divided into three subranges (they are given ranks (1, 2, 3): each season has a common range of values of min ÷ max $Y_\Sigma^{[y]s}$.

The segments with different ranks are shown in different colors: green (rank 1 is the minimum values of Y_Σ), yellow (rank 2), and red (rank 3 is the maximum values of Y_Σ).

In addition, here vulnerability maps for each season are developed for two different ranges. Relative vulnerability maps: for each season there is a unique range of vulnerability (min ÷ max for specific season). Absolute vulnerability maps—for all seasons of the year, there is one, common range of vulnerability (min ÷ max for the year).

3.7. Representation of Water Area Total Vulnerability (the Problem of Classification)

The maps should contain the areas of the highest and lowest vulnerability for planning operations of OSR or for the waters and shores cleaning. Representation of these zones on maps is a result of vulnerability calculations (performed with the use of GIS programs). It partly depends on the choice of the method for classifying the final range of integrated vulnerability to subranges (the methods of equal intervals, natural breaks, etc.). This choice also affects the overall picture of site vulnerabilities. The classification of subranges may possibly be such that the maps of sea-coastal zones would reflect, along the shoreline, areas with a low, medium, and high vulnerability in an approximately equal manner (proportions). This direction also requires further research.

There are other problems of mapping the vulnerability of sea-coastal areas to oil: whether it is necessary to create separate vulnerability maps for oil with various densities, how to take into account the hydrological situation in the area (for example, the density jump layer), the ice conditions, etc.

4. Conclusions

Vulnerability maps of sea-coastal zones are important elements of oil spill response plans, environmental support for offshore projects, preparation of the Environmental Impact Assessment, and integrated marine environmental management. The compilation of such maps is a complex scientific problem. When developing them, it is necessary to take into account the presence of different biotic and abiotic components in the cartographic area's ecosystem, differences in their specific vulnerability, seasonal variability of components, spills of different types of oil, and other factors.

Coastal sensitivity maps and water vulnerability maps have been developed and are being used in many countries. The brief analysis of several existing international and Russian methods for the mapping of sea-coastal vulnerability to oil allows us to draw the following conclusions. The majority of the reviewed methods are based on arithmetic operations with rank values. For all the simplicity of the approach, it is not acceptable, and leads to incorrect vulnerability maps. Using such maps, in turn, results in erroneous spill liquidators' actions in terms of the minimization of damage to the environment from both oil spills and the operations aimed to eliminate them. It is necessary to reject the use of rank estimates of parameters if the latter are used in arithmetic operations during calculation of vulnerability maps.

The article also briefly describes the main problems of constructing the maps of sea-coastal vulnerability to oil. For some problems, the main solutions are outlined. We suppose that, despite the complexity of these problems, acceptable and rational solutions exist. The necessary conditions are initial rigor of the approach to development of mapping algorithm. There is a need to consider existing mathematical rules and restrictions, processes of oil spreading in water, regularities of biota distribution and behavior, and oil impact on it. Simplifications and assumptions should be done on the later stages of calculations considering the complexity of all processes and the necessity of clearly showing the vulnerable zones on the maps. Only this will allow us to create correct and comprehensible maps of sea coast vulnerabilities to oil for OSR plans. This direction requires further investigation and the joint efforts of researchers, experts, and liquidators of spills from different countries.

Author Contributions: A.S. and A.K. analyzed presented methods and wrote the paper. All authors read and approved the final manuscript.

Funding: This research was funded by FASO number in state assignment of MMBI KSC RAS 01 2013 66838.

Acknowledgments: This paper is based on the results of scientific research performed for the state assignment of MMBI KSC RAS "Evaluation of vulnerability and ecological monitoring of Arctic ecosystems during offshore development" (State Registration No. 01 2013 66838).

Conflicts of Interest: The authors declare no conflict of interest.

Appendix

Appendix A.1 Rank Values in Arithmetic Operations

In the development of vulnerability maps, many algorithms for their calculation assume execution of arithmetic operations. Then there is a question: is it feasible to use the variables presented on ordinal (rank) scales in such calculations? This question is discussed in detail in several publications [19–22]. The sequence of numerical characteristics of a value can be denoted by any method as they increase, for example, by a series of natural numbers. These are ranks. But they are not numerical (metric) values. They are mere markers that reflect the order of objects. They do not reflect the correlation of values. Using ranks (points) is permitted if they are arithmetized. Then all the resulting dimensions on the linear ordinary scale, which have no numeric character, take the form of numerical information [21]. But this operation is usually not performed in constructing vulnerability maps.

Let us give an example of arithmetic operations when the values X and Y are represented on the metric scale with a rational zero (ratio scale) and on the rank (order) scale. Let us normalize the metric values X_m and Y_m and represent them as ranks (X_r and Y_r)—please, see the left part of the example table. The obtained products of the ranks $X_r \times Y_r$ do not reflect the sequence of products of these values $X_m \times Y_m$ on the metric scale (the right part of the example Table A1).

Table A1. Example of arithmetic operations—on the metric and on the rank (order) scales.

Initial Values X and Y on the Metric (m) and Rank (r) Scales						The Product $X \times Y$ of Metric ($X_m \times Y_m$) and Rank ($X_r \times Y_r$) Values		
X	X_m	X_r	Y	Y_m	Y_r	$X \times Y$	$X_m \times Y_m$	$X_r \times Y_r$
A	50	1	P	60	5	$A \times P$	$50 \times 60 = 3000$	$1 \times 5 = 5$
B	60	2	Q	40	4	$C \times Q$	$70 \times 40 = 2800$	$3 \times 4 = 12$
C	70	3	R	30	3	$A \times Q$	$50 \times 40 = 2000$	$1 \times 4 = 4$
D	110	4	S	10	2	$E \times S$	$120 \times 10 = 1200$	$5 \times 2 = 10$
E	120	5	T	4	1	$B \times S$	$60 \times 10 = 600$	$2 \times 2 = 4$

If actual quantities are not known and replaced with ranks, performing operations with them as operations with metric numbers leads to incorrect results. In arithmetic operations, including calculation of sea-coastal vulnerability maps to oil, it is unacceptable to use any values if all or even some of them (biota density distribution, biota vulnerability, etc.) are ranks or points (in case points are not presented on the metric ratio scale). This approach may lead to incorrect results (to incorrect vulnerability maps). The latter means that in reality the most vulnerable areas can be marked as areas having an average or even low vulnerability and vice versa. The use of such vulnerability maps in OSR will not minimize the damage from oil spills and reduce the OSR operations efficacy.

References

1. International Maritime Organization (IMO); International Petroleum Industry Environmental Conservation Association (IPIECA). *Sensitivity Mapping for Oil Spill Response*; IPIECA: London, UK, 1994; Volume 1.
2. International Petroleum Industry Environmental Conservation Association (IPIECA); International Maritime Organization (IMO); International Association of Oil & Gas Producers (OGP). *Sensitivity Mapping for Oil Spill Response*; IPIECA: London, UK, 2012.
3. Gundlach, E.R.; Hayes, M.O. Vulnerability of coastal environments to oil spill impacts. *Mar. Technol. Soc.* **1978**, *12*, 18–27.
4. Petersen, J.; Michel, J.; Zengel, S.; White, M.; Lord, C.; Plank, C. *Environmental Sensitivity Index Guidelines*; Ver. 3.0. Technical Memorandum NOS OR&R 11; NOAA Ocean Service: Seattle, WA, USA, 2002.
5. Introduction to Environmental Sensitivity Index Maps. NOAA, 2008; 56p. Available online: http://response. restoration.noaa.gov/sites/default/files/ESI_Training_Manual.pdf (accessed on 18 September 2018).
6. Statens Forurensningstilsyn (SFT). *Beredskap Mot Akutt Forurensning—Modell for Prioritering av Miljøressurser ved Akutte Oljeutslipp Langs Kysten*; TA-nummer 1765/2000; SFT: Oslo, Norway, 2004; ISBN 82-7655-401-6. (In Norwegian)
7. Pogrebov, V.B. Integral assessment of the environmental sensitivity of the biological resources of the coastal zone to anthropogenic influences. In *Basic Concepts of Modern Coastal Using*; RSHU: St. Petersburg, Russia, 2010; Volume 2, pp. 43–85. (In Russian)
8. World Wildlife Fund (WWF). *Methodological Approaches to Ecologically Sensitive Areas and Areas of Priority Protection Map Development and Coastline of the Russian Federation to Oil Spills*; Vladivostok: Moscow/Murmansk/St. Petersburg, Russia, 2012. (In Russian)

9.   Fiolek, A.; Pikula, L.; Voss, B. Resources on Oil Spills, Response, and Restoration: A Selected Bibliography. Library and Information Services Division, Current References 2010-2, June 2010, revised December 2011. Available online: ftp://ftp.library.noaa.gov/noaa_documents.lib/NESDIS/NODC/LISD/Central_Library/current_references/current_references_2010_2.pdf (accessed on 18 September 2018).

10.   Belter, C. Deepwater Horizon: A Preliminary Bibliography of Published Research and Expert Commentary. NOAA Central Library Current References Series No. 2011-01; First Issued: February 2011; Last Updated: 13 May 2014. Available online: https://repository.library.noaa.gov/view/noaa/10854 (accessed on 18 September 2018).

11.   Albaigés, J.; Bernabeu, A.; Castanedo, S.; Jiménez, N.; Morales-Caselles, C.; Puente, A.; Viñas, L. The Prestige Oil Spill. In *Handbook of Oil Spill Science and Technology*; Fingas, M., Ed.; John Wiley & Sons, Inc.: Hoboken, NJ, USA, 2015; pp. 513–545. ISBN 978-0-470-45551-7.

12.   Sweet, S.T.; Kennicutt, M.C., II; Klein, A.G. The Grounding of the Bahía Paraíso, Arthur Harbor, Antarctica: Distribution and Fate of Oil Spill Related Hydrocarbons. In *Handbook of Oil Spill Science and Technology*; Fingas, M., Ed.; John Wiley & Sons, Inc.: Hoboken, NJ, USA, 2015; pp. 547–556. ISBN 978-0-470-45551-7.

13.   Siddiqi, H.A.; Munshi, A.B. Tasman Spirit Oil Spill at Karachi Coast, Pakistan. In *Handbook of Oil Spill Science and Technology*; Fingas, M., Ed.; John Wiley & Sons, Inc.: Hoboken, NJ, USA, 2015; pp. 557–573. ISBN 978-0-470-45551-7.

14.   The International Tanker Owners Pollution Federation Limited (ITOPF). Oil Tanker Spill Statistics 2017. London, 2018. Available online: http://www.itopf.org/fileadmin/data/Photos/Statistics/Oil_Spill_Stats_2017_web.pdf (accessed on 18 September 2018).

15.   Offringa, H.R.; Låhr, J. *An Integrated Approach to Map Ecologically Vulnerabilities in Marine Waters in the Netherlands (V-Maps)*; RIKZ working document RIKZ 2007-xxx; Ministry of Transport, Public Works and Water Management, Rijkswaterstaat, National Institute for Marine and Coastal Management: Hague, The Netherlands, 2007.

16.   *Sub-Regional Risk of Spill of Oil and Hazardous Substances in the Baltic Sea (BRISK). Environmental Vulnerability*; Doc. No. 3.1.3.3, Ver. 1; Admiral Danish Fleet HQ, National Operations, Maritime Environment: Copenhagen, Denmark, 2012.

17.   Shavykin, A.A.; Ilyn, G.V. *An Assessment of the Integral Vulnerability of the Barents Sea from Oil Contamination*; MMBI KSC RAS: Murmansk, Russia, 2010. (In Russian)

18.   Shavykin, A.A. A method for constructing maps of vulnerability of coastal and marine areas from oil. Example maps for the Kola Bay. *Bull. Kola Sci. Centre RAS* **2015**, *2*, 113–123. (In Russian)

19.   Sachs, L. *Statistische Auswertungsmethoden (Methods of Statistical Analysis)*, 3rd revised and expanded ed.; Springer: Berlin/Heidelberg, Germany, 1972.

20.   Glantz, S.A. *Primer of Biostatistics*, 7th ed.; The McGraw-Hill Companies: New York, NY, USA, 2012; ISBN 978-0071781503.

21.   Khovanov, N.V. *Analysis and Synthesis of Indicators in Information Deficit*; SPbU: St. Petersburg, Russia, 1996. (In Russian)

22.   Orlov, A.I. *Organizational-Economic Modeling: In 3 Parts. Part 2: Expert Evaluation*; Bauman MSTU: Moscow, Russia, 2011. (In Russian)

23.   Shavykin, A.A.; Kalinka, O.P.; Vashchenko, P.S.; Karnatov, A.N. Method of Vulnerability Mapping of Sea-Coastal Zones to Oil, Oil Products and Other Chemical Substances. RF. Patent 2613572, 17 March 2017. (In Russian)

24.   Shavykin, A.A.; Matishov, G.G.; Karnatov, A.N. A Procedure for mapping vulnerability of sea-coastal zones to oil. *Dokl. Earth Sci.* **2017**, *475*, 907–910. [CrossRef]

25.   Shavykin, A.A.; Karnatov, A.N. Method of Vulnerability Mapping of Sea-Coastal Zones to Oil, Oil Products and Other Chemical Substances Based on Calculations with Metric Values. RF. Patent 2648005, 21 March 2018.

26.   International Maritime Organization (IMO); International Petroleum Industry Environmental Conservation Association (IPIECA). *Sensitivity Mapping for Oil Spill Response*; IPIECA: London, UK, 2010.

27.   Rice, S.D.; Moles, D.A.; Karinen, J.F.; Kern, S.; CarIs, M.G.; Brodersen, C.C.; Gharrett, J.A.M.M. *Effects of Petroleum Hydrocarbons on Alaskan Aquatic Organisms: A Comprehensive Review of All Oil-Effects Research on Alaskan Fish and Invertebrates Conducted by Theauke Bay Laboratory, 1970–81*; NOAA: Silver Spring, MD, USA, 1984.

28. Markarian, R.K.; Nicolette, J.P.; Barber, T.R.; Giese, L.H. *A Critical Review of Toxicity Values and an Evaluation of the Persistence of Petroleum Products for Use in Natural Resource Damage Assessments*; API Publication Number 4594; American Petroleum Institute: Washington, DC, USA, 1995.

29. French-McCay, D.P. Development and application of an oil toxicity and exposure model, OILTOXEX. *Environ. Toxicol. Chem.* **2002**, *21*, 2080–2094. [CrossRef] [PubMed]

30. Environmental Impacts of Arctic Oil Spills and Arctic Spill Response Technologies. Literature Review and Recommendations December 2014. Arctic Oil Spill Response Technology Joint Industry Programme. Available online: http://neba.arcticresponsetechnology.org/assets/files/Environmental%20Impacts%20of%20Arctic%20Oil%20Spills%20-%20report.pdf (accessed on 18 September 2018).

31. Gardiner, W.W.; Word, J.Q.; Word, J.D.; Perkins, R.A.; McFarlin, K.M.; Hester, B.W.; Word, L.S.; Ray, C.M. The acute toxicity of chemically and physically dispersed crude oil to key Arctic species under Arctic conditions during the open water season. *Environ. Toxicol. Chem.* **2013**, *32*, 2284–2300. [CrossRef] [PubMed]

32. Dupuis, A.; Ucan-Marin, F. *A Literature Review on the Aquatic Toxicology of Petroleum Oil: An overview of Oil Properties and Effects to Aquatic Biota*; Canadian Science Advisory Secretariat: Vancouver, BC, USA, 2015.

Journal of
Marine Science and Engineering

MDPI

Article

Multi-Criteria Analysis of Different Approaches to Protect the Marine and Coastal Environment from Oil Spills

Antigoni Zafirakou *, Stefania Themeli, Eythymia Tsami and Georgios Aretoulis

Department Civil Engineering, Aristotle University of Thessaloniki, 54124 Thessaloniki, Greece;
stefthemeli@gmail.com (S.T.); efits1993@gmail.com (E.T.); garet@civil.auth.gr (G.A.)
* Correspondence: azafir@civil.auth.gr; Tel.: +30-2310-994371

Received: 17 September 2018; Accepted: 13 October 2018; Published: 24 October 2018

check for updates

Abstract: Marine pollution has many different sources. This study focuses on oil spills that may occur after a ship collision or during oil extraction and other oil tanker activities. The most critical oil spill accidents are presented, followed by the regulatory framework on maritime oil spill management. Among the measures taken towards the protection of the marine and coastal environment from oil pollution are floating booms and barriers, oil collecting materials and vessels, absorbent materials, chemical dispersants, other chemicals, physical degradation, biodegradation, on-site oil burning. These measures may assist coastal facilities and local authorities in their strategic development of oil spill mitigation planning and response towards coastal and marine protection from oil spills. In the present paper, the aim is to rank the approaches of dealing with the oil spill by means of a multicriteria method. The theoretical background of the selected multicriteria method, called PROMETHEE, is briefly presented; necessary to understand the ranking of the treatment approaches as well as the subsequent findings of the possible criteria for the analysis. Almost all of the scenarios evaluated rank floating booms and barriers as the most suitable methods to deal with oil spill containment, followed by oil collecting materials and vessels.

Keywords: marine pollution; oil spill pollution; oil spill accidents; oil spill mitigation plans; coastal protection; statistical analysis; PROMETHEE methodology

1. Introduction

Pollution that comes from a single source, like an oil or chemical spill, is known as *point source*. In addition, nitrates and point-source pollution near estuaries and other water outlets can be catastrophic. Figure 1 depicts clearly how pollution sources are distributed. Urban, agricultural and industrial runoff coincide with air pollution (namely non-point sources) to a very high 77% of marine pollution, whereas point sources such as marine transport, dumbing and oil extraction (when focusing only on oil pollution) contribute only by 23% [1,2]. In addition to that, discharge from malfunctioned or damaged factories, wastewater treatment plants and desalination facilities, is also considered point source pollution [3].

Ship collision or malfunction, or simply cleaning and sailing, can contribute to marine pollution by spreading garbage, black or grey water, sludge, water ballast, coatings or even air emissions, to the deep sea or at seashore. Oil spills occur after a collision and/or sinking of oil tankers, under bad weather conditions and are extremely hazardous in ports with dense maritime traffic. Ship malfunctions and accidents on-board are the main causes of ships running aground, or colliding contributing to the spreading of the spills. The most common nautical accidents occur due to sinking or foundering, grounding, structural failure, scuttling, by contact or collision, explosion or fire, or after disappearance or abandonment [2]. The environmental impact of oil accidents is immense on both the water ecosystems and the coastal environment, including the urban and economic growth of the affected coastal zones (Figure 2) but only for a set period of time, as some areas have natural oil cleaning ability.

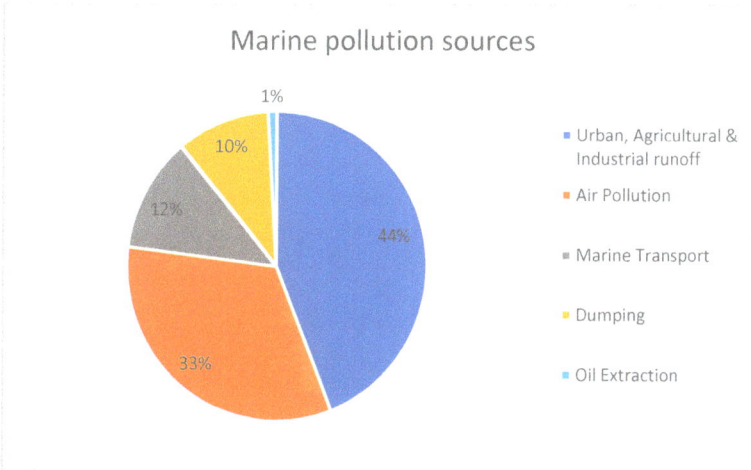

Figure 1. Distribution of marine pollution sources (point and non-point), data from [1]. (Reproduced from [2], with permission from Themeli, S. and Tsami, E., 2017)

Figure 2. Oil stranded on the shoreline adjacent to a fishing farm. Reproduced from [4], with permission from ITOPF, 2018.

According to the International Tanker Owners Pollution Federation (ITOPF), oil spills can be attributed to allision/collision, grounding, hull or equipment failure, fire/explosion, while ships are at anchor (inland/restricted or in open waters), underway (inland/restricted or in open waters), or during loading/discharging or other operations, in the following percentages (Figure 3).

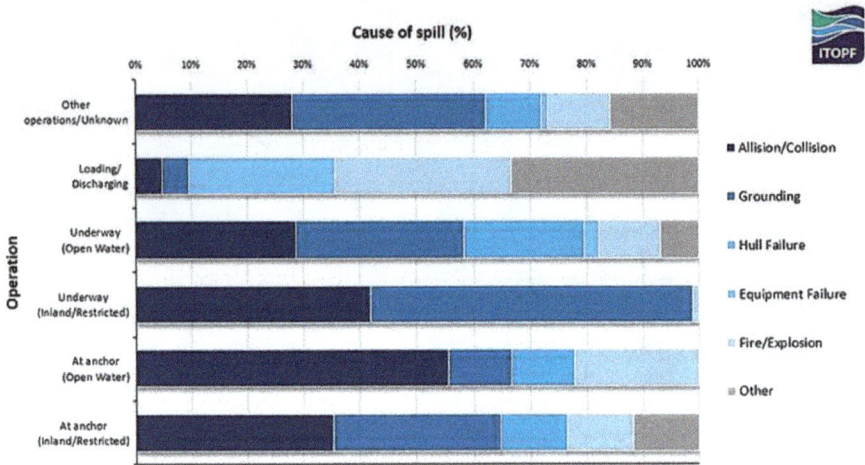

Figure 3. Primary causes of spills > 700 tn by operation at time of incident, 1970–2017. Reproduced from [5], with permission from ITOPF, 2018.

For historical reasons, spills are generally categorized by size: below 7 tn, between 7–700 tn and above 700 tn. Information is now available on over 10,000 incidents; luckily most of the spills are below 7 tn. According to the International Tanker Owners Pollution Federation (ITOPF) the average number of recorded small and large-scale oil spills worldwide is remarkably decreasing [2], as graphically depicted in Figure 4. This is mainly due to the fact that illegal discharges are being increasingly monitored and ship owners punished by doing such discharges.

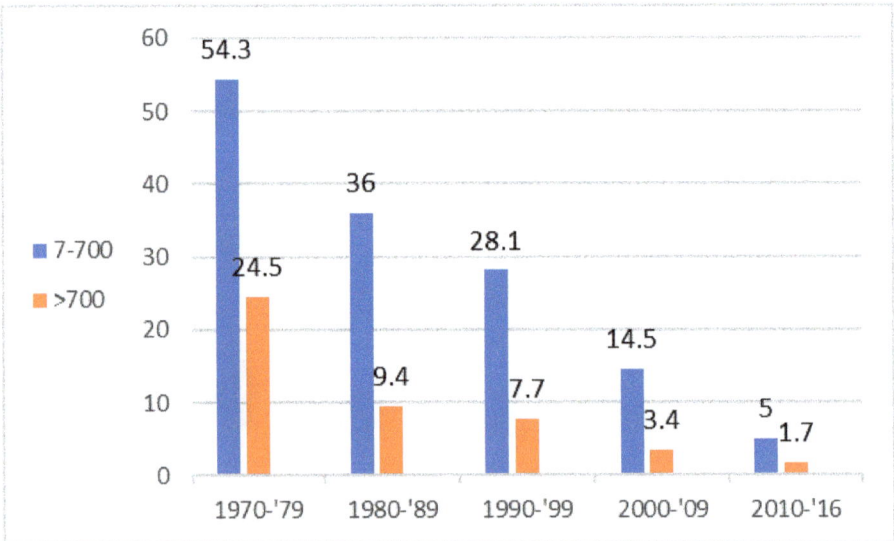

Figure 4. Recorded oil spills on international scale, annually averaged by decade. Reproduced from [2], with permission from Themeli, S. and Tsami, E., 2017.

When looking at the frequency and quantities of oil spilled, it should be noted that a few very large spills are responsible for a high percentage of the oil spilled. For example, in more recent decades the following can be seen [5]:

- In the 1990s, there were 358 spills of more than 7 tn, resulting in 1,134,000 tn of oil lost; 73% of this amount was spilled in just 10 incidents.
- In the 2000s, there were 181 spills of more than 7 tn, resulting in 196,000 tn of oil lost; 75% of this amount was spilled in just 10 incidents.
- In the period 2010–2017 there have been 53 spills of more than 7 tn, resulting in 47,000 tn of oil lost; 80% of this amount was spilled in just 10 incidents.

In terms of the oil volume spilled, the statistics for a particular year may be severely distorted by a single large incident. This is clearly illustrated in Figure 5 [5].

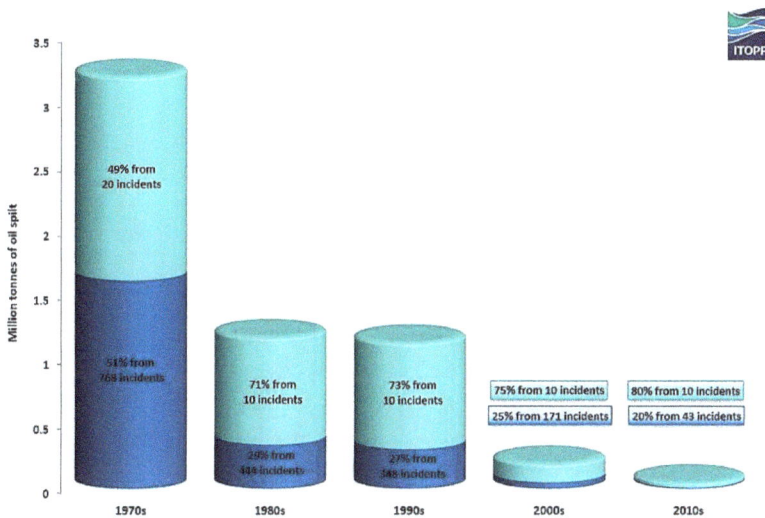

Figure 5. Oil spills per decade (1970s–2010s), of over 7 tn, showing the influence of a relatively small number of comparatively large spills on the overall figure. Reproduced from [5], with permission from ITOPF, 2018.

Table 1 summarizes the top 10 oil spills accidents in history, in descending order of oil volume spilled, which were caused due to collision, fire, explosion or sinking. Two of the biggest oil spills in history, caused after a wellhead blowout, were Deepwater Horizon, on April 21st, 2010, in the Gulf of Mexico, with 633,116 tn, and IXTOC 1, on June 3rd, 1979, in the Gulf of Mexico, with 470,000 tn.

Figure 6 demonstrates two of the most renowned oil spills in history. The 1989 Exxon Valdez, the 1991 Gulf war, the 2010 Deepwater Horizon are few of the most renowned oil spills in history that alarmed the scientific community and directed all towards the management of anthropogenic environmental disasters in the marine environment. The TORREY CANYON (1967) was the first major tanker disaster to be brought to the notice of the general public due to enormous media coverage, and drew universal attention to the dangers of dispersants. The spill triggered the international Conventions, which form the basis for compensation for damage caused by tanker spills, and interim voluntary agreements to bridge the gap before the Conventions entered into force and became widely accepted [5].

Table 1. The 10 largest oil spills in history in descending order. Reproduced from [5], with permission from ITOPF, 2018.

Rank	Oil Spill/Case Name	Date (MM/DD/YYYY)	Location	Oil Volume (tn)	Cause
1	Atlantic Empress	07/19/1979	Off Tobago, West Indies	287,000	Collision
2	ABT Summer	05/28/1991	700 n.mi off the coast of Angola	260,000	Fire/explosion
3	Castillo de Bellver	08/06/1983	Off Saldanha Bay, South Africa	252,000	Fire
4	Amoco Cadiz	03/16/1978	Brittany, France	223,000	Collision
5	M/T Heaven Tanker	04/11/1991	Genoa, Italy	144,000	Explosion
6	ODYSSEY	11/10/1988	700 nautical miles off Nova Scotia, Canada	132,000	Sinking/Fire
7	TORREY CANYON	03/18/1967	Scilly Isles, UK	119,000	Collision on reef
8	SEA STAR	12/19/1972	Gulf of Oman	115,000	Collision
9	IRENES SERENADE	02/23/1980	Navarino Bay, Greece	100,000	Explosion
10	URQUIOLA	05/12/1976	La Coruna, Spain	100,000	Collision on bottom

(a) Atlantic Empress (1979)

(b) Exxon Valdez Oil spill clean-up efforts (1989)

Figure 6. Oil spills and clean-up efforts

2. Oil Spill Contingency Plans and Relative Legislation

In the wake of global concerns, the Marine Environment Protection Committee of the International Marine Organization (IMO) developed the International Convention on Oil Pollution Preparedness, Response and Cooperation (OPRC) in 1990, to provide a framework for international cooperation for combating oil pollution incidents. It came into effect only after the ratification from the majority of the countries-members, in 1995. The OPRC Convention has 19 Articles and 10 Resolutions, covering both administrative and technical aspects. These call, among others, for parties' oil exploration and production activities, shipyards, oil refineries, terminals and depot, ports, harbours and marinas, manufacturing plants and other establishments using oil, to develop and maintain oil spill response plans. Ships and vessels are also required to develop and maintain on board a Shipboard Oil Pollution Emergency Plan (SOPEP), while countries are required to develop and maintain a National Oil Spill Contingency Plan for major incidents. The parties to the OPRC Convention are required to establish measures for dealing with pollution incidents, either nationally or in cooperation with other countries [6]. A contingency plan should comprise three parts:

1. A strategy/preparedness section to get ready for real action, which should describe the scope of the plan, the geographical coverage, the perceived risks and hazard assessment, the roles and responsibilities of those charged with implementing the plan, the available equipment,

scenarios of oil spill incidents accompanied by oil slick simulation and identification of combating options.

2. A response action, to deal with the spilled oil, which should set out the emergency procedures that will allow rapid assessment of the spill and the mobilization of appropriate response resources, as well as application of models upon request for prediction of oil spill movement.

3. A data directory, which should contain all relevant maps, resource lists and data sheets required to support an oil spill response effort and conduct the response according to an agreed strategy.

All countries with extremely important maritime traffic, are obliged, according to the legislation, to establish an oil contingency plan and next to them, every port authority, company and facility that handles oil products in coastal areas. Methodologically the development of a Port Contingency Plan (PCP) should include the following stages [7]:

a. Recording of general information about the area and description of its basic morphological, economic, social and environmental characteristics.

b. Establishment of a catalogue of all the hazardous substances (name, characteristics, properties) handled by the port facilities and mapping of the handling locations in the port area.

c. Identification of all the incidents/scenarios that could lead to emergency situations.

d. Qualitative and quantitative risk assessment for the marine environment in any case of incident involving any of the dangerous substances.

e. Description of the equipment and the sequence of actions for the collection and disposal of the dangerous waste.

f. Definition and description of the preparedness exercises that the authorized-to handle emergency situations personnel should take.

g. Definition of rules and measures to be followed to preserve the personnel's safety and health.

h. Prediction of the necessary procedures in the case of a plan's revision.

It is easily derived that stages (e) and (f) are of utmost importance. The recognition of actions that should be taken in case of an accidental oil spill and the tools required to achieve the best marine protection comprise the operational activity needed to handle successfully an oil pollution emergency.

Besides the OPRC, the management of the maritime pollution from oil is further supported by a series of regulatory and legislative acts, such as the International Convention on Civil Liability for Oil Pollution Damage (1969), the International Convention for Marine Pollution MARPOL73/78, Bonn Agreement (1983) for cooperation in dealing with pollution by oil and other substances in the North Sea, Helsinki Convention (1992) on the protection of the marine environment of the Baltic Sea area, OSPAR Convention (1992) for the protection of the marine environment of the North-East Atlantic sea, the Protocol on Preparedness, Response and Co-operation to Pollution Incidents by Hazardous and Noxious Substances (2000), the International Convention on Civil Liability for Bunker Oil Pollution Damage (2001), the Directive 2013/30/EU on safety for offshore oil and gas operations and amending Directive 2004/35 EC, other regional/European initiatives, national laws and port regulations. The NEREIDS project [8] provides an integrated vision of maritime policy and surveillance, aimed to enhance automatic and unsupervised ship monitoring capabilities for Maritime Situational Awareness (MSA) and to support advanced and efficient decision-making tools.

International research has focused on modelling the oil spills. The trajectory and the transport mechanism of the M/V Marathassa oil spill have been studied. The research results showed that the fraction of the oil on the water surface and on the shoreline, as well as the amount of oil recovered were affected by the time of the initial release, the overall duration of the discharge, wind and recovery actions [9].

Another study compares the Department of Energy's (DOE's) National Energy Technology Laboratory's (NETL's) Blowout and Spill Occurrence Model (BLOSOM), with the National Oceanic and Atmospheric Administration's (NOAA's) General NOAA Operational Modelling Environment

(GNOME). A complex approach was used for the comparison of the two models. This proposed methodology could be used to illustrate the approach an oil spill modeller would typically follow when trying to hindcast or forecast an oil spill, including detailed technical information on basic aspects [10].

Detection of oil spills is another research challenge. In this context, an algorithm is developed to effectively analyse large-scale oil spill areas in SAR images. Furthermore, an ANN algorithm was used to generate probability maps of oil spills [11]. In the same scientific area of the oil detection is also focusing the next study. Authors developed algorithms for oil spill detection using radar remote sensing. The algorithms take into account both the mathematical and the physical modelling of the sea surface covered by oil slicks [12].

Modelling the "fate" of oil in shallow waters is extremely important. This is the aim of a study, where a model for the dynamics of oil in suspension, appropriate for shallow waters, including the nearshore environment is presented. The proposed model is capable of oil mass conservation and does so by evolving the oil on the sea surface as well as the oil in the subsurface [13]. The theme of the following papers focuses on the treatment approaches of oil spills. This is field of research of the next study that describes a two-stage method for optimizing the location of marine oil spill combat forces and assessing the costs related to this action at the sea. Response time, cost and effectiveness of the means to treat multiple oil spills is the aim of the current paper. This translates into an optimization problem that relates to positioning the oil pollution combat ships in ports, in such a way that they are able combat the anticipated number of oil spills in certain positions in the Polish coast of the Baltic Sea area in the shortest possible time [14].

Another step towards the treatment of oil spills is the use of novel methods and materials to prevent or slow the advancement of oil spills and remove them from the sea. Materials and techniques environmentally friendly are being developed. This is the goal of the following study. An environmentally friendly and degradable material, Poly (lactic acid) (PLA) ultrafine fibres is introduced, for the removal of oil from water. It is emphasized that this work is expected to promote the mass production and application of biodegradable PLA fibres in the treatment of marine oil spill pollution [15].

Last but not least, is the contribution of the Aristotle University of Thessaloniki, to the development of oil spill transport models that can predict the fate and evolution of an oil spill and provide useful information to the authorities in order to apply the abovementioned measures [16–20].

Research has focused on preventing and detecting oil spills, developing methods for treating the oil spills and removing them from the natural environment, assessing the impact that such accidents cause on the various forms of life and especially in the human health and nutrition. The current paper focuses on the treatment stage and highlights a methodological approach for prioritizing/ranking available treatment methods for oil spills based on the unique characteristics of the affected location.

3. Methodological Approach and Findings

In the present paper, the aim is to rank the alternative treatment approaches for dealing with the oil spill by means of a multicriteria method. The methodological steps included the following:

- Consideration of the alternative treatment approaches for the oil containment (please see Table 2)
- Definition of selection parameters/scenarios for the alternative treatment approaches, namely:

  - ✓ Coastal description
  - ✓ Weather conditions
  - ✓ Oil type
  - ✓ Approach characterization

- Definition of corresponding selection criteria per parameter/scenario, for the alternative treatment approaches

- Creation and dissemination of a structured questionnaire to experts in order to assign weights to the considered criteria (please see Tables 3–6)
- Creation and dissemination of a structured questionnaire to experts for assessing the performance-effectiveness of each alternative treatment approach against each criterion and within each scenario (please see Tables 7–10)
- Application of Visual PROMETHEE per parameter/scenario
- Presentation of results and comparison among parameters/scenarios (please see Tables 11–14)

Table 2. Selected alternatives for the multi-criteria analysis.

Codes	Alternatives
A_1	Floating booms and barriers
A_2	Oil collecting materials
A_3	Oil collecting vessels
A_4	Absorbent materials
A_5	Chemical dispersants
A_6	Other Chemicals
A_7	Physical Degradation
A_8	Biodegradation
A_9	On-site oil burning

Table 3. Weighting factors for the 1st scenario—Coastal description.

	Scenario 1: Coastal Description		Weights
C_1	Port	W_1	2.98
C_2	Beach (Bathing waters—Touristic zones)	W_2	2.73
C_3	Natura-Ramsar protected wetlands	W_3	2.92
C_4	Fish/mussel-cultures	W_4	2.81
C_5	Waterfront-urban area	W_5	2.87

Table 4. Weighting factors for the 2nd scenario—Weather conditions.

	Scenario 2: Weather Conditions		Weights
C_1	Sunshine	W_1	2.95
C_2	Rain	W_2	3.13
C_3	Wind	W_3	2.60
C_4	Snow	W_4	2.36
C_5	Fog	W_5	2.48

Table 5. Weighting factors for the 3rd scenario—Oil type.

	Scenario 3: Oil type		Weights
C_1	Light fractions	W_1	3.30
C_2	Medium fractions	W_2	3.09
C_3	Heavy fractions	W_3	2.90

Table 6. Weighting factors for the 4th scenario—Approach characterization.

	Scenario 4: Approach Characterization		Weights
C_1	Effective	W_1	3.30
C_2	Required (to apply)	W_2	2.97
C_3	Time consuming	W_3	3.05
C_4	Economical	W_4	3.06
C_5	Well-known	W_5	3.09
C_6	Environmentally friendly	W_6	3.13

Table 7. Treatment Effectiveness per Coastal Description Criteria.

Approach/Coastal Description	Port	Beach (Bathing Waters—Touristic Zones)	Natura-Ramsar Protected Wetlands	Fish/Mussel-Cultures	Waterfront-Urban Area
Floating Booms and Barriers	5	5	5	5	5
Oil Collecting Materials	3	3	3	3	3
Oil Collecting Vessels	3	3	3	3	3
Absorbent Materials	2	2	2	2	2
Chemical Dispersants	1	1	1	1	2
Other Chemicals	1	1	1	1	2
Physical Degradation	1	1	1	1	1
Biodegradation	1	1	1	1	1
On-Site Oil Burning	1	1	1	1	1

Table 8. Treatment Effectiveness per Weather Condition Criteria.

Approach/Weather Condition	Sunshine	Rain	Wind	Snow	Fog
Floating Booms and Barriers	5	5	2	2	2
Oil Collecting Materials	3	1	1	2	2
Oil Collecting Vessels	4	3	1	1	2
Absorbent Materials	2	2	3	3	2
Chemical Dispersants	1	3	2	2	2
Other Chemicals	1	2	2	2	2
Physical Degradation	2	2	1	1	1
Biodegradation	2	2	1	1	1
On-Site Oil Burning	1	1	1	1	1

Table 9. Treatment Effectiveness per Oil Types Criteria.

Approach/Oil Types	Light Fractions	Medium Fractions	Heavy Fractions
Floating Booms and Barriers	2	3	4
Oil Collecting Materials	2	3	4
Oil Collecting Vessels	1	3	4
Absorbent Materials	3	2	1
Chemical Dispersants	3	1	1
Other Chemicals	3	1	1
Physical Degradation	2	1	1
Biodegradation	2	1	1
On-Site Oil Burning	4	2	1

Table 10. Treatment Effectiveness per Approach Characterization Criteria.

Approach/Approach Characterization	Effective	Required	Time Consuming	Economical	Well-Known	Environmentally Friendly
Floating Booms and Barriers	4	4	4	4	4	4
Oil Collecting Materials	4	4	4	4	2	4
Oil Collecting Vessels	4	4	3	2	3	3
Absorbent Materials	3	2	2	2	2	3
Chemical Dispersants	2	2	3	2	2	1
Other Chemicals	2	2	3	2	2	1
Physical Degradation	1	2	1	2	2	1
Biodegradation	1	2	1	2	2	2
On-Site Oil Burning	1	2	1	2	2	1

Table 11. Alternatives' ranking with respect to the coastal description (Scenario 1).

Ranking	Alternatives	Phi	Phi+	Phi-
1	Floating booms and barriers	1.000	1.000	0.0000
2	Oil collecting materials	0.6250	0.7500	0.1250
3	Oil collecting vessels	0.6250	0.7500	0.1250
4	Absorbent materials	0.1999	0.5749	0.3750
5	Chemical dispersants	−0.3997	0.0752	0.4749
6	Other chemicals	−0.3997	0.0752	0.4749
7	Natural degradation	−0.5501	0.0000	0.5501
8	Biodegradation	−0.5501	0.0000	0.5501
9	On-site oil burning	−0.5501	0.0000	0.5501

Table 12. Alternatives' ranking with respect to the weather conditions (Scenario 2).

Ranking	Alternatives	Phi	Phi+	Phi-
1	Floating booms and barriers	0.6801	0.7260	0.0459
2	Absorbent materials	0.4067	0.5754	0.1686
3	Chemical dispersants	0.2114	0.4499	0.2384
4	Oil collecting vessels	0.1719	0.4333	0.2615
5	Other chemicals	0.0378	0.3341	0.2963
6	Oil collecting materials	−0.0554	0.3197	0.3751
7	Natural degradation	−0.3718	0.1397	0.5115
8	Biodegradation	−0.3718	0.1397	0.5115
9	On-site oil burning	−0.7090	0.0000	0.7090

Table 13. Alternatives' ranking with respect to the oil type (Scenario 3).

Ranking	Alternatives	Phi	Phi+	Phi-
1	Floating booms and barriers	0.3504	0.5280	0.1776
2	Oil collecting materials	0.3504	0.5280	0.1776
3	On-site oil burning	0.2797	0.5215	0.2418
4	Oil collecting vessels	0.1284	0.4836	0.3552
5	Absorbent materials	0.1021	0.3883	0.2862
6	Chemical dispersants	−0.1473	0.2220	0.3693
7	Other chemicals	−0.1473	0.2220	0.3693
8	Natural degradation	−0.4582	0.0444	0.5026
9	Biodegradation	−0.4582	0.0444	0.5026

Table 14. Alternatives' ranking with respect to the approaches' characterization (Scenario 4).

Ranking	Alternatives	Phi	Phi+	Phi-
1	Floating booms and barriers	0.5667	0.7101	0.1435
2	Oil collecting materials	0.3590	0.5440	0.1850
3	Oil collecting vessels	0.3584	0.5444	0.1860
4	Absorbent materials	0.0059	0.3185	0.3126
5	Biodegradation	−0.1526	0.2071	0.3597
6	Natural degradation	−0.2578	0.1230	0.3808
7	On-site oil burning	−0.2578	0.1230	0.3808
8	Chemical dispersants	−0.3109	0.1075	0.4184
9	Other chemicals	−0.3109	0.1075	0.4184

The theoretical background of multicriteria methods was sought, necessary for understanding the ranking of the treatment approaches as well as for the subsequent findings based on the considered scenarios and criteria of the analysis.

The selected method for this evaluation is PROMETHEE (Preference Ranking Organization METHod for the Enrichment of Evaluations), by Brans and Marechal [21]. In PROMETHEE methodology there are no specific instructions about how the weights will be assigned but instead each decision maker is capable of assigning priorities depending on the criteria [22]. In this context, various approaches existed in international literature and Macharis et al. [23] were the first to investigate potential synergies between European and American multicriteria methods. To understand PROMETHEE method, one should consider a decision problem with n alternatives and k criteria. For each criterion f_j ($j = 1, \ldots, k$), a preference function $P_j(a,b)$ shall be adopted to translate the deviation between two alternatives a and b into Preference degree, with a range between 0 and 1. This function describes the difference d = $f_j(a) - f_j(b)$ between the evaluations of the alternatives on each criterion.

$$P_j(a,b) = G_j \{f_j(a) - f_j(b)\} \tag{1}$$

A different preference function corresponds to each criterion, regardless if it is qualitative or quantitative. In Reference [24] the following, six possible functions are proposed: U-Shape, V-Shape, Linear and Gaussian for quantitative criteria and Usual and Level for qualitative criteria. For qualitative criteria with no large rating scale, such as the 5-scale rating, Usual function appears as a decent choice, although for larger scales the most suitable function is Level. For some of the remaining functions, it is necessary to choose thresholds of Indifference (Q) or Preference (P) [21,25–31]. After choosing the most suitable function, criteria weights w_j are used to calculate the multicriteria preference index $\pi(a,b)$ taking into consideration all the criteria.

$$\pi(a, b) = \sum_{j=1}^{k} w_j P_j(a, b) \tag{2}$$

That index is used to calculate the positive preference flow (*Phi*$^+$, $\varphi^+(\alpha)$) and the negative preference flow (*Phi*$^-$, $\varphi^-(\alpha)$), where

$$\varphi^+(a) = \frac{1}{n-1} \sum_b \pi(a, b) \tag{3}$$

$$\varphi^-(a) = \frac{1}{n-1} \sum_b \pi(a, b) \tag{4}$$

The difference between the preference flows is the Net Preference flow (*Phi*, φ). Higher value of net preference flow represents higher appeal of the alternative solution. Therefore, in this paper the highest value of a net preference flow assigned to a treatment approach/alternative identifies the latter as the best choice among the available treatment methods for the examined scenario.

3.1. Choice of Alternative Approaches for the Oil Containment

In this particular study, the focus is on oil spill pollution and prevention. The alternative approaches in dealing with oil collection and removal can be ranked in terms of their effectiveness for the prevention and treatment of marine pollution from the oil spilled in case of an accident. The most common tools and methods for the spilled oil containment and removal in the sea are: (1) floating booms and barriers, (2) oil collecting materials, (3) oil collecting vessels, (4) absorbent materials, (5) chemical dispersants, (6) other chemicals, (7) physical degradation, (8) biodegradation and (9) on-site oil burning. These are recognized as the 'alternatives' (treatment approaches) in the multi-criteria analysis.

3.2. Selection of Scenarios and Corresponding Criteria

For the optimal choice among the alternatives some 'criteria' should be defined. Parameters (in the context of Visual PROMETHEE, they are called "scenarios") that are critical for proper selection of responses to the oil spill, include: (a) coastal description (location - morphology), (b) weather conditions and (c) oil type. Coastal location and morphology, play a significant role in terms of economic, ecological or touristic value, as an even small oil spill would create multiple environmental and economic effects. The oil type that is spread after an accident and the weather conditions at the time of the accident, contribute correspondingly to the problem's dimensions and its solution. For instance, light, medium and heavy fractions of oil, require different cleaning measures. More specifically, light products (e.g., gasoline, diesel) do not require costly and time-consuming applications to deal with, quite the opposite with heavy oil products. Lastly, the weather conditions, such as strong winds, can be prohibitive and disastrous during cleaning operations if they direct the spill towards a coast, or can be beneficial if they direct it towards the open seas where a mitigation plan can be applied.

The considered scenarios are assigned specific criteria. Therefore, the proposed scenarios along with the corresponding criteria include the following:

- Scenario 1: Coastal description

  - ✓ Port
  - ✓ Beach (Bathing waters—Touristic zones)
  - ✓ Natura-Ramsar protected wetlands
  - ✓ Fish/mussel-cultures
  - ✓ Waterfront-urban area

- Scenario 2: Weather conditions

  - ✓ Sunshine
  - ✓ Rain
  - ✓ Wind
  - ✓ Snow
  - ✓ Fog

- Scenario 3: Oil type

  - ✓ Light fractions
  - ✓ Medium fractions
  - ✓ Heavy fractions

- Scenario 4: Approach characterization

  - ✓ Effective
  - ✓ Required (to apply)
  - ✓ Time consuming
  - ✓ Economical
  - ✓ Well-known
  - ✓ Environmentally friendly

3.3. Determination of Weights for the Criteria

PROMETHEE multi-criteria analysis method was chosen to rank the approaches than can be used to deal with oil spills, based on the criteria already mentioned. The Visual PROMETHEE Academic Edition program was used. As with any multicriteria method, it is imperative to define alternatives as well as the evaluation criteria. The following tables show in detail the alternatives (A_i) and the criteria (C_j) to be used along with their codification to facilitate data input in the Visual PROMETHEE application.

Weights are required according to the PROMETHEE methodology. This was pursued by collecting questionnaires and following a statistical analysis through the SPSS program (Statistical Package for Social Sciences). These questionnaires were distributed to scientists of various disciplines, who were asked to rank, according to their knowledge, the criteria that should play a significant role in the decision making of an oil spill containment. The scale was defined in such a way that 1 denotes the minimum significance of each criterion examined and 5 the maximum significance, respectively. The average weights for each criterion were ultimately used as the final weight-value for each criterion. Tables 3–6 below, present the weights assigned to the criteria per scenario and serve as the entry data in the Visual PROMETHEE application.

Visual PROMETHEE software used the average weights as seen in the criteria tables and the performance/effectiveness assessment according to experts for each alternative per criterion. The application succeeded in assessing which alternatives are considered best with respect to the aforementioned criteria per scenario.

3.4. Determination of Treatment Approaches' Effectiveness Against Each Criterion

The Visual PROMETHEE application additionally needs as inputs the effectiveness/performance of each treatment approach against each criterion. This was made possible through a structured questionnaire survey towards a number of selected experts. The questionnaires were completed through interviews. The following Tables 7–10, present the mean values of the effectiveness for each treatment approach per criterion as recorded through the survey.

4. Results

The results of Visual PROMETHEE software for the four scenarios (coast description, weather conditions, oil type and approach characterization), are presented herein based on net, positive and negative flows. Higher value of net preference flow, as mentioned before, represents higher appeal of the alternative solution. The net preference flow (Phi) is calculated by adding the positive (Phi+) and negative (Phi-) flows.

Table 11 presents the ranking of the alternative solutions, based on net, positive and negative flows. It seems that the best four approaches to deal with oil spills, with respect to the coastal description, are, in descending order: floating booms and barriers, oil collecting materials and vessels and absorbent materials. The remaining five techniques were not ranked as important.

Table 12 shows the ranking of alternative solutions to deal with oil spills with net, positive and negative flows. It depicts that the best five ways to deal with oil spills, with respect to the weather conditions, are, in descending order: floating booms and barriers, absorbent materials, chemical dispersants, oil collecting materials and other chemicals. The remaining four show a negative flow, so they are not considered as best approaches, when dealing with the weather conditions.

Table 13 shows the ranking of approaches with net, positive and negative flows. It seems that the best five ways to deal with oil spills, with respect to the oil type spread, are in descending order: floating booms and barriers, oil collecting materials, on-site oil burning, oil collecting vessels and absorbent materials. The remaining four show a negative flow, so they are not accounted for best approach, when the oil type is considered.

Table 14 presents the ranking of approaches with net, positive and negative flows for each alternative in relation to the features examined. It depicts that the best four ways to deal with oil spills are, in descending order: floating booms and barriers, oil collecting materials and vessels and absorbent materials. The remaining five show a negative flow, so they are not considered as best approaches, with respect to the characterization of these approaches.

Taking into consideration all results, it is noticed that for the four examined scenarios, the alternative methods preferred by the participants of the survey, are the floating booms and barriers, the oil collecting vessels and the absorbent materials, which all show a positive net flow. Next, the oil collecting materials show a positive net flow in three of the four scenarios (only in the case of the weather conditions they show a negative net flow). The opposite is noticed for the treatment approaches of natural degradation and biodegradation, which always present a negative net flow. In addition, chemical dispersants and other chemicals exhibit a negative net flow in three out of four cases (only in the scenario of weather conditions a positive net flow is presented) and the on-site oil burning exhibits a negative net flow in three out of the four cases (only because of the criteria for the type of oil present a positive net flow).

Finally, Table 15 was created to illustrate and summarize the ranking of the alternatives. Table 15 highlights that in all four cases, the floating booms and barriers are in the 1st place and in one of them (criteria related to the scenario: type of oil) together with the oil collecting materials are ranked 1st. In general, oil collecting vessels and absorbent materials occupy high positions in the ranking (2nd–5th) while other chemicals, chemical dispersants, natural degradation, biodegradation and on-site oil burning occupy mainly the last positions (with the exception of on-site burning which in one case occupies the 2nd position).

Table 15. Frequency of occurrence of approaches for each position in the ranking.

RANKING APPROACH	1	2	3	4	5	6	7	8	9
Floating Booms and Barriers	4	1	0	0	0	0	0	0	0
Oil Collecting Materials	1	3	1	0	0	1	0	0	0
Oil Collecting Vessels	0	1	2	2	0	0	0	0	0
Absorbent Materials	0	1	0	2	1	0	0	0	0
Chemical Dispersants	0	0	1	0	1	2	1	1	1
Other Chemicals	0	0	0	0	2	2	1	1	1
Natural Degradation	0	0	0	0	0	1	3	3	2
Biodegradation	0	0	0	0	1	0	2	3	2
On-Site Oil Burning	0	1	0	0	0	1	2	1	2

5. Conclusions

In the present study, the aim was to prioritize the actions (treatment approaches) that can be taken from local authorities and other relevant authorities, in the case of an oil spill accident, through a multicriteria methodology. Initially reference was made to point sources of marine pollution, such as oil accidents, which even though they contribute much less than the non-point sources of land runoff, their effect on the marine environment is direct, eminent and long-lasting. Both the environment and the economy of the affected areas suffer the consequences. Therefore, strict legislation exists, which requires mitigation plans readily available by the relevant authorities. The prioritization of these plans is very crucial, depending on the coastal characterization (port, beach, wetland etc.), the weather conditions at the time of the accident and the cleaning efforts and the oil type. With the use of Visual PROMETHEE application, the alternative cleaning methods were ranked with respect to the abovementioned criteria. The weight of each criterion and the performance of each treatment approach per criterion, were the subject of a questionnaire survey that took place, with the participation of expert scientists from various disciplines. The theoretical background of the multicriteria method is briefly presented, necessary for understanding the ranking of the treatment approaches. The alternative approaches to deal with an accidental oil spill were selected from among a wide variety. The most prominent are floating booms and barriers, oil collecting materials and vessels, absorbent materials, chemical dispersants and other chemicals, physical degradation and biodegradation and, finally, on-site oil burning.

Based on the weighting factors, provided by the questionnaires and the multicriteria analysis, the alternatives were ranked. The results have identified the floating booms and barriers, as the best oil spill containment approach, followed by the oil collecting vessels and the absorbent materials. Given the scenarios of oil type, weather conditions and the coastal characterization, as well as the opinion of the participants to the questionnaire on each alternative method, those three approaches are nominated as the most popular and effective. Even though the data sample was relatively small, PROMETHEE methodology is capable of completing the analysis and provide reliable results.

However, regarding future work, the same questionnaires could be distributed to a narrower range of disciplines on oil spill or marine pollution related fields, and/or to a greater number of participants. Moreover, additional criteria and scenarios could be considered and alternative multicriteria methods could be applied. Furthermore, the combined simultaneous consideration of all scenarios could be examined, in order to identify the optimum treatment approach or examine relevant combinations of treatment approaches. The criteria weights could be evaluated based on a more extensive body of experts. Finally, as part of the future research, the proposed model will be applied in real case studies and examine the effectiveness of a simulated application of the identified optimum treatment method. Limitations of the current research focus on the criteria weights and treatment approaches' effectiveness, because these are evaluated by experts, bearing in mind the special conditions of Greece.

Conclusively, marine pollution created by oil spills is eminent and requires immediate action. Local authorities must obey the relevant legislation and provide the proper equipment and mitigation plans to deal with the containment of the spills. The needs and requirements for taking measures against pollution vary according to the nature and utilization of each site under consideration. Lack of pollution detection due to extreme weather conditions leads to unpleasant consequences. This study verifies that the allocation of oil spill combating stations, in the form of containers, where the necessary booms, pumps and dispersants are stored, to be transported to the oil spill accident site by special vessels and well-trained crews, in the least time, is an effective approach. Among other tools are the oil spill transport models that can predict the fate and evolution of an oil spill and provide useful information to the authorities in order to apply the abovementioned measures. These tools could be evaluated in another similar analysis.

Author Contributions: For the realization of this research the following people have contributed. Conceptualization, A.Z. and G.A.; methodology, A.Z.; software, G.A.; validation, A.Z. and G.A.; formal analysis, S.T.; investigation, S.T.; resources, E.T.; data curation, S.T. and E.T.; writing—original draft preparation, A.Z.; writing—review and editing, A.Z.; visualization, G.A.; supervision, A.Z.

Funding: This research received no external funding.

Conflicts of Interest: The authors declare no conflict of interest.

References

1. NOAA (National Oceanic and Atmospheric Administration). What Is the Biggest Source of Pollution in the Ocean? Available online: https://oceanservice.noaa.gov/facts/pollution.html (accessed on 17 June 2018).
2. Themeli, S.; Tsami, E. Multi-Parametric Analysis of the Mitigation Plans for the Oil Spill Pollution in the Sea. Bachelor's Thesis, Aristotle University of Thessaloniki, Thessaloniki, Greece, 2017.
3. NOAA (National Oceanic and Atmospheric Administration). Ocean Pollution. Available online: www.noaa.gov/resource-collections/ocean-pollution (accessed on 17 June 2018).
4. ITOPF (International Tank Owners Pollution Federation). Environmental Effects. Available online: http://www.itopf.org/knowledge-resources/documents-guides/environmental-effects/ (accessed on 20 July 2018).
5. ITOPF (International Tank Owners Pollution Federation). Oil Tanker Spill Statistics 2017. Available online: http://www.itopf.org/knowledge-resources/data-statistics/statistics/ (accessed on 18 July 2018).
6. IMO (International Maritime Organization). List of IMO. Available online: http://www.imo.org/en/About/Conventions/ListOfConventions/Pages/Default.aspx (accessed on 18 June 2018).
7. Palantzas, G.; Koutitas, C.; Naniopoulos, A. Methodology of development of a hazardous substances contingency plan. Port of Thessaloniki case study. In Proceedings of the 5th International Exhibition and Conference on Environmental Technology (HELECO'05), Athens, Greece, 3–6 February 2005.
8. NEREIDS (New Service Capabilities for Integrated and Advanced Maritime Surveillance). NEREIDS Report Summary. Available online: https://cordis.europa.eu/result/rcn/159365_en.html (accessed on 1 October 2018).
9. Zhong, X.; Niu, H.; Wu, Y.; Hannah, C.; Li, S.; King, T. A modeling study on the oil spill of M/V Marathassa in Vancouver harbour. *J. Mar. Sci. Eng.* **2018**, *6*, 106. [CrossRef]

10. Duran, R.; Romeo, L.; Whiting, J.; Vielma, J.; Rose, K.; Bunn, A.; Bauer, J. Simulation of the 2003 foss barge-point wells oil spill: A comparison between BLOSOM and GNOME oil spill models. *J. Mar. Sci. Eng.* **2018**, *6*, 104. [CrossRef]

11. Kim, D.; Jung, H.S. Mapping oil spills from dual-polarized SAR images using an artificial neural network: Application to oil spill in the Kerch Strait in November 2007. *Sensors* **2018**, *18*, 2237. [CrossRef] [PubMed]

12. Hammoud, B.; Faour, G.; Ayad, H.; Ndagijimana, F.; Jomaah, J. Performance analysis of detector algorithms using drone-based radar systems for oil spill detection. *Proceedings* **2018**, *2*, 370. [CrossRef]

13. Restrepo, J.M.; Ramírez, J.M.; Venkataramani, S. An oil fate model for shallow-waters. *J. Mar. Sci. Eng.* **2015**, *3*, 1504–1543. [CrossRef]

14. Łazuga, K.; Gucma, L.; Perkovic, M. The model of optimal allocation of maritime oil spill combat ships. *Sustainability* **2018**, *10*, 2321. [CrossRef]

15. Li, H.; Li, Y.; Yang, W.; Cheng, L.; Tan, J. Needleless melt-electrospinning of biodegradable poly (lactic acid) ultrafine fibers for the removal of oil from water. *Polymers* **2017**, *9*, 3. [CrossRef]

16. Sofianos, S.; Kallos, G.; Mantziafou, A.; Tzali, M.; Zafirakou, A.; Dermisis, V.; Koutitas, C.; Zervakis, V. Oil spill dispersion forecasting system for the region of installation of the Burgas Alexandroulopis pipeline outlet (N.E. Aegean) in the framework of "DIAVLOS" Project. In Proceedings of the 9th National Symposium of Oceanography & Fisheries, Patras, Greece, 13–16 May 2009.

17. Zafirakou, A.; Koutitas, C.; Sofianos, S.; Mantziafou, A.; Tzali, M.; Dermissis, V.; Dermisi, S.C. Modeling the evolution and fate of an oil slick with a 3-D simulation model. In Proceedings of the 3rd IC-EpsMsO, Athens, Greece, 8–11 July 2009.

18. Zafirakou, A.; Palantzas, G.; Samaras, A.; Koutitas, C. Oil spill modeling aiming at the protection of ports and coastal areas. *Environ. Process.* **2015**, *2*, S41–S53. [CrossRef]

19. Zafirakou, A.; Palantzas, G.; Samaras, A.; Koutitas, C. The use of oil spill simulation in developing and applying oil pollution contingency plans in ports and coastal areas. In Proceedings of the 12th International Conference on Protection and Restoration of the Environment (PRE-XII), Skiathos Island, Greece, 29 June–4 July 2014.

20. Zafirakou, A. The contribution of oil spill dispersion forecasting models to contingency planning. In *Monitoring of Marine Pollution*; Fouzia, H.B., Ed.; InTech Open Access Book: London, UK, 2018; under review.

21. Brans, J.P.; Mareschal, B. Promethee methods in multiple criteria decision analysis: State of the art surveys. In *International Series in Operations Research & Management Science*; Figueira, J., Greco, S., Ehrgott, M., Eds.; Springer: New York, NY, USA, 2005; pp. 163–186.

22. Macharis, C.; Turcksin, L.; Lebeau, K. Multi actor multi criteria analysis (MAMCA) as a tool to support sustainable decisions: State of use. *Decis. Support Syst.* **2012**, *54*, 610–620. [CrossRef]

23. Macharis, C.; Springael, J.; De Brucker, K.; Verbeke, A. PROMETHEE and AHP: The design of operational synergies in multicriteria analysis: Strengthening PROMETHEE with ideas of AHP. *Eur. J. Oper. Res.* **2004**, *153*, 307–317. [CrossRef]

24. Brans, J.P.; Mareschal, B.; Vincke, P. How to select and how to rank projects: The PROMETHEE method. *Eur. J. Oper. Res.* **1986**, *24*, 228–238. [CrossRef]

25. Vavatsikos, A. Development of a Decision Support System in GIS Environment, Using Fuzzy Multicriteria Methods. Ph.D. Thesis, Department of Production and Management Engineering, Democritus University of Thrace, Xanthi, Greece, 2008.

26. Podvezko, V.; Podviezko, A. Dependence of multi-criteria evaluation result on choice of preference functions and their parameters. *Ukio Technologinis ir Ekonominis Vystymas* **2010**, *16*, 143–158.

27. Macharis, C.; Mareschal, B.; Waaub, J.P.; Milan, L. PROMETHEE-GDSS revisited: Applications so far and new developments. *Int. J. Multicriteria Decis. Mak.* **2015**, *5*, 129–151. [CrossRef]

28. Roukounis, C.; Karambas, T.; Aretoulis, G. Multicriteria decision making for waterdromes allocation in Greece. In Proceedings of the Transport Research Arena (TRA) 2018, Vienna, Austria, 16–19 April 2018.

29. Aretoulis, G.N.; Triantafyllidis, C.H.; Papathanasiou, J.B.; Anagnostopoulos, I.K. Selection of the most competent project designer based on multi-criteria and cluster analysis. *Int. J. Data Anal. Tech. Strateg.* **2015**, *7*, 172–186. [CrossRef]

30. Antoniou, F.; Aretoulis, G.N.; Konstantinidis, D.; Papathanasiou, J. Choosing the most appropriate contract type for compensating major highway project contractors. *J. Comput. Optim. Econ. Financ.* **2014**, *6*, 77–95.
31. Antoniou, F.; Aretoulis, G.N. Comparative analysis of multi-criteria decision making methods in choosing contract type for highway construction in Greece. *Int. J. Manag. Decis. Mak.* **2018**, *17*, 1–28.

Journal of
Marine Science and Engineering

MDPI

Article

Estimating the Usefulness of Chemical Dispersant to Treat Surface Spills of Oil Sands Products

Thomas King [1,*], Brian Robinson [1], Scott Ryan [1], Kenneth Lee [1], Michel Boufadel [2] and Jason Clyburne [3]

1   Fisheries and Oceans Canada, Bedford Institute of Oceanography, Dartmouth, NS B2Y 4A2, Canada;
     brian.robinson@dfo-mpo.gc.ca (B.R.); scott.ryan@dfo-mpo.gc.ca (S.R.); ken.lee@dfo-mpo.gc.ca (K.L.)
2   Center for Natural Resources, Department of Civil and Environmental Engineering, The New Jersey
     Institute of Technology, Newark, NJ 07102, USA; boufadel@gmail.com
3   Atlantic Centre for Green Chemistry, Department of Chemistry and Environmental Science,
     Saint Mary's University, Halifax, NS B3H 3C3, Canada; jason.clyburne@smu.ca
*   Correspondence: thomas.king@dfo-mpo.gc.ca; Tel.: +1-902-426-4172

Received: 22 October 2018; Accepted: 2 November 2018; Published: 6 November 2018

check for
updates

Abstract: This study examines the use of chemical dispersant to treat an oil spill after the initial release. The natural and chemically enhanced dispersion of four oil products (dilbit, dilynbit, synbit and conventional crude) were investigated in a wave tank. Experiments were conducted in spring and summer to capture the impact of temperature, and the conditions in the tank were of breaking waves with a wave height of 0.4 m. The results showed that natural dispersion effectiveness (DE) was less than 10%. But the application of dispersant increased the DE by an order of magnitude with a statistically significant level ($p < 0.05$). Season (spring versus summer) had an effect on chemical DE of all oils, except for the conventional oil. Thus, the DE of dilbit products is highly dependent on the season/temperature. A model was fitted to the DE as a function of oil viscosity for the chemically dispersed oil, and the correlation was found to be very good. The model was then combined with a previous model compiled by the author predicting oil viscosity as a function of time, to produce a model that predicts the DE as function of time. Such a relation could be used for responders tackling oil spills.

Keywords: Access Western Blend (condensate/bitumen-dilbit); Western Canadian Select (condensate mixed with synthetic crude/bitumen-dilsynbit); Synthetic Bitumen (synthetic crude/bitumen-Synbit); Heidrun; dispersant; wave tank; dispersion effectiveness (DE)

1. Introduction

Crude bitumen, produced in Alberta, Canada, is a highly viscous crude oil and semi-solid at room temperature. The majority of the oil produced is shipped via pipeline and railcars outside the province for refinement or export. In order to meet conventional oil pipeline specifications, the crude bitumen is diluted with a lighter hydrocarbon oil to reduce its viscosity and subsequently improve flow. The blending process for crude bitumen is at the discretion of the oil producer, so a wide variety of products of varying chemical composition is produced [1]. Heavy oil sands (blended bitumen) represent ca. two million barrels per day (b/d) of the four million b/d of crude oil produced and transported in Canada [2]. From pipelines, oil products may be transferred to tankers for shipment to global markets. Canada's production, transport, and sale of these products are expected to increase by a million barrels per day in the next decade [2]. The anticipated growth in oil production and transport increases the risk of oil spills in aquatic areas, and places greater demands on oil spill transport routes and capabilities to respond to spills.

In July of 2016, Environment Canada released new regulations and a list of approved oil spill treating agents that included COREXIT®EC9500A as an alternative measure, to recovery, to mitigate oil spills that occur in waters offshore Canada [3]. The use of a spill treating agent in offshore areas, by the responding parties, is to reduce damage to shoreline areas that are highly productive and sensitive, and also costly to clean. The application of conventional techniques, such as spill treating agents, are limited to studies showing significant, but incomplete effectiveness of COREXIT®EC9500A on a Cold Lake bitumen blend (e.g., Cold Lake crude bitumen blend with 30% condensate, dilbit) spilled under different environmental conditions [4,5]. However, there is no information in the literature to support the use of chemical dispersant to treat surface spills of various other oil sands products (e.g., Access Western Blend (AWB) or dilbit, Western Canadian Select (WCS) of dilsynbit and synthetic bitumen or synbit) that have weathered at sea, post spill. In addition, a science-based tool to estimate the window of opportunity to treat such spills is highly desirable. According to a Royal Society of Canada report on the behaviour and environmental impacts of crude oil released into aquatic environments [6], more research is required on the natural and chemically enhanced dispersion of bitumen blends under a variety of oceanographic conditions.

Chemical dispersants have been shown to be effective in treating heavy fuel oil, but water temperature can be a limiting factor [7,8]. Therefore, seasonal temperature variations are considered in this study, since the blended bitumen products are classified as heavy oils. Also, weathering of oil can increase its viscosity, which reduces the effectiveness of chemical dispersant to treat spills [9]. The chemical dispersant, COREXIT®EC9500A (as the only listed dispersant for offshore use in Canada), is tested at a dispersant-to-oil ratio (DOR) of 1:20 (manufacturer's recommended dose) to determine its effectiveness at treating surface spills of bitumen blends and readily dispersible conventional oil spilled on seawater in a flow-through wave tank during spring and summer of 2016 and 2017 in Atlantic Canada.

Attempts are made to address these gaps in knowledge by evaluating (1) dispersant effectiveness by oil type, including fresh and weathered products and seasonal effects (i.e., water temperature) to generate a new dispersant model based on empirical data; and (2) a previously generated viscosity weathering model [9] integrated with the newly generated dispersion effectiveness model to provide a means to estimate the effectiveness of dispersant to treat weathered oil. The information generated will aid oil spill responders and decision-makers on the appropriate conditions, where dispersant might be applicable to treat oil spills that have weathered at sea.

2. Materials and Methods

2.1. Oil Types and Characterizing the Chemical Composition and Physical Properties of the Oil Products

Access Western Blend (a dilbit comprised of crude bitumen blended at 30% with condensate), synthetic bitumen (a synbit made up of 50% synthetic crude oil blended with crude bitumen) and Western Canadian Select (a dilsynbit consisting of 50% synthetic crude oil/condensate blended with crude bitumen) were selected, because they represent the highest volume of oil sands products transported throughout Canada. Heidrun was also selected as the reference conventional crude, since it physical properties are reasonably close to the blended bitumen products. Similar to the technique used by Li et al. [10], the bitumen oil products were artificially weathered by purging them with nitrogen for 48 h at ~20 °C. Weathering the products prior to placing the oil in the tank for dispersion effectiveness testing is a key step as weathering generally increases oil viscosity and is likely to limit chemical dispersant effectiveness.

Samples of the unweathered oils were evaluated for saturates, aromatics, resins and asphaltenes (SARAs) using thin-layer chromatography coupled with flame ionization detection (TLC-FID) [1,11]. To monitor changes in the physical properties of the oil at various seawater temperatures; recovered oil samples were analyzed by an Anton Paar SVM 3000 Analyzer to quantify viscosity [12] and density [13].

2.2. Wave Tank Facility

The wave tank facility is located at the Bedford Institute of Oceanography (BIO) in Dartmouth, Nova Scotia. The tank dimensions are 30 m long, 0.6 m wide and 2.0 m high, with a typical water level of 1.5 m (Figure 1). The tank is equipped with a series of manifolds to generate a more or less uniform current along the wave propagation direction; hence, the label flow-through system has been used to evaluate dispersant effectiveness of fresh and weathered crude oils [4,5,8,14–16].

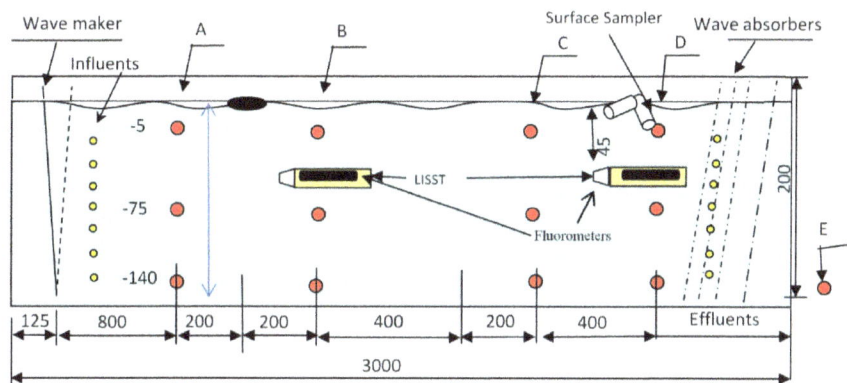

Figure 1. Schematic diagram (not to scale, all units in cm) illustrating the location of the oil source (black ellipse between A and B), Laser In-situ Scattering and Transmissometry (LISST) particle counters, sampling locations at A, B, C, D (3 depths), the effluent port E (1 port) and surface (near sample location D). LISST#1 is at location B (ca. 1.2 m downstream at a 0.45 m depth) and LISST#2 is at location D (ca. 12 m downstream at a 0.45 m depth) of oil release.

The hydrodynamics of the various wave types generated in the wave tank facility has been characterized in prior works [17,18]. Each experiment was conducted for one hour during which each wave cycle (four breakers) lasts for 15 s followed by a quiescence period that lasts for 25 s.

2.3. Oil and Dispersant Application during Wave Tank Tests

The experimental factorial design involves testing of four oils with two treatments (without and with dispersant) in triplicate over two seasons. Therefore, the total number of runs was 24, conducted in random order for spring and summer experiments. Briefly, for each experiment, quiescent conditions were achieved in the tank (i.e., no waves). Next, ca. 240 g of oil product was gently poured onto the filtered seawater surface within a 40 cm diameter ring located 10 m downstream from the wave-maker and ~12 g of the dispersant COREXIT®EC9500A (Nalco, active surfactant is dioctyl sodium sulfosuccinate; U.S. Patent No. 614285) was sprayed gently onto the oil slick through a pressurized nozzle (60 psi, 0.635 mm i.d.). This resulted in a DOR of 1:20. The wave-maker was started, and produced a sequence of waves. The ring was promptly lifted prior to the arrival of the first breaking wave on the location of the ring. The sequence of waves; generated a 0.4 m high plunging breaker (where the water curls and re-enters the water surface downstream) every 40 s at the same location using the dispersive focusing technique [19]. In this study, only breaking waves were investigated with the use of chemical dispersant to treat oil spills, since an earlier study [20] revealed that spill treating agents were ineffective in the dispersion of condensate bitumen blends when no wave breaking occurs.

2.4. Wave Tank In Situ Measuring Devices

Two particle size counters (Laser In-situ Scattering Transmissometry (LISST)-100X, Sequoia Scientific, Inc., Bellevue, WA, USA) were employed during the experiments, one at 1.2 m and another at 12 m downstream of the oil release point and both at a depth of 0.45 m (Figure 1). Particle size (2 to 500 μm) distributions were recorded at 2.0 s intervals for 1 h per experiment as in previous studies [4,8,10,14–16].

The Sauter mean diameter D_{Sauter} was estimated based on the LISST measurement [21]. It is obtained as:

$$D_{Sauter} = D_{32} = \frac{\sum\limits_{i}^{M} c_i D_i^3}{\sum\limits_{i}^{M} c_i D_i^2} \tag{1}$$

where c_i is the concentration of particles, calculated from the volume concentration as obtained from the LISST measurements ($c_i = \frac{V_i}{(\frac{\pi}{6})D_i^3}$); D is the particle diameter; the subscript i refers to the size class, and M is the total size bins (32 intervals herein).

2.5. Laboratory Analysis of Seawater Samples from Wave Tank Studies

The experimental and sampling procedures were consistent with the crude oil dispersant efficacy testing in the flow-through wave tank reported previously [4]. Four water sampling devices were deployed, one at 2.0 m upstream from the oil release point and the other three downstream at 2.0 m, 8.0 m and 12 m from the oil release point (Figure 1). Each of the four samplers, collect water (~100 mL) at three depths (0.05, 0.75 and 1.4 m) in the tank at the time points: 5, 15, 30, 45, and 60 min. In addition, effluent samples (from the side opposite the wave-maker) were taken (Figure 1). Four time-zero samples (prior to oil release, to check background levels) were selected at arbitrary sampling locations (Figure 1).

The collected water samples were extracted and analysed for total petroleum hydrocarbons (TPH) using a gas chromatograph equipped with flame ionization detection (GC-FID) [4,22]. The method is a modified version of EPA 3500C, whereby the sample container is the extraction vessel. Briefly, 12 mLs of dichloromethane (DCM) were added to a 125 mL amber glass sample bottle containing ~80 mLs of seawater collected during the experiments. Next, the sample was placed on a Wheaton R$_2$P roller (VWR, Canada) for 18 h. The roller has been modified to accommodate a 3-inch (internal diameter) PVC pipe into each roller slot. This modification permits sample containers of different sizes to be used in the apparatus. Once extraction was complete, the sample bottles were removed and the DCM was recovered. The recovered DCM was placed in a pre-weighed 15 mL centrifuge tube, and the solvent was removed using a nitrogen evaporator until the final volume reached 1.0 mL graduation on the centrifuge tube. The extracts were then analysed by GC-FID. Calibration standards prepared from the test oils were used to develop calibration curves for evaluating the oil concentration in the seawater extracts. The method detection limit is <0.5 mg/L. The benefit of this procedure is that 240 samples can be extracted simultaneously; thus increasing productivity with acceptable accuracy and precision.

3. Results and Discussion

3.1. Composition and Physical Properties of Test Oils

In their unweathered state, bitumen blends have viscosities >200 cSt @ 15 °C and are classified as heavy oils (Table 1). The bitumen blends contain a greater percentage of resin and asphaltenes compared to Heidrun crude oil, which is the medium conventional crude. Depending on the rate of diluent released and seawater temperatures during a spill, these high molecular weight (>500 atomic mass units) chemicals can greatly affect the physical properties (e.g., density and viscosity) of the oils that are relevant to responding to spills. The source of these chemicals in blended bitumen products is most likely from the crude bitumen. Oil sands products are expected to significantly weather within a few hours post-spill, thus significantly altering their viscosities, which would limit dispersant effectiveness [9]. Heidrun is not expected to weather to the extent that limits its treatment with chemical dispersant after a spill. The viscosities and densities of the four oils, at the recorded experimental seawater temperatures are found in Table S1 (Supplementary Materials) where one notes a viscosity range varying from 60 up to 10,000 cSt.

Table 1. Saturates, aromatics, resins and asphaltenes (SARAs) contribution and physical properties for the three oils (unweathered).

Oil Type	Chemical Composition				Physical Properties			Oil Class
	Sat	Aro	Resin	Asph	Viscosity	Density	API°	
	%Contribution				(cSt)	(g/cm^3)		
Access Western Blend (AWB)	14	23	46	17	244	0.9189	22.3	Heavy
Heidrun	38	40	27	4	68.9	0.9132	23.3	Medium
Synbit	20	10	52	18	205	0.9304	20.4	Heavy
Western Canadian Select (WCS)	20	10	57	13	211	0.9214	21.9	Heavy

3.2. Test Conditions during Wave Tank Studies

To capture the effect of water temperature on the chemical dispersion of the test oil products, experiments were conducted consecutively during the spring and summer of 2016 and 2017 in Atlantic Canada and the physical measurements of the seawater obtained are recorded in Table S1 (Supplementary Materials). The water was obtained directly from the Bedford Basin, Dartmouth Nova Scotia, Canada and its temperature ranged from 3.7 °C to 19.7 °C for the entire study. Water temperature can affect dispersant effectiveness when treating heavy conventional oils such as Intermediate Fuel Oil (IFO) 180 [8]. Salinity is also an important factor to consider, since it can affect the efficacy of dispersants, such as COREXITEC®9500A, that are formulated for saltwater environments [23]. In our experiments, the salinity, over spring and summer months, ranged from 25.5 to 30.4 parts-per-thousand (ppth). The small difference between these values suggests that salinity variation would not make a measureable impact on the behavior of the test oils during the experiment.

3.3. Laser In-Situ Scattering Transmissometry (LISST)-100x

Since natural dispersion was very poor and due to transport and dilution, information collected from the first LISST (1.2 m from oil release) was placed in the Supplementary Materials (Figures S1, S3 and S5). Figure 2 reports contour plots of oil droplet volume concentration (μL/L) obtained from the second LISST-100x (12 m from the oil application) for the natural dispersion case as a function of time. The vertical axis (*y*-axis) represents particle or oil droplet size (μm). Additional information on particle sizes, volume concentrations and Sauter values can be found in the Supplementary Materials (Figures S1–S4). Without any treatment, the four oils showed poor natural dispersion under spring and summer conditions, where very little oil (in dispersed form or as small droplets) was in the water column; only the largest size (i.e., >100 μm) droplets had a non-negligible concentration, but still low. The findings in Figure 2 are consistent with the literature of conventional oils; oil droplets that have been produced by breaking waves in the absence of dispersant are typically larger than 100 μm, have a unimodal distribution, and tend to rise to the surface where they are likely to coalesce [8,10,14]. The Sauter mean diameter values of Figure 2 varied between 150 to 350 μm, which is in agreement with prior studies on heavy oil dispersion.

Figure 3 reports contour plots of oil droplet volume concentrations (μL/L) obtained from the LISST-100x (12 m from the oil application) for the chemical dispersion case as a function of time. The vertical axis (*y*-axis) represents particle or oil droplet size (μm). The oil detected by the first LISST (1.2 m from oil release, Supplementary Materials) remained dispersed in the water.

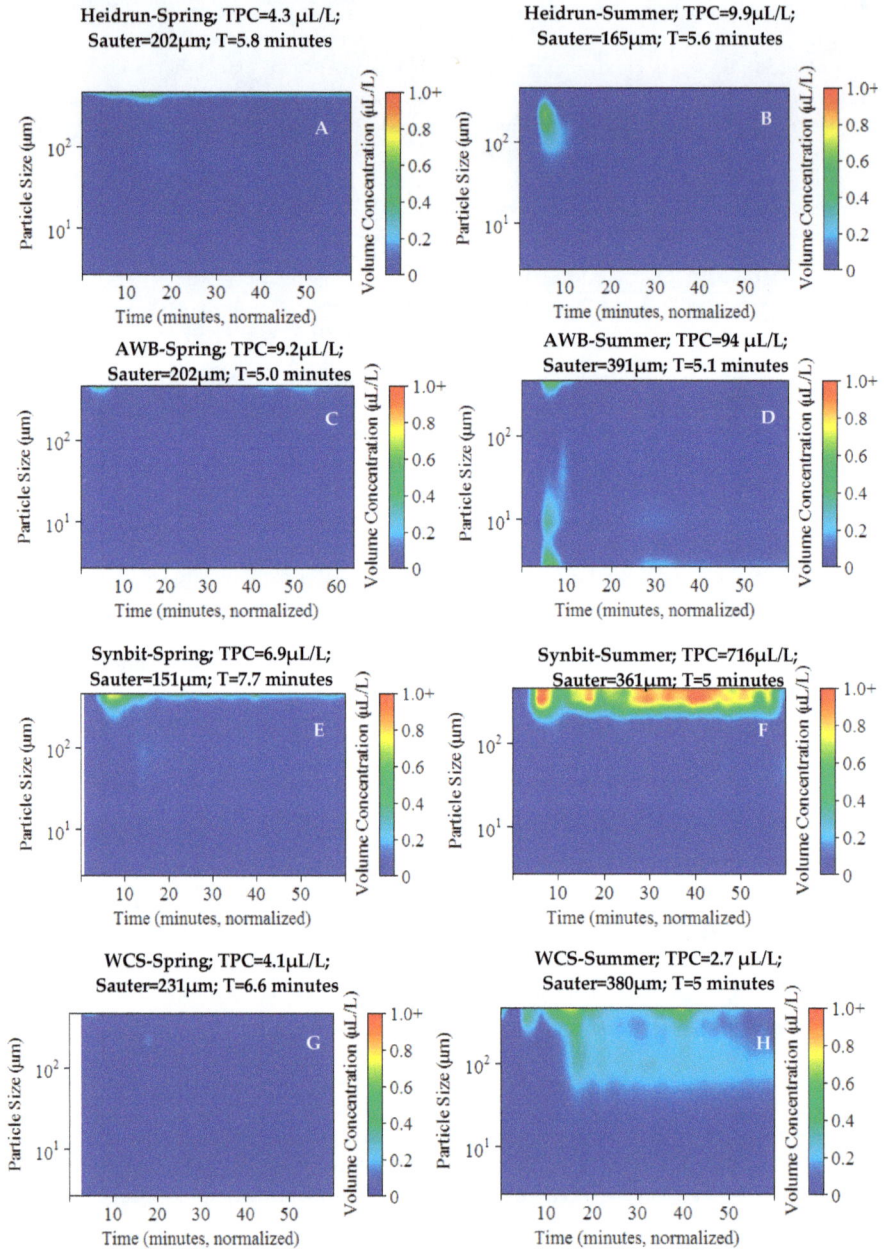

Figure 2. Contour plots (LISST data, 12 m from oil release) showing seasonal effects on the concentration of oil particle size simulated in the wave tank for the natural dispersion of: (**A**) Heidrun-spring, (**B**) Heidrun-Summer, (**C**) AWB-Spring, (**D**) AWB-Summer, (**E**) Synbit-Spring, (**F**) Synbit-Summer, (**G**) WCS-Spring, and (**H**) WCS-summer. The Sauter mean diameter values varied between 150 to 350 μm and low concentration of small particles (<100 μm) were detected in all cases.

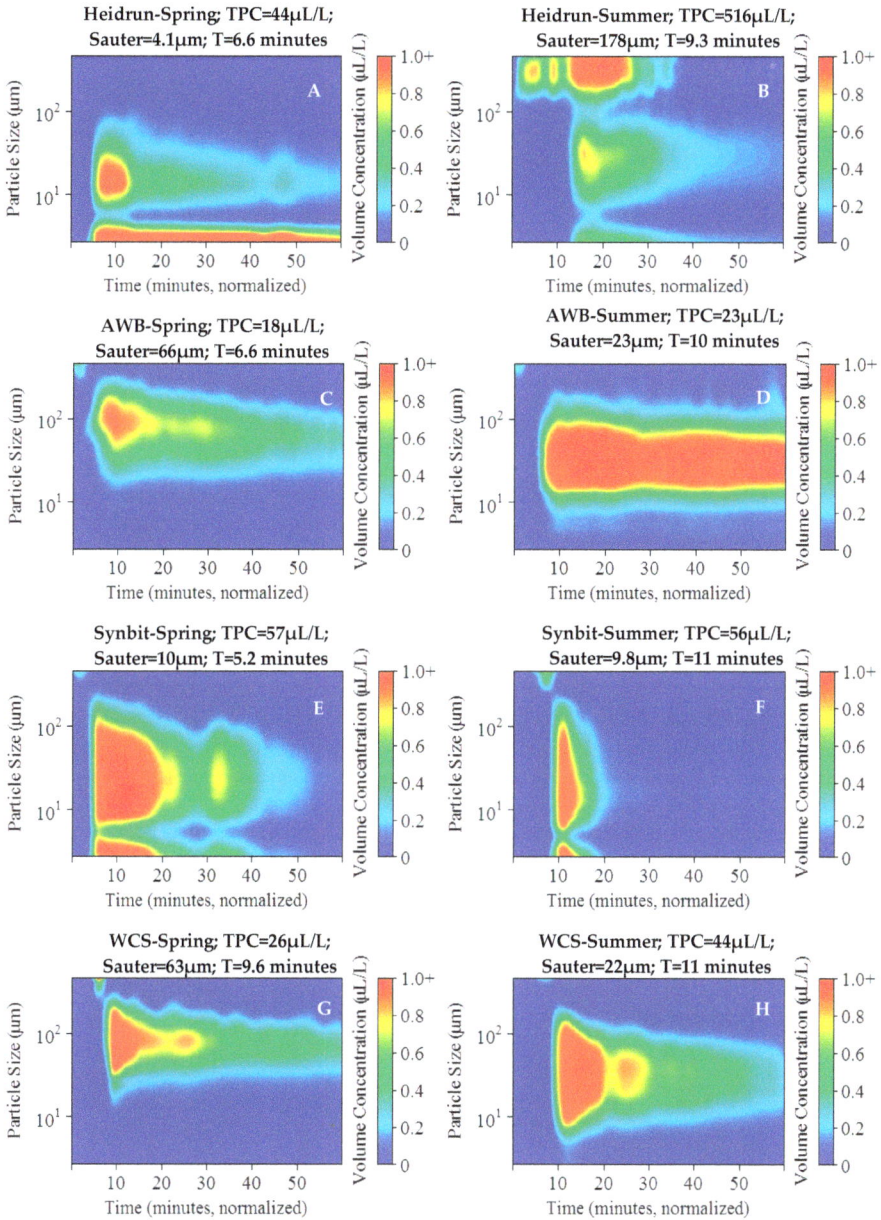

Figure 3. Contour plots (data from LISST, 12 m from oil release) illustrating seasonal effects on the concentration of oil particle size for chemically enhanced dispersion of: (**A**) Heidrun-spring, (**B**) Heidrun-Summer, (**C**) AWB-Spring, (**D**) AWB-Summer, (**E**) Synbit-Spring, (**F**) Synbit-Summer, (**G**) WCS-Spring, and (**H**) WCS-summer. The Sauter mean diameter values varied between 4 to 170 μm and high concentration of small particles (<100 μm) were detected in all cases.

Figure 3 displayed higher volume concentrations and smaller particle sizes (<50 μm) in the water column compared to natural dispersion (without dispersant) of all oil types for spring and summer conditions. In general, higher concentrations of small particles (<50 μm) were detected in summer conditions. The chemical dispersion of Heidrun and synbit (both in spring and summer) produced large concentrations of very small droplets (<10 μm). This occurred also for the chemical dispersion of WCS, but only in summer conditions.

Our recent investigation [21] elucidated two major aspects of the LISST that should be considered when evaluating the droplet size distribution. The first is the impact of high concentrations, and it was found that if the optical transmission drops below 30%, the measured peak value of the LISST tended to underestimate the true peak by up to 50%, and the instrument accuracy decreased by up to ~30%. Fortunately, all LISST measurements in this study had an optical transmission that was larger than 45%. The out-of-range sizes of particles affected the LISST measurements especially near the limits of the range (but also slightly within the mid-range) when very high concentrations were detected. However, the impact of the out-of-range values decreases sharply as the size associated with that concentration is farther from the limit. In Zhao et al. [21], concentrations of 1.0 micron droplets increased the readings of the 2.3 micron concentrations by 20% of the 1.0 micron concentration. Thus, unless the out of range concentrations are 10 times or larger than those within range, the LISST should be viewed as capturing the totality of the mass of the droplets within range.

3.4. Total Petroleum Hydrocarbons (TPH) in the Water Column

Averaged TPH concentrations at all depths (0.05, 0.75, and 1.4 m) for location D (10 m from oil release point; Figure 1) are plotted in Figure 4 as function of time for the four oils under natural and chemically enhanced dispersion conditions in spring and summer. Each curve represents the average of a triplicate. Low TPH concentrations were observed under natural dispersion conditions, which have been reported by others [8,10]. The concentration during the summer was slightly higher than spring (warmer temperatures decrease the viscosity thus affecting dispersion), but remained an order of magnitude smaller than the chemically dispersed TPH for both seasons. For all four oil types, TPH concentrations reached at maximum and gradually declined with dilution and transport by waves and currents in the tank. These trends were similar for the natural and chemically enhanced dispersion of other oils in spring and summer from previous studies [4,8].

During dispersant application, the increased oil concentration in the water column creates controversy from a policy point of view, since it makes the oil more bioavailable to aquatic species, but reduces the amount of oil reaching the sensitive habitats in shoreline areas. Through natural dilution and transport the TPH concentrations in the water column dropped to near background levels for each of the four oil types. Sufficient mixing and water currents to transport dispersed oil are critical components when assessing not only dispersant effectiveness to treat oil products, but also the rate of dilution and transport to ensure minimal impacts to aquatic species and their habitats.

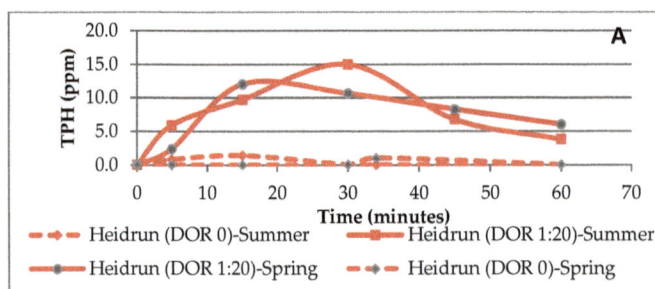

Figure 4. *Cont.*

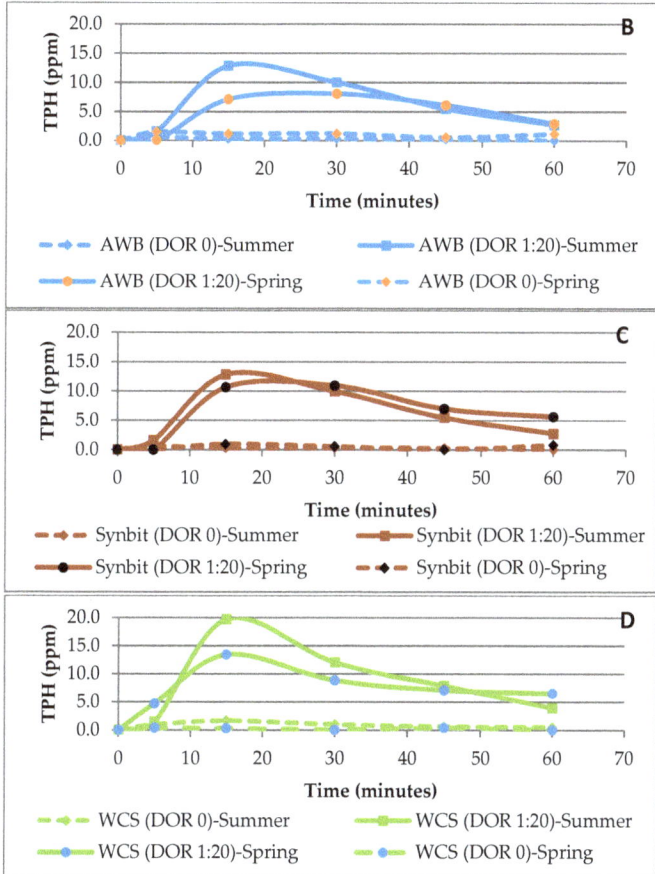

Figure 4. Average total petroleum hydrocarbon (TPH) concentrations in the water column over a depth of 0.05 to 1.4 m located 12 m downstream from the oil release point for treated and untreated oils: (**A**) Heidrun, (**B**) AWB, (**C**) Synbit and (**D**) WCS under spring and summer conditions. TPH concentrations were an order magnitude higher for oils treated with dispersant.

3.5. Dispersant Effectiveness (DE)

The averaged TPH concentrations for all depths at each sampling location (A, B, C, D, and E) were used to generate DE (%) values, which was obtained over the duration of the entire experiment by computing the fraction of dispersed oil in the effluent from the wave tank and the residual dispersed oil in the water column at the end of each experiment. Details on calculating DE (%) values during wave tank studies can be found in King et al. [4,5,8].

Table 2 (analysis of variance [ANOVA] single factor, Excel) shows the natural and chemically enhanced dispersion of the four oils under breaking waves and spring and summer environmental conditions. For each untreated and treated oil type, DE testing was performed in triplicate covering a range of seawater temperatures over two seasons and the experiments were conducted in random order. Natural dispersion effectiveness ranged from 0 to 7% for all four oils (Table S1). The application of a chemical dispersant had a significant ($p < 0.05$) effect on dispersion of all oil types. In the discussions to follow reference is made to viscosities in units of centipoise (cP) when taken from the literature. To get viscosity in cSt, divide cP by the density of the oil. In this case, cSt values would be approximately 10% higher than cP. Oil viscosity is critical in studying dispersants, because thin, medium-viscosity oils

(<2000 cP) are readily dispersible, but heavy, highly-viscous oils (>10,000 cP) are not [24,25]. Lewis [26] reported that a 2000 cP oil treated with COREXIT®EC9500A dispersed quickly and completely, but a more viscous 7000 cP fuel oil did not. This implies that viscosity has an effect on dispersion of oil; however in that study the time window of opportunity to treat weathered oil is not considered. Since, Heidrun's viscosity is <2000 cSt even with seasonal temperature factored in, the performance of the chemical dispersant was not affected when treating it. For the bitumen blends (e.g., AWB, synbit and WCS), DE was notably lower (~20 to 30% less) most likely due to the fact that these products are more viscous, in their pre-weathered state, than the fresh Heidrun crude oil. Also, the dispersant had an effectiveness of 40 to 50% and low as 30% for bitumen blends with viscosities of 4000 and near 10,000 cSt, respectively.

Table 2. Analysis of variance of the randomization tests showing the dispersion effectiveness of dispersant to the natural dispersion of four oils under breaking waves. The value of "p" provides information on the probability of the observation (i.e., difference) to be due to randomness. The smaller the value of "p" the less likely the difference is due to randomness. Average ± Standard Deviation (Ave ± std).

Treatment	n *	Spring Value (%) (Ave ± std)	Difference (%)	p	Summer Value (%) (Ave ± std)	Difference (%)	p
No Treatment	6	1.7 ± 1.6	-	-	1.8 ± 1.1	-	-
Corexit/AWB	6	30.6 ± 2.8	−28.9	1.0×10^{-4}	53.2 ± 3.3	−51.4	1.0×10^{-5}
No Treatment	6	2.7 ± 1.2	-	-	1.8 ± 1.1	-	-
COREXIT/Heidrun	6	70.6 ± 1.7	−67.9	5.0×10^{-7}	76.0 ± 7.4	−74.2	7.9×10^{-5}
No Treatment	6	4.2 ± 3.0	-	-	4.0 ± 1.6	-	-
COREXIT/Synbit	6	48.4 ± 4.8	−44.2	1.7×10^{-4}	59.2 ± 2.9	−55.2	9.0×10^{-6}
No Treatment	6	2.7 ± 1.6	-	-	3.5 ± 1.0	-	-
COREXIT/WCS	6	41.3 ± 4.2	−38.6	1.2×10^{-4}	53.5 ± 4.0	−50.0	3.1×10^{-5}

* $n = n_1 + n_2$ observations.

Significantly ($p < 0.05$) higher DE values were recorded during summer than spring conditions, since seasonal temperature affects the viscosity of these heavy oil products (Table 3). This is consistent with a study on heavy conventional oil products such as IFO 180, which was effectively dispersed with a DE of 90% at high temperature (16 °C) and had low DE (<10%) at low temperatures (<10 °C) using the same test facility [8]. Also, laboratory studies showed a 20% difference in DE of heavy oils between 16 and 5 °C [7]. The seasonal effects (% difference) on DE for the bitumen blends were greatest (22.3%) for AWB (dilbit) and the least (10.8%) for synbit (Table 3). This is most likely due to the fact that synthetic crude as the diluent portion of synbit is less volatile than the condensate in AWB (dilbit). Its composition contains a greater portion of chemicals including saturates in the range of C_{17} to C_{35} and alkylated polycyclic aromatics [27] that are less susceptible to natural attenuation by evaporation than condensate (primarily of low molecular weight aromatics and aliphatics in the range of n-C_5 to C_{10}) [28] when dispensed in a dynamic state in spring and summer. Both diluents (condensate and synthetic crude) are present in WCS (dilsynbit), so the seasonal effect (% difference) on DE falls between the other two blends. Also, one can note that the bitumen blends prior to treatment have different viscosities, since they were pre-weathered (7% w/w) under similar conditions. Heidrun crude oil was readily dispersible when treated with chemical dispersant over the entire temperature range with DE values >70% for both spring (7.1 ± 0.8 °C) and summer (16.8 ± 1.6 °C) conditions. With this medium crude oil, the reported seawater temperature range did not have a significant ($p = 0.28$) influence on the effectiveness of the chemical dispersant (Table 3).

Table 3. Analysis of Variance of the randomization tests show the DE of oils affected by seasonal water temperatures under breaking waves. The value of "*p*" provides information on the probability of the observation (i.e., difference) to be due to randomness. The smaller the value of "*p*" the less likely the difference is due to randomness. Average ± standard deviation (Ave ± std).

The Effect of Seasonal Water Temperature on DE				
Treatment	*n* *	Value (%) (Ave ± std)	Difference (%)	*p*
COREXIT/AWB-Spring	6	30.6 ± 2.8	-	-
COREXIT/AWB-Summer	6	53.2 ± 3.3	−22.6	0.00088
COREXIT/Heidrun-Spring	6	70.6 ± 1.7	-	-
COREXIT/Heidrun-Summer	6	76.0 ± 7.4	−5.4	0.28
COREXIT/Synbit-Spring	6	48.4 ± 4.8	-	-
COREXIT/Synbit-Summer	6	59.2 ± 2.9	−10.8	0.029
COREXIT/WCS-Spring	6	41.3 ± 4.2	-	-
COREXIT/WCS-Summer	6	53.5 ± 4.0	−12.2	0.023

3.6. Modelling Dispersion Effectiveness

The four different oil types, consisting of fresh and artificial weathered products, selected for this study cover a board range of viscosities, over two seasons, with measured DE values (Table S1). The untreated (naturally dispersed) oil DE values were plotted as a function of oil viscosity (Figure 5). The plot revealed that natural DE was very similar or changes were minimal for all four oils dispersed under spring and summer conditions. Figure 6 reports the DE as function of the viscosity for the chemically enhanced dispersion of the four oils. A linear model was fitted to the plot of chemical DE as a function of oil viscosity (Figure 6). Therefore one would write the equation:

$$DE = m\ln(v) + b \tag{2}$$

where *m* is slope, *v* is the viscosity of the oil and *b* is the y-intercept.

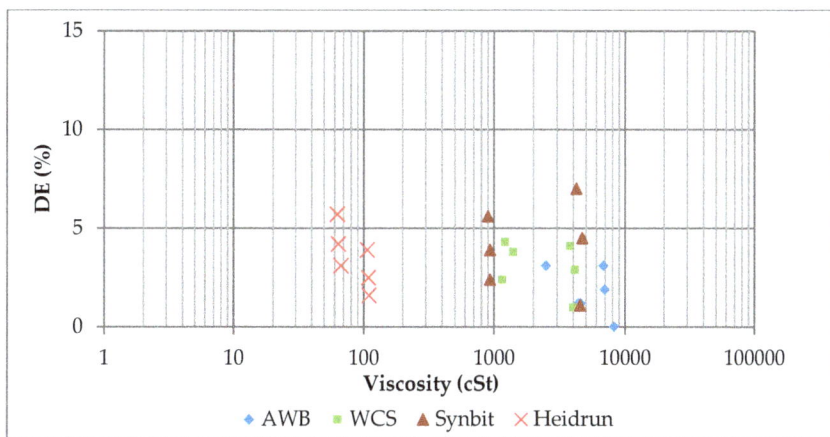

Figure 5. A plot of effectiveness of natural dispersion (DE %) as a function of oil viscosity. Experiments were conducted in spring and summer (Table S1). DE is <10% for all oils tested.

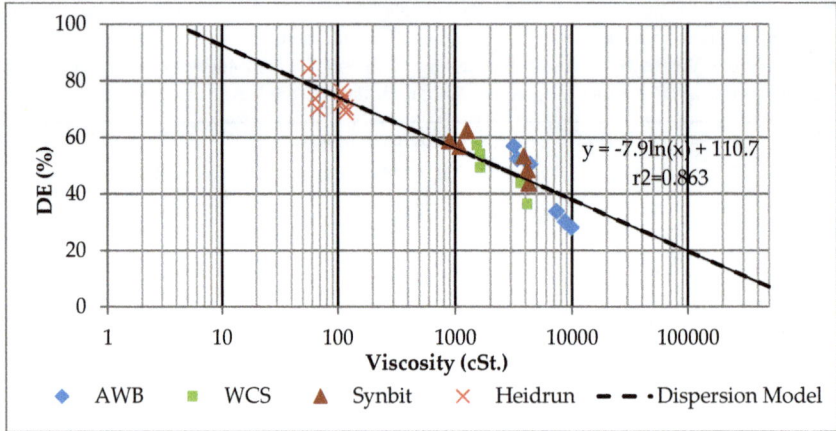

Figure 6. A plot of the effectiveness of chemically enhanced dispersion (DE %) as a function of oil viscosity. Experiments were conducted in spring and summer with fresh and weathered oils (Table S1). A dispersion model was fitted to the data points that represent four different oil types. DE (%) decreases with the increasing viscosities of oils, but effectiveness better than natural dispersion.

The fit was generally good, as one notes visually through the absence of any systematic bias (undershooting or overshooting), and the large coefficient of determination, $R^2 > 0.86$ and ANOVA showed a significant ($p < 0.001$) curve fit. If the viscosity of oil, at a specific temperature is known, then the model can be used to predict the chemically enhanced DE (%) of spilled oil at sea.

3.7. Combining Dispersant Effectiveness (DE) and Viscosity Functions to Estimate the Window of Opportunity to Treat Surface Spills of Oil after the Initial Release

King et al. [9] developed a model for the temporal evolution of viscosity due to weathering:

$$v = (v_0 + (v_f - v_0)\left(\frac{t}{T+t}\right)^n \tag{3}$$

where v represents the viscosity (cSt) of weathered oil, v_0 and v_f are the initial and final oil viscosities, respectively, and t represents time in hours. The parameter "T" represents the "half-weathering rate constant"; the value of t when $v/v_{max} = 0.5$, and the parameter "n" reflects the slope of the curve to reach the maximum value. The advantage of Equation (3) is that it allows for the rapid increase in the early hours, and then for the plateaus in the data.

Equation (3) can be substituted for v into Equation (4) to determine DE as function of time for various oils:

$$DE = a \ln((v_0 + (v_f - v_0)\left(\frac{t}{T+t}\right)^n + b) \tag{4}$$

Table S2 (Supplementary Materials) shows DE values generated using Equation (4) for the four oil products considered herein that were weathered under different conditions. Thus, assuming weathering conditions similar to those reported in King et al. [9], calm seas in either spring or summer, one can predict the DE based simply on time, if conditions change to a more energetic state where dispersants are applicable. The predicted data presented (Table S2) shows changes in oil viscosity with time of weathering and the DE to treat a spill on water after the initial release in spring (ca. 6 °C) and summer (ca. 15 °C) conditions. In this case, the bitumen blends' viscosities exceeded 2000 cSt in 24 h for spring and summer, thus having a great effect on the predicted DE to treat these weathered oil products. For the conventional crude (Heidrun), changes in its viscosity were <500 cSt in 360 h after weathering on water and the predicted DE to treat the weathered oil was good. However, similar to

King et al. [9] the data was based on an oil slick thickness of 4 mm and the effect of slick thickness on the rate of weathering of oil is considered in the discussion to follow.

A previous report [29] suggested that temperature and time were greater factors in oil evaporation than surface wind speed or oil slick thickness for a wide range of crude oils. In contrast, Gros et al. [30] determined that wind speed strongly affected the thickness of the oil slick and thus evaporation very early in an experimental spill of Norwegian crude in the North Sea. So one would assume that the thickness of the oil slick can affect the rate of weathering of condensate bitumen blends (where 30% of the product is gas condensate). This was substantiated by King et al. [9], where the thickness of the oil slick affected the rate of weathering of condensate bitumen blends. However, the results of that study showed that for a 4 mm thick slick of AWB, the viscosity exceeded 10,000 cSt within three hours of weathering on temperate (22 °C) water. Under such conditions there was a rapid change in the viscosity of the oil regardless of oil slick thickness, where the time window for dispersant use would be closed, as indicated from studies by others [24–26]. Temperature and the thickness of the oil slick affects the rate of weathering of oil [27], so these factors could affect the use of Equation (4) to estimate DE. The effects of temperature on DE have been substantiated by Li et al. [8], but not the weathering of oil at different temperatures and its inherent effect on DE to treat a spill at various points in time after the initial release as shown in Table S2. The function (Equation (4)) proposed here may be applicable to other oil types with the limitations mentioned above.

4. Conclusions

The natural and chemically enhanced dispersion of four oil products were investigated in the wave tank of the Center for Offshore Oil and Gas Research (COOGER), placed outdoor in Halifax Nova Scotia, Canada. The products were: Access Western Blend, Heidrun (conventional heavy crude), synbit, and Western Canadian Select, and the dispersant was COREXIT9500A. Experiments were conducted in spring and summer to capture the impact of temperature, and the hydrodynamic conditions in the tank were of breaking waves with a wave height of 0.4 m. The results showed that the natural (or physical) dispersion of these oils was less than 10%, and there was essentially no difference in behavior between seasons. The application of dispersant increased the DE by an order of magnitude within a significant statistical level ($p < 0.05$). Also, temperature (summer versus spring) resulted in larger chemical DE for all oils, except for the conventional oil (Heidrun). For this study, oil type, fresh and weathered oil, and seasonal effect data produced a broad range of oil viscosities with measured DE values that were fitted to a linear regression model. The approach could, therefore, be readily used to estimate the chemical DE values of released oil.

Equation (4) has the potential to predict the DE based simply on the time of weathering of oil. For this study it was applied to various bitumen blends and the conventional crude, Heidrun, but may be applicable to other oils as well. However, some limitations of the function to consider are the oil type, thickness of the oil slick, temperature, and the fact that the weathering of oil initially occurs in calm waters prior to more energetic sea states where dispersant is applicable. Although the dispersion model has only been considered in its application to oil spills in Canadian waters, it may be applicable to predict the use of dispersant to treat spills in international waters as well.

J. Mar. Sci. Eng. **2018**, *6*, 128

Supplementary Materials: The following are available online at http://www.mdpi.com/2077-1312/6/4/128/s1; Table S1: Summary of test conditions, physical properties of test oils, and dispersion effectiveness measurements; Table S2: Predicted viscosity data generated from oil weathering (Equation (2)) and dispersion effectiveness (Equation (3)) models; Figure S1: The plot represents the seasonal effect on particle size distribution data (LISST 100X-#1; 1.2 m from oil release) obtained at the point in time of maximum total particle concentration during the natural dispersion of four oils; Figure S2: The plot represents the seasonal effect on particle size distribution data (LISST 100X-#2; 12 m from oil release) obtained at the point in time of maximum total particle concentration during the natural dispersion of four oils; Figure S3: The plot represents the seasonal effect on particle size distribution data (LISST 100X-#1; 1.2 m from oil release) obtained at the point in time of maximum total particle concentration during the chemically enhanced dispersion of four oils; Figure S4: The plot represents the seasonal effect on particle size distribution data (LISST 100X-#2; 12 m from oil release) obtained at the point in time of maximum total particle concentration during the chemically enhanced dispersion of four oils; Figure S5: Contour plots (LISST 100X-#1; 1.2 m from oil release) illustrating seasonal effect on the concentration of oil particle sizes simulated in the wave tank for the natural dispersion of four oil types.

Author Contributions: The paper was conceived and written by T.K. and J.C. M.B. provided guidance on the model components and editing. B.R. and S.R. provided processing and interpretation of chemistry and oil droplet size data. K.L. provided guidance on the experimental design, oil dispersion work and editing.

Funding: This research was funding by the Department of Fisheries and Ocean Canada under the Government of Canada's Oceans Protection Plan.

Acknowledgments: The authors acknowledge the technical support provided by Patrick Toole, Jennifer Mason and Graeme Soper of Fisheries and Oceans Canada.

Conflicts of Interest: The authors declare no conflict of interest.

References

1. King, T.; Mason, J.; Thamer, P.; Wohlgeschaffen, G.; Lee, K.; Clyburne, J. Composition of Bitumen Blends Relevant to Ecological Impacts and Spill Response. In Proceedings of the 40th Arctic and Marine Oil Spill Program Technical Seminar on Environmental Contamination and Response, Calgary, AB, Canada, 3–5 October 2017; pp. 463–475.

2. CAPP (Canadian Association of Petroleum Producers). Crude Oil Forecasts, Markets and Transportation. Available online: https://www.capp.ca/publications-and-statistics/publications/320294 (accessed on 10 October 2018).

3. Canada Gazette. Regulations Establishing a List of Spill-Treating Agents (Canada Oil and Gas Operations Act). Available online: http://www.gazette.gc.ca/rp-pr/p2/2016/2016-06-15/html/sor-dors108-eng.html (accessed on 10 October 2018).

4. King, T.; Robinson, B.; McIntyre, C.; Ryan, S.; Saleh, F.; Boufadel, M.; Lee, K. Fate of surface spills of Cold Lake Blend diluted bitumen treated with dispersant and mineral fines in a wave tank. *Environ. Eng. Sci.* **2015**, *32*, 250–261. [CrossRef]

5. King, T.; Robinson, B.; Ryan, S.; Lu, Y.; Lee, K.; Zhou, Q.; Ju, L.; Li, J.; Peiyan, S. The Fate of Chinese and Canadain Oil Treated with Dispersants in a Wave Tank. In Proceedings of the 38th Arctic and Marine Oil Spill Program Technical Seminar on Environmental Contamination and Response, Halifax, NS, Canada, 2–4 June 2015; pp. 470–494.

6. Lee, K.; Boufadel, M.; Chen, B.; Foght, J.; Hodson, P.; Swanson, S.; Venosa, A. *Expert Panel Report on the Behaviour and Environmental Impacts of Crude Oil Released into Aqueous Environments*; Royal Society of Canada: Ottawa, ON, Canada, October 2015; ISBN 978–1–928140–02–3.

7. Srinivasan, R.; Lu, Q.; Sorial, G.; Venosa, A.; Mullin, J. Dispersant effectiveness of heavy fuel oils using the baffled flask test. *Environ. Eng. Sci.* **2007**, *24*, 1307–1320. [CrossRef]

8. Li, Z.; Lee, K.; King, T.; Boufadel, M.; Venosa, A. Effects of temperature and wave conditions on chemical dispersant efficacy of heavy fuel oil in an experimental flow-through wave tank. *Mar. Pollut. Bull.* **2010**, *60*, 1550–1559. [CrossRef] [PubMed]

9. King, T.; Robinson, B.; Cui, F.; Boufadel, M.; Lee, K.; Clyburne, J. An oil spill decision matrix in response to surface spills of various bitumen blends. *Environ. Sci. Proc. Impacts* **2017**, *19*, 928–938. [CrossRef] [PubMed]

10. Li, Z.; Lee, K.; King, T.; Boufadel, M.; Venosa, A. Evaluating crude oil chemical dispersant efficacy in a flow-through wave tank under regular non-breaking and breaking wave conditions. *Mar. Pollut. Bull.* **2009**, *58*, 735–744. [CrossRef] [PubMed]

11. Maki, H.; Sasaki, T. Analytical method of crude oil and characterization of spilled heavy oil. *J. Jpn. Soc. Water Environ.* **1997**, *20*, 639–642.

12. ASTM D341–09. *Standard Practice for Viscosity-Temperature Charts for Liquid Petroleum Products*; ASTM International: West Conshohocken, PA, USA, 2015.

13. ASTM D 5002. *Standard Test Method for Density and Relative Density of Crude Oils by Digital Density Analyzer*; ASTM International: West Conshohocken, PA, USA, 2010.

14. Li, Z.; Lee, K.; King, T.; Boufadel, M.; Venosa, A. Assessment of chemical dispersant effectiveness in a wave tank under regular non-breaking wave conditions. *Mar. Pollut. Bull.* **2008**, *56*, 903–912. [CrossRef] [PubMed]

15. Li, Z.; Lee, K.; King, T.; Kepkay, P.; Boufadel, M.; Venosa, A. Evaluation chemically dispersant efficacy in a wave tank: 1-Dispersant effectiveness as a function of energy dissipation rate. *Environ. Eng. Sci.* **2009**, *26*, 1139–1148. [CrossRef]

16. Li, Z.; Lee, K.; King, T.; Boufadel, M.; Venosa, A. Evaluation chemically dispersant efficacy in a wave tank: 2-Significant factors determining in situ oil droplet size distribution. *Environ. Eng. Sci.* **2009**, *26*, 1407–1418. [CrossRef]

17. Venosa, A.; Lee, K.; Boufadel, M.; Li, Z.; King, T.; Wickely-Olsen, E. Dispersant Effectiveness as a Function of Energy Dissipation Rate in an Experimental Wave Tank. In Proceedings of the International Oil Spill Conference, Savannah, GA, USA, 4–8 May 2008; pp. 777–784.

18. Wickley-Olsen, E.; Boufadel, M.; King, T.; Li, Z.; Lee, K.; Venosa, A. Regular and Breaking Waves in Wave Tank for Dispersion Effectiveness Testing. In Proceedings of the International Oil Spill Conference, Savannah, GA, USA, 4–8 May 2008; pp. 499–508.

19. Botrus, D.; Boufadel, M.; Wickley-Olsen, E.; Weaver, J.; Weggel, R.; Lee, K.; Venosa, A. Wave tank to Simulate the Movement of Oil under Breaking Waves. In Proceedings of the 31th Arctic and Marine Oil Spill Program Technical Seminar on Environmental Contamination and Response, Calgary, AB, Canada, 3–5 June 2008; pp. 53–59.

20. Environment Canada, Fisheries and Oceans Canada, and Natural Resources Canada. *Properties, Composition and Marine Spill Behaviour, Fate and Transport of Two Diluted Bitumen Products from the Canadian Oil Sands*; Federal Government Technical Report: Ottawa, ON, Canada, November 2013; pp. 1–85.

21. Zhao, L.; Boufadel, M.; King, T.; Robinson, B.; Conmy, R.; Lee, K. Impact of particle concentration and out-of-range sizes on the measurement of the LISST. *Meas. Sci. Technol.* **2018**, *29*, 1–15. [CrossRef]

22. Cole, M.; King, T.; Lee, K. Analytical technique for extracting hydrocarbons from water using sample container as extraction vessel in combination with a roller apparatus. *Can. Tech. Rep. Fish. Aquat. Sci.* **2007**, *2733*, 1–12.

23. Chandrasekar, S.; Sorial, G.; Weaver, J. Dispersant effectiveness on oil spills—Impact of salinity. *J. Mar. Sci.* **2006**, *63*, 1418–1430. [CrossRef]

24. National Research Council (US). *Using Oil Spill Dispersant on The Sea*; National Academy Press: Washington, DC, USA, 1989; p. 54.

25. GENIVAR. *Risk Assessment for Marine Spills in Canadian Waters: Phase 1, Oil Spills South of the 60th Parallel*; WSP Canada Inc.: Montreal, QC, Canada, 2013; pp. 1–254.

26. Lewis, A. *Determination of the Limiting oil Viscosity for Chemical Dispersion At-Sea. (MCA Project MSA 10/9/180)*; Final Report for DEFRA, ITOPF, MCA and OSRL, Maritime and Coastguard Agency Project MSA 10/0/180; MCA Project MSA: Middlesex, UK, 2004; pp. 1–86.

27. Yang, C.; Wang, Z.; Yang, Z.; Hollebone, B.; Brown, C.; Landriault, M.; Fieldhouse, B. Chemical fingerprints of Alberta oil sands and related petroleum products. *Environ. Forensics* **2011**, *12*, 173–188. [CrossRef]

28. Environment and Climate Change Canada. Risk Management Approach for Natural Gas Condensate. Available online: http://www.ec.gc.ca/ese-ees/default.asp?lang=En&n=BBE4B27E-1 (accessed on 24 August 2018).

29. Fingas, M. Modeling oil and petroleum evaporation. *J. Pet. Sci. Res.* **2014**, 2, 104–115.
30. Gros, J.; Nabi, D.; Würz, B.; Wick, L.; Brussaard, C.; Huisman, J.; van der Meer, L.; Reddy, C.; Arey, J. First day of an oil spill on the open sea: Early mass transfers of hydrocarbons to air and water. *Environ. Sci. Technol.* **2014**, *48*, 9400–9411. [CrossRef] [PubMed]

Journal of
Marine Science and Engineering

MDPI

Article

Refined Analysis of RADARSAT-2 Measurements to Discriminate Two Petrogenic Oil-Slick Categories: Seeps versus Spills

Gustavo de Araújo Carvalho [1,*], Peter J. Minnett [2], Eduardo Tavares Paes [3], Fernando Pellon de Miranda [1] and Luiz Landau [1]

[1] LabSAR—Laboratório de Sensoriamento Remoto por Radar Aplicado à Indústria do Petróleo, LAMCE—Laboratório de Métodos Computacionais em Engenharia, PEC—Programa de Engenharia Civil, COPPE—Instituto Alberto Luiz Coimbra de Pós-Graduação e Pesquisa de Engenharia, UFRJ—Universidade Federal do Rio de Janeiro, Rio de Janeiro 21941-909, Brazil; pellon@labsar.coppe.ufrj.br (F.P.M.); landau@lamce.coppe.ufrj.br (L.L.)
[2] OCE—Department of Ocean Sciences, RSMAS—Rosenstiel School of Marine and Atmospheric Science, UM—University of Miami, Miami, FL 33145, USA; pminnett@rsmas.miami.edu (P.J.M.)
[3] LEMOPA—Laboratório de Ecologia Marinha e Oceanografia Pesqueira da Amazônia, ISARH—Instituto Socioambiental e dos Recursos Hídricos, UFRA—Universidade Federal Rural da Amazônia, Belém 66077-830, Brazil; etpaes@gmail.com (E.T.P.)
* Correspondence: ggus.ocn@gmail.com (G.A.C.)

Received: 1 November 2018; Accepted: 30 November 2018; Published: 11 December 2018

check for updates

Abstract: Our research focuses on refining the ability to discriminate two petrogenic oil-slick categories: the sea surface expression of naturally-occurring oil seeps and man-made oil spills. For that, a long-term RADARSAT-2 dataset (244 scenes imaged between 2008 and 2012) is analyzed to investigate oil slicks (4562) observed in the Gulf of Mexico (Campeche Bay, Mexico). As the scientific literature on the use of satellite-derived measurements to discriminate the oil-slick category is sparse, our research addresses this gap by extending our previous investigations aimed at discriminating seeps from spills. To reveal hidden traits of the available satellite information and to evaluate an existing Oil-Slick Discrimination Algorithm, distinct processing segments methodically inspect the data at several levels: input data repository, data transformation, attribute selection, and multivariate data analysis. Different attribute selection strategies similarly excel at the seep-spill differentiation. The combination of different Oil-Slick Information Descriptors presents comparable discrimination accuracies. Among 8 non-linear transformations, the Logarithm and Cube Root normalizations disclose the most effective discrimination power of almost 70%. Our refined analysis corroborates and consolidates our earlier findings, providing a firmer basis and useful accuracies of the seep-spill discrimination practice using information acquired with space-borne surveillance systems based on Synthetic Aperture Radars.

Keywords: oil-slick discrimination algorithm; petrogenic oil-slick category; naturally-occurring oil seeps; man-made oil spills; exploratory data analysis; remote sensing; synthetic aperture radar; RADARSAT; Gulf of Mexico; Campeche Bay

1. Introduction

The impact of mineral oil pollution is a widely spread source of environmental concern in various ecosystems [1,2]. The detection of the sea surface expression of oil using space-borne surveillance systems is an extensively studied subject [3–5]. Oil floating on the surface of the ocean can be located, to some extent, with different types of remote sensing sensors—e.g., thermal infrared (AVHRR: Advanced Very High Resolution Radiometer [6]), visible/near infrared (MODIS: Moderate Resolution Imaging Spectroradiometer [7]), etc.—but generally, most attempts concentrate on using satellite-derived measurements from active microwave-imaging instruments (SAR: Synthetic Aperture Radars [8–10]), e.g., RADARSAT [11,12].

Research projects using SAR measurements to study petrogenic oil slicks usually focus on understanding two major processes: (1) Identification of smoother regions observed at the sea surface with reduced radar backscattering signal, i.e., classification and segmentation for dark spot detection (e.g., [13]); and (2) Differentiation of radar signature of mineral oil slicks from what is commonly referred to as "radar look-alikes" (e.g., [14])—for instance, surface natural oil produced by plants or animals (i.e., biogenic oil films), atmospheric conditions (e.g., low wind and rain cells), oceanographic features (e.g., upwelling regions and internal gravitational waves), etc. [15]. Apart from the scientific effort studying these two processes [16], few investigations are directed at using remote sensing systems to differentiate the mineral oil-slick type—i.e., differences among types of anthropogenic oil slicks observed at the sea surface, for instance: oil slicks formed from heavy versus light oil [17]; or oil slicks from production oil tests (i.e., oil released at the surface of the ocean in the process of evaluating new drilling wells) versus oily water (i.e., oil slicks from leakages occurring during the exploration or production phases) [18].

The available literature covering the subject of identifying oil slicks at the surface of the ocean using space-borne surveillance systems, for the most part, does not address the petrogenic oil-slick category discrimination: telling apart the oil-slick sea surface expression in relation to their source, thus considering oil seeps (i.e., natural oil seepages from a hydrocarbon reservoirs) versus oil spills (i.e., mineral oil spillages from man-made activities) [19–22]. The seep-spill discrimination mostly regards two points of view: economic and environmental. While the former deals with the discovery of new oil exploration frontiers in finding the presence of active petroleum systems, the latter is capable of improving the relationship between the oil- and gas-related industry and environmental organizations (and society as a whole) by reducing any origin uncertainty about the oil slick source (i.e., naturally-occurring seeps versus man-made spills). A third point of view is the one of the remote sensing community, in which if a certain methodology is capable of discriminating oil from oil using microwave measurements acquired from space [19–22], it might be plausible to say that such methodology can also be applied to differentiate oil from look-alike features in SAR imagery. This framework scientifically strengthens the other two points of view.

Notwithstanding the relative neglect of research projects on the use of satellite sensors for the discrimination of the oil-slick category, Carvalho [19] showed it is feasible to use SAR-derived measurements for seep-spill discrimination—see also [20–22]. These authors have used a series of Multivariate Data Analysis Techniques to devise a novel idea to discriminate the oil-slick category while studying seeps and spills observed on the surface of the ocean in the Gulf of Mexico off the Mexican coast in the Campeche Bay region (Figure 1). They have proposed a simple Oil-Slick Discrimination Algorithm based on SAR backscatter signature, i.e., sigma-naught (σ°), beta-naught (β°), and gamma-naught (γ°) [23–25], along with the geometry, shape, and dimension of the oil slicks. Their best outcome is reached with optimal Overall Accuracies of approximately 70%, based on the oil slicks' areas and perimeters.

We report on analyses to refine the ability to discriminate the petrogenic oil-slick category (seeps versus spills) proposed in our previous investigations [19–22]. Exploiting the same dataset, but with expanded Data Processing Segments, we extend our earlier studies onto a firmer basis. Based on our methodical data mining exercise, we seek to improve the seep-spill discrimination accuracy, as well as to answer three scientific questions:

1. Among the several Data Transformation Approaches we tested, which one provides the most accurate oil-slick category discrimination?
2. Is there a specific Attribute Selection Process that excels at choosing variables to discriminate seeps from spills?
3. Which combination of Oil-Slick Information Descriptors promotes the best discrimination between seeps and spills?

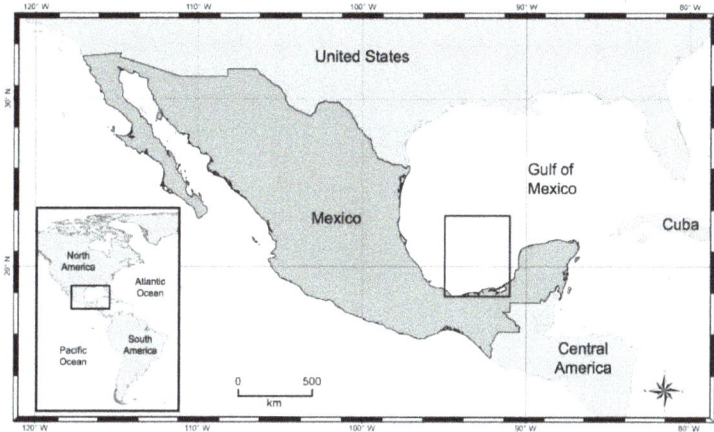

Figure 1. Campeche Bay located off the Mexican coast on the southernmost bight of the Gulf of Mexico. The highlighted region shows the location of the analyzed oil slicks. Courtesy of Adriano Vasconcelos (LabSAR/UFRJ).

2. Methods

We developed a comprehensive Exploratory Data Analysis (EDA) to reveal hidden information contained in the satellite-derived measurements and to refine the analysis to discriminate slicks by category, as proposed in our earlier studies [19–22]. The design of our EDA focuses on a data-driven scheme to investigate possible ways to improve the seep-spill discrimination with the simplest possible analysis and the lowest satellite-imaging cost. The research strategy employed herein is a development of our previous investigations [19–22], and consists of four distinct Data Processing Segments (i.e., A, B, C, and D in Figure 2)—devised in eight individual Phases—separately described in detail and introduced in a complete manner easily enabling replicability of our data mining exercise. A summary of our EDA design is depicted in Figure 2. While in-house Python codes are used to run the oil slick RADARSAT-2 related analyses (i.e., Phases 1–4), PAST (PAleontological STatistics: version 3.20, Oslo, Norway [26]) is used in the implementation of Phases 5–8.

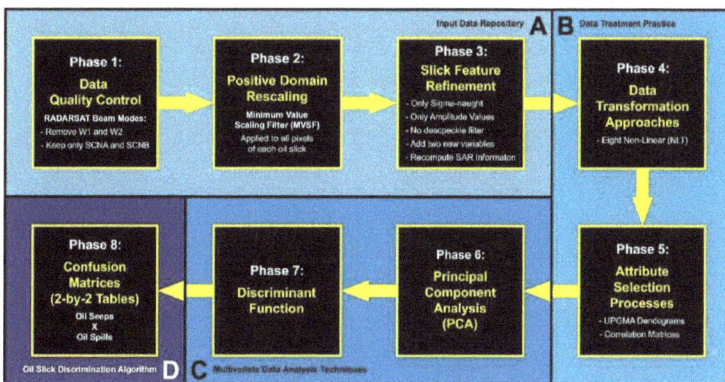

Figure 2. Research strategy developed to refine the ability to discriminate between two petrogenic oil-slick categories (i.e., seeps versus spills), as proposed in our previous studies [19–22]. The proposed Exploratory Data Analysis (EDA) has four distinct Data Processing Segments defined as: (**A**) Input Data Repository (Phases 1–3); (**B**) Data Treatment Practice (Phases 4–5); (**C**) Multivariate Data Analysis Techniques (Phases 6–7); and (**D**) Oil-Slick Discrimination Algorithm (Phase 8).

A multi-year dataset of RADARSAT-2 scenes imaged between 2008 and 2012 gave rise to the oil slick data archive analyzed in our earlier investigations [19–22]. This data archive consists of polygons representative of oil slicks that had been identified and field validated as seeps and spills by domain experts. For more information about this dataset, see [19–22]. The workable dataset explored herein is defined after fine-tuning this data archive along the 1st Data Processing Segment (Figure 2A: Input Data Repository—Phases 1–3).

2.1. Phase 1: Data Quality Control

The initial oil slick data archive from our previous studies [19–22] is sorted by the satellite scene-imaging configuration (i.e., beam modes determining the acquisition swath width and ground resolution), thus establishing the amount of RADARSAT-2 imagery and the seeps and spills of our workable dataset.

2.2. Phase 2: Positive Domain Rescaling

The initially available oil slick data archive analyzed in our earlier investigations [19–22] had undergone a linear scaling action (Negative Values Scaling Filter: NVSF) that is comprised of a two-fold procedure applied to individual oil slicks: the subtraction of the minimum negative pixel value within each oil slick from every single pixel of such oil slick, followed by the addition of 1 to every single pixel—the minimum pixel value becomes 1. This brings all pixel values to the positive domain, which is a requirement of data normalization procedures that cannot be applied to negative values, e.g., $\log_{10}$. The NSVF is applied at the pixel level, i.e., taking into account all pixels of each oil slick to provide a single measure representative of all pixels of such oil slick (see below: Section 2.3.2). Nevertheless, previously, the NVSF was only applied to certain oil slicks: those having at least one negative pixel value—for instance, oil slicks that had spurious negative SAR backscatter signature caused by intrinsic multiplicative random granular speckle noise destructive imprecision in the range-dependent gain calculation [27,28].

Although we also conduct this filtering strategy, we apply it in the present research to all oil slicks. In essence, hereafter, for our purpose, the NVSF is referred to as Minimum Values Scaling Filter (MVSF), such that: PIXpos = (PIX-PIXmin) + 1, in which PIXpos corresponds to the new positive pixel value, PIX is the original pixel value, PIXmin is the minimum pixel value of all pixels of each oil slick. Therefore, this is a dissimilarity between our previous investigations and the current EDA: NVSF versus MVSF. The reason for applying the MVSF to all oil slicks is three-fold: (1) To avoid possible biases caused by gradient differences among oil slicks with and without NVSF; (2) To circumvent the application of despeckle filtering (e.g., Frost Filter: FFrost [29]; see also Phase 3) that eventually would eliminate negative values, but would alter (e.g., smoothing) the SAR backscatter signature values—the lack of such filter is justifiable to preserve the data-driven design of our EDA; and (3) To exploit data transformations that do not accept negative values (see below: Phase 4).

2.3. Phase 3: Slick Feature Refinement

2.3.1. SAR Backscatter Signature

Previously, we explored twelve SAR backscatter signatures: SAR backscatter coefficients corresponding to the radar cross-section (RCS: σ) normalized by the unit area calculated in three different surface planes (i.e., σ^0, β^0, and γ^0 [30–34]) computed in four radiometric-calibrated image products—i.e., the amplitude (1st) of the received radar beam and its dimensionless physical quantity form that represents power expressed in dB (2nd), both with (3rd) and without (4th) despeckle filtering (FFrost: 3-by-3 window). However, herein we perform a simplification for a more controlled EDA solely using σ^0 given in amplitude without despeckle filtering. As such, from this point onwards, unless otherwise stated, any reference to SAR backscatter signature synonymously refers to this simplification.

2.3.2. Oil-Slick Information Descriptors

As before [19–22], we start our research analyzing the same ten attributes describing the oil slicks' geometry, shape, and dimension (these are collectively referred to as Size Information Descriptors) derived from two basic morphological features characterizing the oil slicks—i.e., area (Area) and perimeter (Per):

- **AtoP**: Area to Per ratio;
- **PtoA**: Per to Area ratio [35];
- **PtoAnor**: Normalized Per to Area ratio = $Per/[(2.(Pi.Area))^{1/2}]$ [36];
- **Complex Index** = $[Per^2]/Area$ [37];
- **Compact Index** = $[4.Pi.Area]/[Per^2]$ [18];
- **Shape Index** = $[Per/4]/[Area^{1/2}]$ [38];
- **Fractal Index** = $[2.Ln(Per/4)]/[Ln(Area)]$ [39];
- **LEN**: Number of pixels of each oil slick polygon.

Analogously, we also exploit the same 36 basic descriptive statistics metrics experimentally explored to characterize the oil slicks' SAR backscatter signature as in our previous investigations [19–22]. These metrics are calculated based on all pixels inside individual oil slick polygons:

- Four central tendency measures: Average (AVG), Median (MED), Mode (MOD), and Mid-mean (MDM: mean of the values between the 2nd and 3rd interquartiles, i.e., it trims off 25% of both ends);
- Six measures of dispersion: Range (RNG), Coefficient of Dispersion (COD: the subtraction of the 1st interquartile from the 3rd interquartile and the division by their sum), Standard Deviation (STD), Variance (VAR), Average Absolute Deviation (AAD: mean of the absolute difference of each value to the mean), and Median Absolute Deviation (MAD: median of the absolute difference of each value minus the median);
- 24 pair-values of Coefficients of Variation (COV: ratio between STD and AVG [18], such that each of the six dispersion measures are individually divided by the four central tendencies);
- The Minimum (MIN) and Maximum (MAX) pixel values of each oil slick.

Herein we introduce two new variables that describe the distribution patterns of the pixels within each oil slick: Skewness (SKW) and Kurtosis (KUR). As such, this collection of 38 basic descriptive statistics metrics characterizing the oil slick's SAR backscatter signature is henceforth referred to as SAR Information Descriptors. Together, these two types of Oil-Slick Information Descriptors (i.e., Size and SAR) determine the initial number of variables (48) accounted in our workable dataset.

2.4. Phase 4: Data Transformation Approaches

In contrast with our previous investigations [19–22], which implemented only a single non-linear normalization (log_{10}) and one linear standardization (Ranging [40]), we exploit several Non-Linear Transformations (NLTs [41–44]):

- **NLT.0**: No Transformation (x);
- **NLT.1**: Reciprocal (1/x);
- **NLT.2**: Logarithm Base 10 ($log_{10}(x)$);
- **NLT.3**: Napierian Logarithm (Ln(x));
- **NLT.4**: Square Root ($x^{1/2}$);
- **NLT.5**: Square Power (x^2);
- **NLT.6**: Cube Root ($x^{1/3}$);
- **NLT.7**: Third Power (x^3).

In which x corresponds to the actual value of each oil slick variable (i.e., Oil-Slick Information Descriptors—see Phase 3). Half of these (i.e., NLT.1, NLT.2, NLT.3, and NLT.4) do not accept negative values (x). To simplify our analyses, we do not perform linear standardizations.

2.5. Phase 5: Attribute Selection Processes

The processes of selecting relevant attributes deals with the complex matter of reducing dimensionality in the variable-hyperspace domain (see also Phase 6); this generally helps to elucidate the problem solution of numerical ecology assessments and to improve the performance of classification algorithms [42,45]. As such, another difference from our earlier studies is the number of explored attributes: before, we investigated 44 data sub-divisions with 502, 433, 423, 151, 141, 35, 10, and 2 variables [19–22]. Indeed, we considerably reduce these numbers with the SAR backscatter signature simplification (see Phase 3: Section 2.3.1). Additionally, we start with 48 Oil-Slick Information Descriptors (see Phase 3: Section 2.3.2) but use even fewer variables upon the completion of the Attribute Selection Processes (see below: Section 2.5.1).

2.5.1. Unweighted Pair Group Method with Arithmetic Mean (UPGMA)

Two attribute selection strategies (i.e., R-mode) have been performed in our previous investigations [19–22]: UPGMA [42,43,46] and CFS (Correlation-Based Feature Selection [47,48]). Based on our earlier results, we only implement the former as it allows a user-defined strategy to select relevant variables: the choice of the similarity index (Pearson's r correlation coefficient) used in the UPGMA dendrogram as cut-off to form groups of similar variables, i.e., phenon line [49,50]. See also [19–22] for further information about analyses and interpretations of rooted tree UPGMA dendrograms.

Moreover, an imperative distinction from our earlier investigations is that herein we are experimenting the use of a strict cut-off level, i.e., a fixed similarity value of 0.3, in relation to the previous fixed value of 0.5 and varying one ranging around 0.9 [19–22]. The selection of the 0.3 similarity cut-off is enlightened by the Bonferroni Adjustment as the level of minimum significance (p value) for large datasets ($n > 100$); below this there is no statistically significant correlation and variables are considered different from one another [51].

2.5.2. Histograms and Correlation Matrices

Histograms and correlation matrices assist in the verification of residual inter-variable correlation and to help with the decision of which variables to select on the groups formed on the UPGMA analyses.

2.6. Phase 6: Principal Component Analysis (PCA)

PCAs reduce the large correlated variables set into a smaller set of uncorrelated hypothetical variables—Principal Components (PCs)—containing most of the relevant information of the initial larger set [42,43]. The rotation of the original axes to the new orthogonal coordinate system is implemented in the same manner as our earlier work: square symmetric correlation matrix and 1000 bootstraps [52]. However, the approach to select relevant axes (i.e., PCs) is a departure from our earlier investigations. While, herein we use only the Kaiser Cut, i.e., Kaiser-Guttman criterion (eigenvalues > 1 [53]), previously we explored several PC-selection practices, e.g., Jolliffe, Scree Plot (Knee/Elbow), and a combined strategy using the Scree Plot (broken stick) with Kaiser [54–57].

2.7. Phase 7: Discriminant Function

Discriminant Analysis differs from Clustering Analysis as it is not meant to determine to which group each object belongs [43]. Instead, Discriminant Functions use a priori measured information (Oil-Slick Information Descriptors) and knowledge of the object's (oil slick) group membership (seep or spill), to obtain the maximum discriminating power that minimizes the probability of erroneous discrimination: $[DF(X) = (W_1X_1 + W_2X_2 + \ldots + W_nX_n) - C_{off}]$; in which DF(X) corresponds to the dependent variable (i.e., Discriminant Function); X_n to the independent variables (i.e., Oil-Slick Information Descriptor value); W_n to the independent variables' weight; and C_{off} to the constant offset [58–61].

The use of uncorrelated attributes (selected PCs from Phase 6), or at least with the lowest possible degree of dependence (UPGMA selected variables from Phase 5), is a pressing need for Discriminant Functions [62], and as such, this concerns a crucial development of the current EDA from our previous investigations [19–22]: herein, we are not only using the PCA scores (PCs) as input to the Discriminant Functions, we are also testing the use of UPGMA dendrogram selected variables (see Phase 5: Section 2.5.1) without passing through the PCA.

2.8. Phase 8: Confusion Matrices (2-by-2 Tables)

The Oil-Slick Discrimination Algorithm accuracy is reported based on the Discriminant Function results by means of the complete understanding of adapted 2-by-2 Tables (Confusion Matrices: CMs). See also [19–22,63–65] for information on how to analyze and to better interpret 2-by-2 Tables. The conjunct interpretation of five metrics [66] is essential to fully evaluate the algorithm's effectiveness. Table 1 gives a picture of these metrics that are color-coded for clarity:

- **CM.1**: Overall Accuracy (shown in Green);
- **CM.2**: Producer's Accuracy (i.e., Sensitivity and Specificity—shown in Yellow);
- **CM.2**: Commission Error (i.e., False Negative and False Positive);
- **CM.3**: User's Accuracy (i.e., Positive and Negative Predictive Values—shown in Purple);
- **CM.3**: Omission Error (i.e., Inverse of the Positive and Negative Predictive Values).

Table 1. Adapted 2-by-2 Tables (Confusion Matrix: CM [19–22,63–65]) illustrating the various metrics explored to evaluate the Oil-Slick Discrimination Algorithm accuracy, i.e., Discriminant Function (DF) results.

CM.1	DF Oil Seep	DF Oil Spill	Total Real
Oil Seep Real	A	B	A+B
Oil Spill Real	C	D	C+D
Total DF	A+C	B+D	A+B+C+D
			$\frac{(A+D)}{(A+B+C+D)}$ *

CM.2**	DF Oil Seep	DF Oil Spill	Total Real
Oil Seep Real	A/(A+B)	B/(A+B)	100%
Oil Spill Real	C/(C+D)	C/(C+D)	100%

CM.3**	DF Oil Seep	DF Oil Spill
Oil Seep Real	A/(A+C)	B/(B+D)
Oil Spill Real	C/(A+C)	D/(B+D)
Total DF	100%	100%

	Number of Samples (i.e. oil slicks)	
CM.1	A = Oil Seeps: Correct Identification	
	B = Oil Seeps: Miss-Identification	
	C = Oil Spills: Miss-Identification	
	D = Oil Spills: Correct Identification	
	A+B+C+D = Oil Slicks (i.e. 4,562)	
	* Overal Accuracy	
CM.2	Producer's Acuraccy	A/(A+B) = Sensitivity
		C/(C+D) = Specificity
	Comission Error	B/(A+B) = False Negative
		C/(C+D) = False Positive
	A+B = Oil Seeps (i.e. 1,994)	
	C+D = Oil Spills (i.e. 2,568)	
CM.3	User's Accuracy	A/(A+C) = Positive Predictive Value
		D/(B+D) = Negative Predictive Value
	Omission Error	B/(B+D) = Inverse of the Pos. Predict. Val.
		C/(A+C) = Inverse of the Neg. Predict. Val.
	A+C = DF Oil Seeps	
	B+D = DF Oil Spills	

** Values given in pecentage.

3. Results and Discussion

3.1. Phase 1: Data Quality Control

The initially available oil slick data archive is composed of 4,916 oil slick polygons—2021 oil seeps (41%) and 2895 oil spills (59%)—imaged with 277 RADARSAT-2 scenes (Table 2 I), all of which are 16-bit and VV polarized [19–22]. These include two different RADARSAT beam modes—Wide [W1 and W2: 354 oil slicks (7%)] and ScanSAR Narrow [SCNA and SCNB: 4562 oil slicks (93%)]—that own two fundamental imaging differences: (1) W1 and W2 are Single Beam Modes (i.e., a strip-map SAR mode with certain imaging aspects constant along the entire scene), whereas SCNA and SCNB are ScanSAR Modes (i.e., combine two or more of the Single Beam Modes) [67]; the latter provides larger area coverage: swath width of 300 km—almost twice that of W1 and W2: 170 km and 150 km, respectively; and (2) Wide has a finer ground resolution of 25 m, which is $\frac{1}{4}$ of the ScanSAR Narrow one: 50 m.

Table 2. Number (and percentage) of explored oil slicks (seeps and spills) and satellite images.

Oil Category	I: Previous Work	II: Our Reseach	III: Difference
Oil Seeps	2,021 (41%)	1,994 (44%)	-27 (-1%)
Oil Spills	2,895 (59%)	2,568 (56%)	-327 (-11%)
Oil Slicks	4,916 (18%)*	4,562 (13%)*	-354 (-7%)
RADARSAT-2 scenes	277	244	-33 (-12%)

I: Initially available oil slick data archive [19-20].

II: Our workable dataset, i.e. large collection of oil slick polygons imaged with RADARSAT-2 scenes of the two ScanSAR Narrow beam modes: SCNA and SCNB.

III: Removal of RADARSAT-2 beam mode (Wide: W1 and W2) dunig Phase 1 (Data Quality Control).

* Difference between oil spills and oil seeps.

Regarding their specification differences, eventual inaccuracies may be introduced to beam mode cross-comparisons. Notwithstanding that W1 and W2 provide better delineation of smaller oil slicks with their finer ground resolution, only SCNA and SCNB are kept in our analysis as these represent more than 90% of the available scenes. Furthermore, the ScanSAR Narrow swath width is more appropriate for monitoring applications requiring large-scale coverage such as the one that gave rise to the initially available oil slick data archive [19–22]. In fact, the lower scene cost of using ScanSAR Narrow to monitor larger ocean regions is rather preferable than the smaller area coverage of the Wide images.

Consequently, our workable dataset is composed of the collection of oil slick polygons imaged with the two ScanSAR Narrow beam modes: 4562 oil slicks—1994 oil seeps (44%) and 2568 oil spills (56%)—Table 2 II. Despite the fact that our EDA has 7% (354) fewer oil slicks than our previous study [19–22], representing about 1% (27) fewer seeps and approximately 11% (327) fewer spills (Table 2 III), such data reduction results in a more balanced dataset as compared to the one explored in our previous investigations, i.e., a smaller difference between the number of analyzed spills and seeps: 13% instead of 18% (Table 2: I–II). Indeed, this provides a firmer basis in the oil-slick category discrimination. Moreover, the oil slick polygons imaged with SCNA and SCNB come from 244 RADARSAT-2 scenes imaged between 2008 and 2012—12% (33) fewer images than our earlier investigations (Table 2).

3.2. Phase 2: Positive Domain Rescaling

As the MVSF is applied at the pixel level to all oil slicks in our workable dataset (Table 2 II: 4562), it affects the values of the 38 SAR Information Descriptors but not of the 10 Size Information Descriptors (see Phase 3: Section 2.3.2). The latter is independent of the MVSF application as they are derived from and include the two basic morphological oil slick features: Area and Perimeter.

3.3. Phase 3: Slick Feature Refinement

The consequence of MVSF (see Phase 2) is two-fold: (1) the SAR Information Descriptors are not the same as in our previous investigations and need to be recomputed for all analyzed oil slicks; (2) MIN loses its meaning as its value for all oil slicks becomes 1; accordingly, it is not pursued in our analysis.

3.4. Phase 4: Data Transformation Approaches

Although the NLTs can be independently applied to each attribute, for consistency, during our EDA, all-numeric variables uniformly undergo the same column-wise transformation. Because three Oil-Slick

Information Descriptors—i.e., Fractal, SKW, and KUR—have values that range from negative to positive, they are not used on half of the NLTs that require only positive values: NLT.1; NLT.2; NLT.3; and NLT.4.

3.5. Phase 5: Attribute Selection Processes

Histograms show that the distribution of some Size Information Descriptors is the same as others, sometimes being inverted independent of NLT, meaning that there is no new information revealed. As a result, only one of these variables is selected, for instance: (1) AtoP and PtoA have equal but inverted distributions; (2) PtoAnor, Complex, Compact, and Shape, also have equal distribution but Compact is inverted from the three other. Of these variables, we only keep PtoA and Compact, as Area and Perimeter appear in opposition in their formula: PtoA has area in the denominator, as opposed to Compact, which has area in the numerator; the contrary holds true for the perimeter (see Phase 3: Section 2.3.2).

3.5.1. Unweighted Pair Group Method with Arithmetic Mean (UPGMA)

The combined analysis of dendrograms and correlation matrices show that the 24 COV pair-values have a strong intra-correlation, as well as that they are highly correlated with most of the other variables; hence, they are not further explored. Therefore, out of the 48 initial Oil-Slick Information Descriptors (see Phase 3: Section 2.3.2), only 19 remain for further analyses—Size (6): Area, Per, PtoA, Compact, Fractal, and LEN; and SAR (13): AVG, MED, MOD, MDM, RNG, COD, STD, VAR, AAD, MAD, MAX, SKW, and KUR. However, only half of the NLTs (NLT.0, NLT.5, NLT.6, and NLT.7) utilize these 19 variables; the other half (NLT.1, NLT.2, NLT.3, and NLT.4) explores three fewer Oil-Slick Information Descriptors, i.e., only 16 variables (see Phase 4: Section 3.4).

Figure 3 depicts eight UPGMA dendrograms (one for each of the analyzed NLT), in which it is possible to observe a number of differences, as well as resemblances, between them; mostly regarding inter-variable correlations. An evident characteristic of the two Logarithm functions (NLT.2: Log_{10}; and NLT.3: Ln) is that their dendrograms are equal; the same holds true for their correlation matrices that are also identical.

Prior to the uncorrelated variables selection, we have to identify the groups of correlated variables. The process of defining and/or interpreting groups in UPGMA dendrograms is quite subjective [43], but, at first glance, the global picture of Figure 3 clearly reveals how equivalent are the groups between the several NLTs; these are color-coded for clarity. In the visual analysis of Figure 3, one can note that variables tend to group based on their main characteristics, following the Oil-Slick Information Descriptor features, such that:

- **Green**: Measures of central tendency (AVG, MED, MOD, and MDM);
- **Blue**: Dispersion measures (RNG, COD, STD, VAR, AAD, and MAD);
- **Grey**: Metrics of pixel distribution (SKW and KUR);
- **Yellow**: Basic morphological features (Area and Per) and LEN;
- **Red**: Ratios derived from the morphological features (PtoA, Compact, and Fractal).

An advanced analysis of the UPGMA dendrograms shown in Figure 3 discloses that:

- The three morphological ratios (Red group) are not correlated with any other variable (similarity close to or equal to zero)—PtoA and Compact form an uncorrelated group, and Fractal usually stands alone; the exception is in NLT.0 where Compact is the one by itself;
- The two groups of SAR Information Descriptor, i.e., Green (central tendency) and Blue (dispersion), generally form a larger group—Geen + Blue—the exception is in NLT.7;
- The Grey group (pixel distribution metrics) is usually correlated with the Yellow group (basic morphological features)—Grey + Yellow group—the exception is in NLT.7 where it groups with the Green group (measures of central tendency);
- RNG is an exception in three NLTs (NLT.0, NLT.6, and NLT.7) as it correlates with the central tendency variables (Green group);
- MAX groups among the central tendency variables (Green group) except in NLT.4.

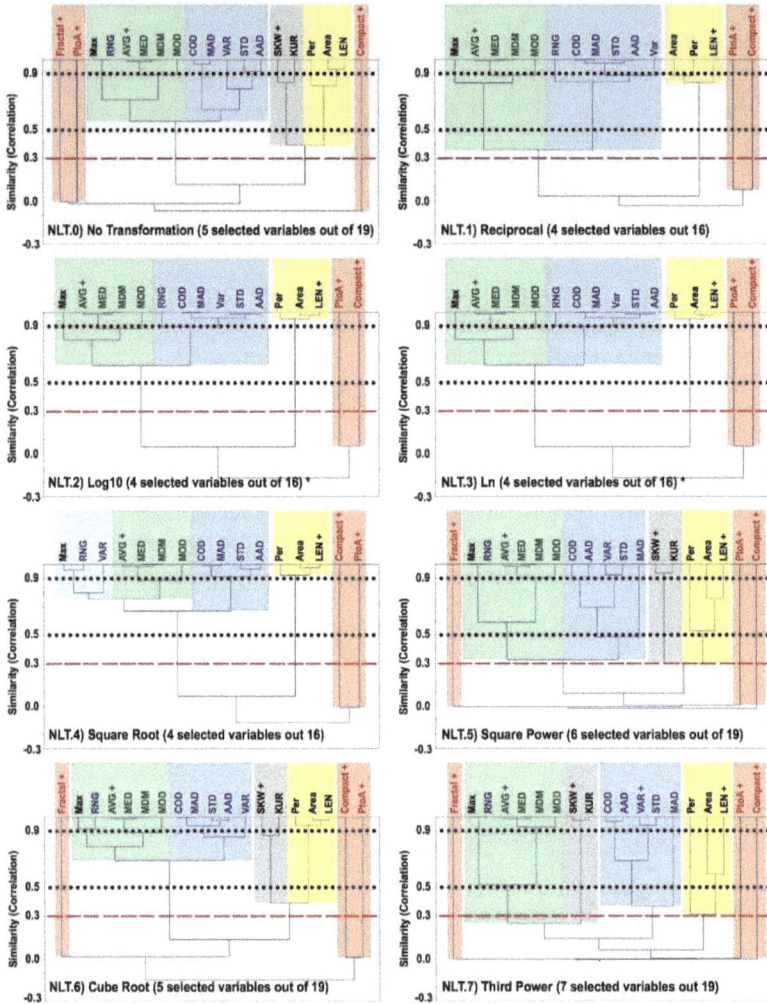

Figure 3. Rooted tree dendrograms (Unweighted Pair Group Method with Arithmetic Mean: UPGMA—see Phase 5: Sections 2.5.1 and 3.5.1) of the several Non-Linear Transformations (NLTs—see Phase 4: Section 2.4). While the horizontal red dashed line represents the phenon line exploited herein to form groups of variables, i.e., similarity value of 0.3 (i.e., Pearson's r correlation coefficient), the two horizontal black dotted lines correspond to the more relaxed thresholds reported in our previous investigations [19–22]. The various color-colored boxes indicate the main groups of variable (see Phase 3: Section 2.3.2). Size Information Descriptors: Yellow [basic morphological oil slick features, i.e., area (Area) and perimeter (Per), and the number of pixels (LEN)] and Red [three ratios derived from the morphological features]. SAR Information Descriptors: Green [measures of central tendency, i.e., average (AVG), median (MED), mode (MOD), and mid-mean (MDM); an exception is the maximum pixel value (MAX)], Blue [dispersion measures, i.e., range (RNG), coefficient of dispersion (COD), standard deviation (STD), variance (VAR); average absolute deviation (AAD), and median absolute deviation (MAD)], and Grey [metrics of the pixel distribution: skewness (SKW) and kurtosis (KUR)]. Selected variables are indicated (+); see also Table 3. * Same outcome: NLT.2 = NLT.3.

The phenon line, represented by the horizontal red dashed line in Figure 3 (i.e., 0.3 Pearson's r correlation coefficient) defines the actual groups from which we select one variable of each—groups are formed when this cut-off line crosses a vertical line (i.e., branch or edge) [49,50]. In fact, the groups formed in this manner match the preliminary visual analysis of the dendrograms:

- Three groups are observed (Green + Blue, Yellow, and one Red) when 16 variables are analyzed, i.e., NLT.1, NLT.2, and NLT.3—NLT.4 is an exception;
- Four other groups are also formed (Green + Blue, Grey + Yellow, and two Red ones) when 19 variables are accounted for, i.e., NLT.0, NLT.5, NLT.6—NLT.7 is an exception;
- Three other groups are formed in NLT.4 (16 variables) in which VAR, RNG, and MAX cluster together forming an extra assemblage (Light Blue)—Light Blue + Green + Blue, Yellow, and one Red;
- Six groups are formed in NLT.7 (19 variables): Green, Grey, Blue, Yellow, and two Red ones.

One should pay close attention to the two Red groups, as from them, three variables are selected—e.g., NLT.0 (Fractal, PtoA, and Compact)—because such variables have no correlation.

The number of selected variables ranges between 4 and 7 variables, depending on the NLT (Table 3), such that:

- **AVG** is selected from the Green + Blue group to maintain the simplest possible analysis;
- **VAR** is selected when the Blue group is alone (only in NLT.7) to keep it simple as possible;
- **SKW** is preferable from the Grey group as it measures asymmetry;
- **LEN** is selected from the Yellow group as Area and Perimeter are both present in the ratios;
- The three morphological ratios (Red group: **PtoA**, **Compact**, and **Fractal**) are always selected when present.

Table 3. Summary of the Attribute Selection Processes (Phase 5).

Attribute Selection Processes (Phase 5) based on the analyses of dendrograms (UPGMA: Unweighted Pair Group Method with Arithmetic Mean)			No Transformation: (X)	Reciprocal: $(1/X)$	Logarithm Base 10: $Log_{10}(x)$	Napierian Logarithm: $Ln(x)$	Square Root: $(x^{1/2})$	Square Power: (x^2)	Cube Root: $(x^{1/3})$	Third Power: (x^3)
Non-Linear Transformations (NLT)			NLT.0	NLT.1	NLT.2*	NLT.3†	NLT.4	NLT.5	NLT.6	NLT.7
Acronym	Variables	n =	19	16	16	16	16	19	19	19
1) **LEN**	Number of Pixels	(n=3) Basic Morphological Oil Slick Features	@	@	@	@	@			@
2) **Area¹**	Area									
3) **Per¹**	Perimeter									
4) **PtoA**	Perimeter to Area Ratio		@	@	@	@	@	@	@	@
5) **Compact**	Compact Index = [4.Pi.Area/(Per²)]		@	@	@	@	@	@	@	@
6) **Fractal**	Fractal Index = [2.ln(Per/4)]/[ln(Area)]		@	~	~	~	~	@	@	@
7) **MAX**	Maximum									
8) **AVG**	Average	(n=4) Central Tendency	@	@	@	@	@	@	@	@
9) **MED**	Median									
10) **MOD**	Mode									
11) **MDM²**	Mid-Mean									
12) **RNG**	Range	(n=6) Dispersion Measures								
13) **COD³**	Coefficient of Dispersion									
14) **VAR**	Variance									@
15) **STD**	Standard Deviation									
16) **AAD⁴**	Average Absolute Deviation									
17) **MAD⁵**	Median Absolute Deviation									
18) **SKW⁶**	Skewness	(n=2) Pixel Distribution	@	~	~	~	~	@	@	@
19) **KUR⁶**	Kurtosis			~	~	~	~			
UPGMA Selected Variables**		n =	5	4	4	4	4	6	5	7

The color-coding relates to the outcomes of the rooted tree UPGMA dendrograms (see Figure 3)

The @ denoted the selected variables.
The ~ denotes the variables that are not accounted for (see Phase 4).

Size Information (n=6): rows 1–6
SAR Information (n=13): rows 7–19

* Same outcome: NLT.2=NLT.3.
** See Figure 3.
¹ Best oil slick discrimination variables (i.e. Area and Perimeter) reported in our previous studies [19-22].
² MDM = Mean of the values between the 2nd and 3rd interquartiles (i.e. trims off 25% of both ends).
³ COD = Subtract the 1st interquartile from the 3rd interquartile and divide by their sum.
⁴ AAD = Mean of the absolute difference of each value minus the mean.
⁵ MAD = Median of the absolute difference of each value minus the median.
⁶ The two new introduced variables (i.e. SKW and KUR) in relation to [19-22].

3.6. Phase 6: Principal Component Analysis (PCA)

The scatterplots show a large overlap between seeps and spills, but their centroids are somehow distinctively independent of NLT. When all variables (16 or 19) are directly input to the PCA, the cumulative variance of the selected PCs (3 to 7) ranges between 80 to 90% for all NLTs. However, when the input is the UPGMA selected variables (4 to 7), the PC-selection (2 to 4 PCs) shows a much lower cumulative variance: from 52% to 70%; the exceptions are the Logarithm functions (NLT.2: Log_{10}; and NLT.3: Ln) with 99.5% (2 PCs). Table 4 reports the number of selected PCs and their cumulative variance per NLT.

Table 4. Outcome of the Principal Component Analysis (PCA: Phase 6) showing the number of selected Principal Components (PCs) and cumulative variance.

Principal Components Analysis (PCA: Phase 6) showing the selected Principal Components (PCs) based on the Kaiser Cut (i.e. Eigenvalue > 1.0)	All Variables[4,5] UPGMA Variable Selection: No[7]			All Variables[4,6] UPGMA Variable Selection: Yes[3]			Separate Analysis[4,7] Area and Perimeter[+]			Separate Analysis[4,7] PtoA and Compact			Separate Analysis[4,7] Average and Skewness		
	[2](n)	PCs	%%	[2](n)	PCs	%%	[2](n)	PCs	%%	[2](n)	PCs	%%	[2](n)	PCs	%%
NLT.0) No Transformation [x]	19	5	80.49%	5	3	69.13%*	2	2	100% (92.21%")	2	2	100% (52.33%")	2	2	100% (67.98%")
NLT.1) Reciprocal [1/x]	16	3	82.42%	4	2	59.95%*									
NLT.2)[1] Logarithm Base 10: [$Log_{10}(x)$]	16	3	90.08%	4	2	99.50%	2	2	100% (97.87%")	2	2	100% (52.76%")			
NLT.3)[1] Logarithm Napierian: [Ln(x)]	16	3	90.08%	4	2	99.50%									
NLT.4) Square Root [$x^{1/2}$]	16	4	90.07%	4	2	68.35%*									
NLT.5) Square Power [x^2]	19	6	80.00%	6	4	70.61%*									
NLT.6) Cube Root [$x^{1/3}$]	19	4	85.05%	5	2	52.32%*	2	2	100% (97.44%")	2	2	100% (50.47%")	2	2	100% (66.27%")
NLT.7) Third Power [x^3]	19	7	81.46%	7	4	59.93%*									

[1] Same outcome: NLT.3=NLT.4
[2] Number of input variables.
[3] See Figure 3 (UPGMA Dendograms).
[4] See Figure 4.
[5] See Table 5.
[6] See Table 6.
[7] See Table 7.
%% Cumulative variance of the selected PCs.
* Low cumulative variance in the selected PCs.
" Cumulative variance of the 1st PC.
+ Best oil slick discrimination variables reported in our earlier investigations [19-22].

3.7. Phase 7: Discriminant Function

As we are comparing the results of using the score values of the selected PCs versus the use of actual values of the Oil-Slick Information Descriptors, both directly input to the Discriminant Analysis, four different Discriminant Function sets are analyzed per NLT:

- **Set.1:** No UPGMA variable selection, i.e., all variables (16 or 19), without PCA;
- **Set.2:** No UPGMA variable selection, i.e., all variables (16 or 19), with PCA (3 to 7 PCs);
- **Set.3:** UPGMA selected variables (4 to 7) without PCA;
- **Set.4:** UPGMA selected variables (4 to 7) with PCA (2 to 4 PCs).

Figure 4 portrays the scheme defining these four input dataset versions for each NLT (8x). Another improvement from our earlier studies is that besides exploring the seep-spill discrimination capabilities of using the PC-scores and values of the variables, as well as the sole use of Area with Perimeter as before [19–22], we also test a separate analysis with a pair of Size Information Descriptors (PtoA with Compact) and with a pair of SAR Information Descriptors (AVG with SKW)—see Figure 4. These are chosen based on the interpretation of the UPGMA dendrograms (Phase 5: Section 3.5.1—see also Figure 3). Although the histograms of the Discriminant Functions' axes show that seep and spill properties overlap, independent of NLT, their centroids are separate.

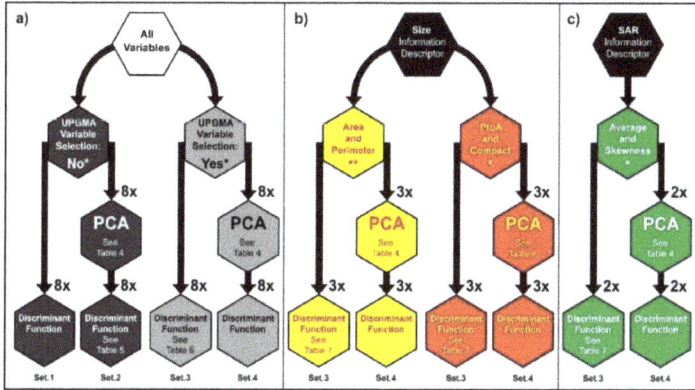

Figure 4. Discriminant Functions explored to discriminate oil seeps from oil spills: (**a**) All variables; (**b**) Separate analysis of Size Information Descriptors (Area with Perimeter and PtoA with Compact); and c) Separate analysis of SAR Information Descriptors (AVG with SKW). The four input dataset versions are shown: all variables (16 or 19—see Phase 5: Section 3.5.1; Figure 3 and Table 3) without (Set.1) and with (Set.2) PCA (Principal Component Analysis—see Phase 6), UPGMA (Unweighted Pair Group Method with Arithmetic Mean—see Phase 5: Section 2.5.1) attribute selection (i.e., 4 to 7 variables—see Phase 5: Section 3.5.1; Figure 3 and Table 3) without (Set.3) and with (Set.4) PCA. 8× refers to the several Non-Linear Transformations (NLT—see Phase 4: Section 2.4); 3x to the best NLT: NLT.0, NLT.2, and NLT.6; and 2× to NLT.0 and NLT.6. * See Figure 3. ** See [19–22].

3.8. Phase 8: Confusion Matrices (2-by-2 Tables)

Each NLT is evaluated with the four input dataset versions (Figure 4), and usually, Set.1 presents the highest discrimination power. However, these variables (16 or 19) are strongly correlated (Figure 3) and do not fulfill a Discriminant Functions requirement to use independent, or the least as correlated as possible, attributes [62]. The second best discrimination accuracy occurs with Set.2, which is closely followed by Set.3. The lowest observed accuracies are from Set.4, as the selected PCs have a very low cumulative variance in the selected PCs; the exceptions are the Logarithm functions (NLT.2: Log_{10}; and NLT.3: Ln—see Table 4).

The global analysis of all 32 Data Transformation Approaches combinations (i.e., eight NLTs versus four input dataset versions) demonstrates the Logarithm functions (NLT.2: Log_{10}; and NLT.3: Ln) and Cube Root (NLT.6) as the most effective NLTs in supporting an accurate Oil-Slick Discrimination Algorithm. The Confusion Matrices evaluating the results of the Discriminant Functions for the several NLTs are shown on the color-coded Table 5 (Pink) and Table 6 (Red): Set.2 and Set.3, respectively. In the examination of these two tables that report the accuracy of the Oil-Slick Discrimination Algorithm, if taking the Log_{10} (NLT.2), for example, one can find that:

- **CM.1**: Overall Accuracies ranging about 69%;
- **CM.2**: Producer's Accuracy, i.e., Sensitivities (65%) or Specificities (71%);
- **CM.2**: Commission Error, i.e., False Negative (35%) and False Positive (29%);
- **CM.3**: User's Accuracy, i.e., Positive (64%) and Negative (73%) Predictive Values;
- **CM.3**: Omission Error, i.e., Inverse of the Positive (36%) and Negative (27%) Predictive Values.

From Tables 5 and 6, one verifies the successful, and similar, results of the Cube Root (NLT.6) in comparison to the Logarithm functions (NLT.2 and NLT.3). Additionally, the cross-comparison of the results from Set.2 (Table 5) and Set.3 (Table 6) indicates that these two attribute selection strategies—i.e., 1) no UPGMA variable selection with PCA; and 2) UPGMA selected variables without PCA—promote comparable seep-spill discrimination accuracies.

A careful analysis of Table 5 (Set.2) discloses that the preferred lowest rate of False Negatives (20.4%) and Inverse of the Positive Predictive Values (23.3%) are observed in NLT.1 (Reciprocal); however, their counterparts, i.e., False Positives (48.0%) and Inverse of the Negative Predictive Values (43.7%), have undesirable high values among all NLTs. As its Overall Accuracy is reasonable (64.07%), this is an example that one needs to look into the conjunct interpretation of the five main metrics shown in Table 1 [19–22,63–65]. Similarly, the cautious analysis of Table 6 (Set.3) reveals that an ideal low rate of False Negatives (11.7%) and Inverse of the Positive Predictive Values (21.8%) are observed in NLT.5 (Square Power), but on the other hand, their counterparts, i.e., False Positives (67.5%) and Inverse of the Negative Predictive Values (49.6%), have unwanted high values. In this case, its Overall Accuracy is quite low (56.90%) though.

When considering the separate analysis of the Oil-Slick Information Descriptors (i.e., Size and SAR), Set.3 and Set.4 (see Figure 4: without PCA and with PCA, respectively) present the same result—these are shown in Table 7. The foremost outcome revealed in Table 7 is that the sole use of SAR Information Descriptors (AVG with SKW) is not as effective as using only Size Information Descriptors (Area with Perimeter and PtoA with Compact). Table 7 also discloses that these two pairs of Size Information Descriptors have the same results in the Logarithm function (NLT.2), and in fact, these results present superior discrimination power than in the other two analyzed NLTs, i.e., NLT.0 (No Transformation) and NLT.6 (Cube Root). Slightly better Overall Accuracies are achieved when using Area with Perimeter than PtoA with Compact; however, one should note that the False Negatives of the former pair are much higher than those of using the second pair: 67.7% against 21.0% (NLT.0), and 43.4% against 28.9% (NLT.6).

We can also evaluate the results of using several variables (Tables 5 and 6) against the use of individual pairs of attributes, i.e. the separate analysis of Size and SAR Information Descriptors (Table 7). If one compares the outcomes of NLT.2 (Log_{10}) in Tables 5–7, it is possible to notice that the sole use of the two Size Information Descriptor pairs (Table 7) has equivalent results as the ones from the other two attribute selection strategies, i.e., no UPGMA variable selection with PCA (Set.2: Table 5) versus UPGMA selected variables without PCA (Set.3: Table 6).

4. Conclusions

Our research addresses a gap in our scientific knowledge regarding the discrimination of the oil-slick category, i.e., sea surface expression of oil seeps versus oil spills observed in Campeche Bay (Figure 1). We report on analyses to refine the ability of using SAR-derived measurements for this task, thus addressing expanded Data Processing Segments (A, B, C, and D in Figure 2) as compared to our previous investigations [19–22]. A firmer basis to discriminate slicks by category has been established with the specific data-driven design of our Exploratory Data Analysis (EDA). An innovative strategy to select uncorrelated attributes based on the Bonferroni Adjustment (i.e., Pearson's r correlation coefficient of 0.3 [51]) has been successfully implemented using rooted tree dendrograms (Unweighted Pair Group Method with Arithmetic Mean: UPGMA—see Figure 3). We investigate several Non-Linear Transformations (NLTs—see Phase 4: Data Transformation Approaches) and various strategies to select uncorrelated attributes: we tested more than 32 combinations of Data Transformation Approaches, i.e., eight NLTs versus four input dataset versions (see Set.1, Set.2, Set.3, and Set.4 in Phase 7: Discriminant Function—Figure 4).

Table 5. Confusion Matrices (CMs) expressing the results of the Discrimination Functions (DFs) of the Oil-Slick Discrimination Algorithm from Set.2—i.e., all variables (16 or 19—see Phase 5: Section 3.5.1—Figure 3 and Table 3) without the dendrogram selection (no UPGMA: Unweighted Pair Group Method with Arithmetic Mean) but with the application of the PCA (Principal Component Analysis—see Phase 6: Section 3.6—Table 4). Note that NLT.2 and NLT.3 have the same outcome.

Block NLT.0

CM.1	DF Oil Seep	DF Oil Spill	Total Real
Oil Seep Real	1,331	663	1,994
Oil Spill Real	980	1,588	2,568
Total DF	2,311	2,251	4,562
NLT.0: No Transformation (x)			2,919 / 63.99%
CM.2	DF Oil Seep	DF Oil Spill	Total Real
Oil Seep Real	66.8%	33.2%	100.0%
Oil Spill Real	38.2%	61.8%	100.0%
CM.3	DF Oil Seep	DF Oil Spill	UPGMA: No
Oil Seep Real	57.6%	29.5%	No
Oil Spill Real	42.4%	70.5%	PCA: Yes
Total DF	100.0%	100.0%	Yes

Block NLT.1

CM.1	DF Oil Seep	DF Oil Spill	Total Real
Oil Seep Real	1588	406	1994
Oil Spill Real	1233	1335	2568
Total DF	2821	1741	4562
NLT.1: Reciprocal (1/x)			2,923 / 64.07%
CM.2	DF Oil Seep	DF Oil Spill	Total Real
Oil Seep Real	79.6%	20.4%	100.0%
Oil Spill Real	48.0%	52.0%	100.0%
CM.3	DF Oil Seep	DF Oil Spill	UPGMA: No
Oil Seep Real	56.3%	23.3%	No
Oil Spill Real	43.7%	76.7%	PCA: Yes
Total DF	100.0%	100.0%	Yes

Block NLT.2

CM.1	DF Oil Seep	DF Oil Spill	Total Real
Oil Seep Real	1299	695	1994
Oil Spill Real	734	1834	2568
Total DF	2033	2529	4562
NLT.2: Logarithm Base 10 ($\log_{10}(x)$)			3,133 / 68.68%
CM.2	DF Oil Seep	DF Oil Spill	Total Real
Oil Seep Real	65.1%	34.9%	100.0%
Oil Spill Real	28.6%	71.4%	100.0%
CM.3	DF Oil Seep	DF Oil Spill	UPGMA: No
Oil Seep Real	63.9%	27.5%	No
Oil Spill Real	36.1%	72.5%	PCA: Yes
Total DF	100.0%	100.0%	Yes

Block NLT.3

CM.1	DF Oil Seep	DF Oil Spill	Total Real
Oil Seep Real	1299	695	1994
Oil Spill Real	734	1834	2568
Total DF	2033	2529	4562
NLT.3: Napierian Logarithm (Ln(x))			3,133 / 68.68%
CM.2	DF Oil Seep	DF Oil Spill	Total Real
Oil Seep Real	65.1%	34.9%	100.0%
Oil Spill Real	28.6%	71.4%	100.0%
CM.3	DF Oil Seep	DF Oil Spill	UPGMA: No
Oil Seep Real	63.9%	27.5%	No
Oil Spill Real	36.1%	72.5%	PCA: Yes
Total DF	100.0%	100.0%	Yes

Block NLT.4

CM.1	DF Oil Seep	DF Oil Spill	Total Real
Oil Seep Real	1269	725	1994
Oil Spill Real	723	1845	2568
Total DF	1992	2570	4562
NLT.4: Square Root ($x^{1/2}$)			3,114 / 68.26%
CM.2	DF Oil Seep	DF Oil Spill	Total Real
Oil Seep Real	63.6%	36.4%	100.0%
Oil Spill Real	28.2%	71.8%	100.0%
CM.3	DF Oil Seep	DF Oil Spill	UPGMA: No
Oil Seep Real	63.7%	28.2%	No
Oil Spill Real	36.3%	71.8%	PCA: Yes
Total DF	100.0%	100.0%	Yes

Block NLT.5

CM.1	DF Oil Seep	DF Oil Spill	Total Real
Oil Seep Real	1100	894	1994
Oil Spill Real	1100	1468	2568
Total DF	2200	2362	4562
NLT.5: Square Power (x^2)			2,568 / 56.29%
CM.2	DF Oil Seep	DF Oil Spill	Total Real
Oil Seep Real	55.2%	44.8%	100.0%
Oil Spill Real	42.8%	57.2%	100.0%
CM.3	DF Oil Seep	DF Oil Spill	UPGMA: No
Oil Seep Real	50.0%	37.8%	No
Oil Spill Real	50.0%	62.2%	PCA: Yes
Total DF	100.0%	100.0%	Yes

Block NLT.6

CM.1	DF Oil Seep	DF Oil Spill	Total Real
Oil Seep Real	1270	724	1994
Oil Spill Real	688	1880	2568
Total DF	1958	2604	4562
NLT.6: Cube Root ($x^{1/3}$)			3,150 / 69.05%
CM.2	DF Oil Seep	DF Oil Spill	Total Real
Oil Seep Real	63.7%	36.3%	100.0%
Oil Spill Real	26.8%	73.2%	100.0%
CM.3	DF Oil Seep	DF Oil Spill	UPGMA: No
Oil Seep Real	64.9%	27.8%	No
Oil Spill Real	35.1%	72.2%	PCA: Yes
Total DF	100.0%	100.0%	Yes

Block NLT.7

CM.1	DF Oil Seep	DF Oil Spill	Total Real
Oil Seep Real	560	1434	1994
Oil Spill Real	451	2117	2568
Total DF	1011	3551	4562
NLT.7: Third Power (x^3)			2,677 / 58.68%
CM.2	DF Oil Seep	DF Oil Spill	Total Real
Oil Seep Real	28.1%	71.9%	100.0%
Oil Spill Real	17.6%	82.4%	100.0%
CM.3	DF Oil Seep	DF Oil Spill	UPGMA: No
Oil Seep Real	55.4%	40.4%	No
Oil Spill Real	44.6%	59.6%	PCA: Yes
Total DF	100.0%	100.0%	Yes

Table 6. Confusion Matrices (CMs) expressing the results of the Discrimination Functions (DFs) of the Oil-Slick Discrimination Algorithm from Set.3—i.e., with the UPGMA (Unweighted Pair Group Method with Arithmetic Mean—see Phase 5: Section 2.5.1) attribute selection (i.e., 4 to 7 variables—see Phase 5: Section 3.5.1; Figure 3 and Table 3) and without the application of the PCA (Principal Component Analysis—see Phase 6). Note that NLT.2 and NLT.3 have the same outcome.

CM.1	DF Oil Seep	DF Oil Spill	Total Real
Oil Seep Real	1,570	424	**1,994**
Oil Spill Real	1,224	1,344	**2,568**
Total DF	**2,794**	**1,768**	**4,562**
NLT.0: No Transformation (x)			2,914 **63.88%**
CM.2	DF Oil Seep	DF Oil Spill	Total Real
Oil Seep Real	**78.7%**	**21.3%**	100.0%
Oil Spill Real	**47.7%**	**52.3%**	100.0%
CM.3	DF Oil Seep	DF Oil Spill	UPGMA:
Oil Seep Real	**56.2%**	**24.0%**	Yes
Oil Spill Real	**43.8%**	**76.0%**	PCA:
Total DF	100.0%	100.0%	No

CM.1	DF Oil Seep	DF Oil Spill	Total Real
Oil Seep Real	1454	540	**1994**
Oil Spill Real	987	1581	**2568**
Total DF	**2441**	**2121**	**4562**
NLT.1: Reciprocal (1/x)			3,035 **66.53%**
CM.2	DF Oil Seep	DF Oil Spill	Total Real
Oil Seep Real	**72.9%**	**27.1%**	100.0%
Oil Spill Real	**38.4%**	**61.6%**	100.0%
CM.3	DF Oil Seep	DF Oil Spill	UPGMA:
Oil Seep Real	**59.6%**	**25.5%**	Yes
Oil Spill Real	**40.4%**	**74.5%**	PCA:
Total DF	100.0%	100.0%	No

CM.1	DF Oil Seep	DF Oil Spill	Total Real
Oil Seep Real	1296	698	**1994**
Oil Spill Real	739	1829	**2568**
Total DF	**2035**	**2527**	**4562**
NLT.2: Logarithm Base 10 ($\log_{10}(x)$)			3,125 **68.50%**
CM.2	DF Oil Seep	DF Oil Spill	Total Real
Oil Seep Real	**65.0%**	**35.0%**	100.0%
Oil Spill Real	**28.8%**	**71.2%**	100.0%
CM.3	DF Oil Seep	DF Oil Spill	UPGMA:
Oil Seep Real	**63.7%**	**27.6%**	Yes
Oil Spill Real	**36.3%**	**72.4%**	PCA:
Total DF	100.0%	100.0%	No

CM.1	DF Oil Seep	DF Oil Spill	Total Real
Oil Seep Real	1296	698	**1994**
Oil Spill Real	739	1829	**2568**
Total DF	**2035**	**2527**	**4562**
NLT.3: Napierian Logarithm (Ln(x))			3,125 **68.50%**
CM.2	DF Oil Seep	DF Oil Spill	Total Real
Oil Seep Real	**65.0%**	**35.0%**	100.0%
Oil Spill Real	**28.8%**	**71.2%**	100.0%
CM.3	DF Oil Seep	DF Oil Spill	UPGMA:
Oil Seep Real	**63.7%**	**27.6%**	Yes
Oil Spill Real	**36.3%**	**72.4%**	PCA:
Total DF	100.0%	100.0%	No

CM.1	DF Oil Seep	DF Oil Spill	Total Real
Oil Seep Real	1426	568	**1994**
Oil Spill Real	947	1621	**2568**
Total DF	**2373**	**2189**	**4562**
NLT.4: Square Root ($x^{1/2}$)			3,047 **66.79%**
CM.2	DF Oil Seep	DF Oil Spill	Total Real
Oil Seep Real	**71.5%**	**28.5%**	100.0%
Oil Spill Real	**36.9%**	**63.1%**	100.0%
CM.3	DF Oil Seep	DF Oil Spill	UPGMA:
Oil Seep Real	**60.1%**	**25.9%**	Yes
Oil Spill Real	**39.9%**	**74.1%**	PCA:
Total DF	100.0%	100.0%	No

CM.1	DF Oil Seep	DF Oil Spill	Total Real
Oil Seep Real	1761	233	**1994**
Oil Spill Real	1733	835	**2568**
Total DF	**3494**	**1068**	**4562**
NLT.5: Square Power (x^2)			2,596 **56.90%**
CM.2	DF Oil Seep	DF Oil Spill	Total Real
Oil Seep Real	**88.3%**	**11.7%**	100.0%
Oil Spill Real	**67.5%**	**32.5%**	100.0%
CM.3	DF Oil Seep	DF Oil Spill	UPGMA:
Oil Seep Real	**50.4%**	**21.8%**	Yes
Oil Spill Real	**49.6%**	**78.2%**	PCA:
Total DF	100.0%	100.0%	No

CM.1	DF Oil Seep	DF Oil Spill	Total Real
Oil Seep Real	1407	587	**1994**
Oil Spill Real	857	1711	**2568**
Total DF	**2264**	**2298**	**4562**
NLT.6: Cube Root ($x^{1/3}$)			3,118 **68.35%**
CM.2	DF Oil Seep	DF Oil Spill	Total Real
Oil Seep Real	**70.6%**	**29.4%**	100.0%
Oil Spill Real	**33.4%**	**66.6%**	100.0%
CM.3	DF Oil Seep	DF Oil Spill	UPGMA:
Oil Seep Real	**62.1%**	**25.5%**	Yes
Oil Spill Real	**37.9%**	**74.5%**	PCA:
Total DF	100.0%	100.0%	No

CM.1	DF Oil Seep	DF Oil Spill	Total Real
Oil Seep Real	892	1102	**1994**
Oil Spill Real	795	1773	**2568**
Total DF	**1687**	**2875**	**4562**
NLT.7: Third Power (x^3)			2,665 **58.42%**
CM.2	DF Oil Seep	DF Oil Spill	Total Real
Oil Seep Real	**44.7%**	**55.3%**	100.0%
Oil Spill Real	**31.0%**	**69.0%**	100.0%
CM.3	DF Oil Seep	DF Oil Spill	UPGMA:
Oil Seep Real	**52.9%**	**38.3%**	Yes
Oil Spill Real	**47.1%**	**61.7%**	PCA:
Total DF	100.0%	100.0%	No

Table 7. Confusion Matrices (CMs) expressing the results of the Discrimination Functions (DFs) of the separate analysis of the Oil-Slick Discrimination Algorithm (see Phase 3: Section 2.3.2): Size Information Descriptors: Area with Perimeter (shown in Orange) and PtoA with Compact (Shown in Blue); and SAR Information Descriptors: AVG with SKW (shown in Black). See also Figure 4.

CM.1	DF Oil Seep	DF Oil Spill	Total Real
Oil Seep Real	645	1349	1994
Oil Spill Real	206	2362	2568
Total DF	851	3711	4562
NLT.0: No Transformation (x)			3,007 / 65.91%
CM.2	DF Oil Seep	DF Oil Spill	Total Real
Oil Seep Real	32.3%	67.7%	100.0%
Oil Spill Real	8.0%	92.0%	100.0%
CM.3	DF Oil Seep	DF Oil Spill	Area and Perimeter
Oil Seep Real	75.8%	36.4%	
Oil Spill Real	24.2%	63.6%	
Total DF	100.0%	100.0%	

CM.1	DF Oil Seep	DF Oil Spill	Total Real
Oil Seep Real	1575	419	1994
Oil Spill Real	1230	1338	2568
Total DF	2805	1757	4562
NLT.0: No Transformation (x)			2,913 / 63.85%
CM.2	DF Oil Seep	DF Oil Spill	Total Real
Oil Seep Real	79.0%	21.0%	100.0%
Oil Spill Real	47.9%	52.1%	100.0%
CM.3	DF Oil Seep	DF Oil Spill	AtoP and Compact
Oil Seep Real	56.1%	23.8%	
Oil Spill Real	43.9%	76.2%	
Total DF	100.0%	100.0%	

CM.1	DF Oil Seep	DF Oil Spill	Total Real
Oil Seep Real	1288	706	1994
Oil Spill Real	727	1841	2568
Total DF	2015	2547	4562
NLT.2: Logarithm Base 10 ($\log_{10}(x)$)			3,129 / 68.59%
CM.2	DF Oil Seep	DF Oil Spill	Total Real
Oil Seep Real	64.6%	35.4%	100.0%
Oil Spill Real	28.3%	71.7%	100.0%
CM.3	DF Oil Seep	DF Oil Spill	Area and Perimeter
Oil Seep Real	63.9%	27.7%	
Oil Spill Real	36.1%	72.3%	
Total DF	100.0%	100.0%	

CM.1	DF Oil Seep	DF Oil Spill	Total Real
Oil Seep Real	1288	706	1994
Oil Spill Real	727	1841	2568
Total DF	2015	2547	4562
NLT.2: Logarithm Base 10 ($\log_{10}(x)$)			3,129 / 68.59%
CM.2	DF Oil Seep	DF Oil Spill	Total Real
Oil Seep Real	64.6%	35.4%	100.0%
Oil Spill Real	28.3%	71.7%	100.0%
CM.3	DF Oil Seep	DF Oil Spill	AtoP and Compact
Oil Seep Real	63.9%	27.7%	
Oil Spill Real	36.1%	72.3%	
Total DF	100.0%	100.0%	

CM.1	DF Oil Seep	DF Oil Spill	Total Real
Oil Seep Real	1128	866	1994
Oil Spill Real	566	2002	2568
Total DF	1694	2868	4562
NLT.6: Cube Root ($x^{1/3}$)			3,130 / 68.61%
CM.2	DF Oil Seep	DF Oil Spill	Total Real
Oil Seep Real	56.6%	43.4%	100.0%
Oil Spill Real	22.0%	78.0%	100.0%
CM.3	DF Oil Seep	DF Oil Spill	Area and Perimeter
Oil Seep Real	66.6%	30.2%	
Oil Spill Real	33.4%	69.8%	
Total DF	100.0%	100.0%	

CM.1	DF Oil Seep	DF Oil Spill	Total Real
Oil Seep Real	1417	577	1994
Oil Spill Real	901	1667	2568
Total DF	2318	2244	4562
NLT.6: Cube Root ($x^{1/3}$)			3,084 / 67.60%
CM.2	DF Oil Seep	DF Oil Spill	Total Real
Oil Seep Real	71.1%	28.9%	100.0%
Oil Spill Real	35.1%	64.9%	100.0%
CM.3	DF Oil Seep	DF Oil Spill	AtoP and Compact
Oil Seep Real	61.1%	25.7%	
Oil Spill Real	38.9%	74.3%	
Total DF	100.0%	100.0%	

CM.1	DF Oil Seep	DF Oil Spill	Total Real
Oil Seep Real	494	1500	1994
Oil Spill Real	456	2112	2568
Total DF	950	3612	4562
NLT.0: No Transformation (x)			2,606 / 57.12%
CM.2	DF Oil Seep	DF Oil Spill	Total Real
Oil Seep Real	24.8%	75.2%	100.0%
Oil Spill Real	17.8%	82.2%	100.0%
CM.3	DF Oil Seep	DF Oil Spill	Average and Skewness
Oil Seep Real	52.0%	41.5%	
Oil Spill Real	48.0%	58.5%	
Total DF	100.0%	100.0%	

CM.1	DF Oil Seep	DF Oil Spill	Total Real
Oil Seep Real	673	1321	1994
Oil Spill Real	614	1954	2568
Total DF	1287	3275	4562
NLT.6: Cube Root ($x^{1/3}$)			2,627 / 57.58%
CM.2	DF Oil Seep	DF Oil Spill	Total Real
Oil Seep Real	33.8%	66.2%	100.0%
Oil Spill Real	23.9%	76.1%	100.0%
CM.3	DF Oil Seep	DF Oil Spill	Average and Skewness
Oil Seep Real	52.3%	40.3%	
Oil Spill Real	47.7%	59.7%	
Total DF	100.0%	100.0%	

Based on our comprehensive approach to find a simple way to discriminate seeps from spills, we are able to answer the three scientific questions:

1. The two Logarithm functions (NLT.2: Log_{10}; and NLT.3: Ln) and Cube Root (NLT.6) have the most accurate seep-spill discrimination among the eight Data Transformation Approaches tested.
2. Of the different strategies tested for selecting relevant attributes (i.e., four input dataset versions—see Phase 7: Section 3.7), two (Set.2 and Set.3) have comparable discrimination power with Overall Accuracies of almost 70%; however, the sole use of UPGMA dendrograms (i.e., Set.3) excels at selecting uncorrelated variables as it provides a simpler form avoiding the implementation of additional Multivariate Data Analysis Techniques (i.e., PCA). This is clearly observed in an inspection of Table 6 (Set.3) and in a comparison with Table 5 [Set.2: the use of all variables (see Phase 5: Section 3.5.1—Figure 3 and Table 3) without the dendrogram selection (i.e., no UPGMA) but with the application of the PCA (see Phase 6: Section 3.6—Table 4)].
3. The use of a collection of variables from two attribute selection strategies, i.e., Set.2 [no UPGMA with PCA (19 or 16 attributes but with 3 to 7 PCs—Table 5)] and Set.3 [UPGMA and no PCA (4 to 7 variables—Table 6)] is equally capable of discriminating seeps from spills. However, these are comparable to the sole use of the two Size Information Descriptor pairs (Area with Perimeter and PtoA with Compact) that outperform the SAR Information Descriptor pair (AVG with SKW)—see Table 7.

Our EDA also demonstrates that using simple and low-cost RADARSAT-2 beam modes (SCNA and SCNB), one can achieve useful seep-spill discrimination accuracies, thus supporting new products for the RADARSAT Constellation Mission (RCM): RADARSAT-2 Mode Selection for Maritime Surveillance (R2MS2).

Author Contributions: The paper was conceived and written by G.A.C. under the supervision of P.J.M., E.T.P., F.P.M., and L.L. All authors participated of the research conceptualization, experiment design, data analysis/interpretation, as well as of the investigation quality improvement, read, edit, and approval of the final manuscript.

Funding: Financial support has been provided by the Programa Nacional de Pós Doutorado (PNPD) of Coordenação de Aperfeiçoamento de Pessoal de Nível Superior (CAPES), Brazil.

Acknowledgments: We thank Pemex and MDA Geospatial Services for the RADARSAT-2 dataset, as well as we are pleased with the support received from COPPE/UFRJ: LabSAR colleagues, LAMCE staff, and PEC employees.

Conflicts of Interest: The authors declare no conflict of interest.

References

1. NRCC (National Research Council Committee). *Oil in the Sea: Inputs, Fates, and Effects*; The National Academies Press: Washington, DC, USA, 1985.
2. NRCC (National Research Council Committee). *Oil in the Sea III: Inputs, Fates, and Effects*; The National Academies Press: Washington, DC, USA, 2003; ISBN 9780309084383.
3. Fingas, M.F.; Brown, C.E. Review of oil spill remote sensing. *Spill Sci. Technol. Bull.* **1997**, *4*, 199–208. [CrossRef]
4. Fingas, M.F.; Brown, C.E. Oil-spill remote sensing—An update. *Sea Technol.* **2000**, *41*, 21–26.
5. Fingas, M.; Brown, C.E. A Review of Oil Spill Remote Sensing. *Sensors* **2018**, *18*, 91. [CrossRef] [PubMed]
6. Asanuma, I.; Muneyama, K.; Sasaki, Y.; Iisaka, J.; Yasuda, Y.; Emori, Y. Satellite thermal observation of oil slicks on the Persian Gulf. *Remote Sens. Environ.* **1986**, *19*, 171–186. [CrossRef]
7. Bulgarelli, B.; Djavidnia, S. On MODIS retrieval of oil spill spectral properties in the marine environment. *IEEE Geosci. Remote Sens. Lett.* **2012**, *9*, 398–402. [CrossRef]
8. Brown, C.E.; Fingas, M. New space-borne sensors for oil spill response. In Proceedings of the International Oil Spill Conference, Tampa, FL, USA, 26–29 March 2001; pp. 911–916.
9. Brown, C.E.; Fingas, M. The latest developments in remote sensing technology for oil spill detection. In Proceedings of the Interspill Conference and Exhibition, Marseille, France, 12–14 May 2009; p. 13.

10. Alpers, W.; Holt, B.; Zeng, K. Oil spill detection by imaging radars: Challenges and pitfalls. *Remote Sens. Environ.* **2017**, *201*, 133–147. [CrossRef]

11. Staples, G.C.; Hodgins, D.O. RADARSAT-1 emergency response for oil spill monitoring. In Proceedings of the 5th International Conference on Remote Sensing for Marine and Coastal Environments, San Diego, CA, USA, 5–7 October 1998; pp. 163–170.

12. Staples, G.; Rodrigues, D.R. Maritime environmental surveillance with RADARSAT-2. In Proceedings of the XVI Brazilian Remote Sensing Symposium (SBSR), Foz do Iguaçu, Brazil, 13–18 April 2013; pp. 8445–8452.

13. Genovez, P.C. Segmentação e Classificação de Imagens SAR Aplicadas à Detecção de Alvos Escuros em Áreas Oceânicas de Exploração e Produção de Petróleo. Ph.D. Dissertation, COPPE, Universidade Federal do Rio de Janeiro (UFRJ), Rio de Janeiro, Brazil, 2010; p. 235.

14. Espedal, H.A. Detection of Oil Spill and Natural Film in the Marine Environment by Spaceborne Synthetic Aperture Radar. Ph.D. Dissertation, Department of Physics, University of Bergen and Nansen Environmental and Remote Sensing Center (NERSC), Bergen, Norway, 1998; p. 200.

15. Johannessen, O.M.; Espedal, H.A.; Jenkins, A.J.; Knulst, J. SAR surveillance of ocean surface slicks. In Proceedings of the 2nd ERS Application Workshop, London, UK, 6–8 December 1995; pp. 187–192.

16. Jackson, C.R.; Apel, J.R. *Synthetic Aperture Radar Marine User's Manual*; NOAA/NESDIS, Office of Research and Applications: Washington, DC, USA, 2004. Freely Available online: http://www.sarusersmanual.com (accessed on 2 December 2018).

17. Wismann, V.; Gade, M.; Alpers, W.; Huehnerfuss, H. Radar signatures of marine mineral oil spills measured by an airborne multi-frequency multi-polarization microwave scatterometer. *Int. J. Remote Sens.* **1998**, *19*, 3607–3623. [CrossRef]

18. Bentz, C.M. Reconhecimento Automático de Eventos Ambientais Costeiros e Oceânicos em Imagens de Radares Orbitais. Ph.D. Dissertation, COPPE, Universidade Federal do Rio de Janeiro (UFRJ), Rio de Janeiro, Brazil, 2006; p. 115.

19. Carvalho, G.A. Multivariate Data Analysis of Satellite-Derived Measurements to Distinguish Natural from Man-Made Oil Slicks on the Sea Surface of Campeche Bay (Mexico). Ph.D. Dissertation, COPPE, Universidade Federal do Rio de Janeiro (UFRJ), Rio de Janeiro, Brazil, 2015; p. 285. Freely Available online: http://www.coc.ufrj.br/index.php?option=com_content&view=article&id=4618:gustavo-de-araujo-carvalho (accessed on 2 December 2018).

20. Carvalho, G.A.; Landau, L.; Miranda, F.P.; Minnett, P.; Moreira, F.; Beisl, C. The use of RADARSAT-derived information to investigate oil slick occurrence in Campeche Bay, Gulf of Mexico. In Proceedings of the XVII Brazilian Remote Sensing Symposium (SBSR), João Pessoa, Brazil, 25–29 April 2015; pp. 1184–1191. Freely Available online: http://www.dsr.inpe.br/sbsr2015/files/p0217.pdf (accessed on 2 December 2018).

21. Carvalho, G.A.; Minnett, P.J.; Miranda, F.P.; Landau, L.; Moreira, F. The use of a RADARSAT-derived long-term dataset to investigate the sea surface expressions of human-related oil spills and naturally-occurring oil seeps in Campeche Bay, Gulf of Mexico. *Can. J. Remote Sens.* **2016**, *42*, 307–321. [CrossRef]

22. Carvalho, G.A.; Minnett, P.J.; de Miranda, F.P.; Landau, L.; Paes, E.T. Exploratory Data Analysis of Synthetic Aperture Radar (SAR) Measurements to Distinguish the Sea Surface Expressions of Naturally-Occurring Oil Seeps from Human-Related Oil Spills in Campeche Bay (Gulf of Mexico). *ISPRS Int. J. Geo-Inf.* **2017**, *6*, 379. Freely Available online: http://www.mdpi.com/2220-9964/6/12/379 (accessed on 2 December 2018). [CrossRef]

23. Freeman, A. Radiometric calibration of SAR image data. In Proceedings of the XVII Congress for Photogrammetry and Remote Sensing, Washington, DC, USA, 2–14 August 1992; pp. 212–222.

24. Laur, H.; Bally, P.; Meadows, P.; Sanchez, J.; Schaettler, B.; Lopinto, E.; Esteban, D. *ERS SAR Calibration: Derivation of the Backscattering Coefficient Sigma-Nought in ESA ERS SAR PRI Products*; Document No.: ES-TN-RS-PM-HL09; ESA (European Space Agency): Paris, France, 1998; p. 51.

25. Shepherd, N. *Extraction of Beta Nought and Sigma Nought from RADARSAT CDPF Products*; Technical Report, Revision 4, AS97-5001; Altrix Systems: Ottawa, ON, Canada, 2000; p. 16.

26. Hammer, Ø.; Harper, D.A.T.; Ryan, P.D. PAST: PAleontological STatistics software package for education and data analysis. *Palaeontol. Electron.* **2001**, *4*, 1–9.

27. Henderson, F.M.; Lewis, A.J. *Principles and Applications of Imaging Radar, Manual of Remote Sensing*, 3rd ed.; Wiley: Hoboken, NJ, USA, 1998; p. 866.

28. Masoomi, A.; Hamzehyan, R.; Shirazi, N.C. Speckle reduction approach for SAR image in satellite communication. *Int. J. Mach. Learn. Comput.* **2012**, *2*, 62–70. [CrossRef]

29. Frost, V.S.; Stiles, J.A.; Shanmugan, K.S.; Holtzman, J.C. A model for radar images and its application to adaptive digital filtering of multiplicative noise. *IEEE Trans. Pattern Anal. Mach. Intell.* **1982**, *4*, 157–166. [CrossRef] [PubMed]

30. AIRBUS (Defense & Space). *Radiometric Calibration of TerraSAR-X Data: Beta Naught and Sigma Naught Coefficient Calculation*; Technical Report TSXXITD-TN-0049; AIRBUS: Friedrichshafen, Germany, 2014; p. 15.

31. AIRBUS (Defense & Space). *TerraSAR-X Value Added Product Specification*; Technical Report TSXX-ITD-SPE-0009, Issue/Revision: 1/3; AIRBUS: Friedrichshafen, Germany, 2014; p. 26.

32. El-Darymli, K.; Mcguire, P.; Gill, E.; Power, D.; Moloney, C. Understanding the significance of radiometric calibration for synthetic aperture radar imagery. In Proceedings of the 27th Canadian Conference on Electrical and Computer Engineering (CCECE), Toronto, ON, Canada, 4–7 May 2014; p. 6. [CrossRef]

33. Thakur, P.K. SAR data processing to extract backscatter response from various features. In Proceedings of the Symposium Tutorials on Polarimetric SAR Data Processing and Applications, International Society for Photogrametry and Remote Sensing (ISPRS), Hyderabad, India, 9–12 December 2014.

34. ASF (Alaska Satellite Facility). *MapReady User Manual Remote Sensing Tool Kit*; Engineering Group Fairbanks: Fairbanks, AK, USA, 2015; p. 120.

35. Fiscella, B.; Giancaspro, A.; Nirchio, F.; Pavese, P.; Trivero, P. Oil spill monitoring in the Mediterranean Sea using ERS SAR data. In Proceedings of the Envisat Symposium (ESA), Göteborg, Sweden, 16–20 October 2010; p. 9.

36. Singha, S.; Bellerby, T.J.; Trieschmann, O. Satellite Oil Spill Detection Using Artificial Neural Networks. *IEEE J. Sel. Top. Appl. Earth Obs. Remote Sens.* **2013**, *6*, 2355–2363. [CrossRef]

37. Solberg, A.H.S.; Storvik, G.; Solberg, R.; Volden, E. Automatic detection of oil spills in ERS SAR images. *IEEE Trans. Geosci. Remote Sens.* **1999**, *37*, 1916–1924. [CrossRef]

38. Pisano, A. Development of Oil Spill Detection Techniques for Satellite Optical Sensors and Their Application to Monitor Oil Spill Discharge in the Mediterranean Sea. Ph.D. Dissertation, Università di Bologna, Bologna, Italy, 2011; p. 146.

39. Mcgarigal, K.; Marks, B.J. *FRAGSTATS: Spatial Pattern Analysis Program for Quantifying Landscape Structure*; General Technical Report Series, PNW-GTR-351; U.S. Department of Agriculture: Portland, OR, USA, 1994; p. 134.

40. Milligan, G.W.; Cooper, M.C. A study of standardization of variables in cluster analysis. *J. Classif.* **1988**, *5*, 181–204. [CrossRef]

41. Moita Neto, J.M.; Moita, G.C. Uma introdução à análise exploratória de dados multivariados. *Química Nova* **1998**, *21*, 467–469. [CrossRef]

42. Legendre, P.; Legendre, L. *Numerical Ecology*, 3rd English ed.; Developments in Environmental Modelling; Elsevier Science B.V.: Amsterdam, The Netherlands, 2012; 990p, ISBN 978-0444538680.

43. Valentin, J.L. *Ecologia Numérica—Uma Introdução à Análise Multivariada de Dados Ecológicos*, 2nd ed.; Editora Interciência: Rio de Janeiro, Brazil, 2012; p. 153, ISBN 978-85-7193-230-2.

44. Lane, D.M.; Scott, D.; Hebl, M.; Guerra, R.; Osherson, D.; Ziemer, H. *Introduction to Statistics*; Online Edition; Rice University: Huston, TX, USA, 2015; p. 695.

45. Guyon, I.; Elisseeff, A. An Introduction to Variable and Feature Selection. *J. Mach. Learn. Res.* **2003**, *3*, 1157–1182.

46. Sneath, P.H.A.; Sokal, R.R. *Numerical Taxonomy—The Principles and Practice of Numerical Classification*; W.H. Freeman and Company: San Francisco, CA, USA, 1973; 573p, ISBN 0-7167-0697-0.

47. Hall, M.A. Correlation-Based Feature Selection for Machine Learning. Ph.D. Dissertation, Department of Computer Science, The University of Waikato, Hamilton, New Zealand, 1999; p. 178.

48. Bouckaert, R.R.; Frank, E.; Hall, M.; Kirby, R.; Reutemann, P.; Seewald, A.; Scuse, D. *WEKA Manual for Version 3-6-0*; The University of Waikato: Hamilton, New Zealand, 2008; p. 212.

49. Sokal, R.R.; Rohlf, F.J. The Comparison of dendrograms by objective methods. *Taxon* **1962**, *11*, 33–40. [CrossRef]

50. NCSS (Number Cruncher Statistical System). *Hierarchical Clustering and Dendrograms*; NCSS Statistical Software: Kaysville, UT, USA, 2015; Chapter 445, p. 15.

51. Zar, H.J. *Biostatistical Analysis*, 5th ed.; Pearson New International Edition; Pearson: Upper Saddle River, NJ, USA, 2014; ISBN 1-292-02404-6.

52. Peres-Neto, P.R.; Jackson, D.A.; Somers, K.M. Giving meaningful interpretation to ordination axes: Assessing loading significance in principal component analysis. *Ecology* **2003**, *84*, 2347–2363. [CrossRef]

53. Kaiser, H.F. A note on Guttman's lower bound for the number of common factors. *Br. J. Stat. Psychol.* **1961**, *14*, 1–2. [CrossRef]

54. Cattell, R.B. The Scree Test for the number of factors. *Multivar. Behav. Res.* **1966**, *1*, 245–276. [CrossRef] [PubMed]

55. Jolliffe, I.T. *Principal Component Analysis*, 2nd ed.; Springer: New York, NY, USA, 2002; p. 487, ISBN 0-387-95442-2.

56. Peres-Neto, P.R.; Jackson, D.A.; Somers, K.M. How many principal components? Stopping rules for determining the number of non-trivial axes revisited. *Comput. Stat. Data Anal.* **2005**, *49*, 974–997. [CrossRef]

57. Hammer, Ø. PAST: Multivariate Statistics. 2015. Freely Available online: http://folk.uio.no/ohammer/past/multivar.html (accessed on 2 December 2018).

58. Lohninger, H. *Teach/Me Data Analysis (Text-Only Light Edition)*; Springer: Berlin, Germany; New York, NY, USA; Tokyo, Japan, 1999; ISBN 3-540-14743-8.

59. Hair, J.F.; Anderson, R.E.; Tatham, R.L.; Black, W.C. *Multivariate Data Analysis*, 5th ed.; Sant'Anna, A.S.; Chaves Neto, A., Translators; (In Portuguese). Análise multivariada de dados, Bookman; Pearson Education, Prentice Hall: Porto Alegre, Brazil, 2005; ISBN 0-13-014406-7.

60. Hammer, Ø. *PAST: PAleontological STatistics, Reference Manual, Version 3.20*; University of Oslo: Oslo, Norway, 2018; p. 264. Freely Available online: http://folk.uio.no/ohammer/past/past3manual.pdf (accessed on 2 December 2018).

61. PUS (Penn State University). *Applied Multivariate Statistical Analysis*; STAT 505; PUS: State College, PA, USA, 2015.

62. McLachlan, G. *Discriminant Analysis and Statistical Pattern Recognition*; A Whiley-Interescience Publication, John Wiley & Sons, Inc.: Queensland, Australia, 1992; ISBN 0-471-61531-5.

63. Carvalho, G.A. The Use of Satellite-Based Ocean Color Measurements for Detecting the Florida Red Tide (Karenia Brevis). Master's Thesis, RSMAS/MPO, University of Miami (UM), Miami, FL, USA, 2008; p. 156. Freely Available online: http://scholarlyrepository.miami.edu/oa_theses/116/ (accessed on 2 December 2018).

64. Carvalho, G.A.; Minnett, P.J.; Fleming, L.E.; Banzon, V.F.; Baringer, W. Satellite remote sensing of harmful algal blooms: A new multi-algorithm method for detecting the Florida Red Tide (Karenia brevis). *Harmful Algae* **2010**, *9*, 440–448. Freely Available online: http://ncbi.nlm.nih.gov/pubmed/21037979 (accessed on 2 December 2018). [CrossRef] [PubMed]

65. Carvalho, G.A.; Minnett, P.J.; Banzon, V.F.; Baringer, W.; Heil, C.A. Long-term evaluation of three satellite ocean color algorithms for identifying harmful algal blooms (Karenia brevis) along the west coast of Florida: A matchup assessment. *Remote Sens. Environ.* **2011**, *115*, 1–18. Freely Available online: http://ncbi.nlm.nih.gov/pubmed/22180667 (accessed on 2 December 2018). [CrossRef] [PubMed]

66. Congalton, R.G. A review of assessing the accuracy of classification of remote sensed data. *Remote Sens. Environ.* **1991**, *37*, 35–46. [CrossRef]

67. MDA (MacDonald, Dettwiler and Associates Ltd.). *RADARSAT-2 Product Description*; Technical Report RN-SP-52-1238, Issue/Revision: 1/13; MDA: Richmond, BC, Canada, 2016; p. 91.

Journal of
Marine Science and Engineering

MDPI

Article

An Operational Marine Oil Spill Forecasting Tool for the Management of Emergencies in the Italian Seas

Alberto Ribotti [1], **Fabio Antognarelli** [1], **Andrea Cucco** [1], **Marcello Francesco Falcieri** [2],
Leopoldo Fazioli [1,*], **Christian Ferrarin** [2], **Antonio Olita** [1], **Gennaro Oliva** [3], **Andrea Pes** [1],
Giovanni Quattrocchi [1], **Andrea Satta** [1], **Simone Simeone** [1], **Costanza Tedesco** [1],
Georg Umgiesser [2] **and Roberto Sorgente** [1]

[1] National Research Council—Institute for the Study of Anthropic impacts and Sustainability in Marine Environment, loc. Sa Mardini snc—Torregrande, 09170 Oristano, Italy; alberto.ribotti@cnr.it (A.R.); fabio.antognarelli@cnr.it (F.A.); andrea.cucco@cnr.it (A.C.); antonio.olita@cnr.it (A.O.); andrea.pes@cnr.it (A.P.); giovanni.quattrocchi@ias.cnr.it (G.Q.); andrea.satta@cnr.it (A.S.); simone.simeone@cnr.it (S.S.); costanza.tedesco@ias.cnr.it (C.T.); roberto.sorgente@cnr.it (R.S.)
[2] National Research Council—Institute of Marine Sciences, Arsenale Castello, 2737/F, 30122 Venezia, Italy; francesco.falcieri@ismar.cnr.it (M.F.F.); christian.ferrarin@ismar.cnr.it (C.F.); georg.umgiesser@ismar.cnr.it (G.U.)
[3] National Research Council—Institute for High Performance Computing and Networking, via Pietro Castellino 111, 80131 Napoli, Italy; gennaro.oliva@cnr.it
* Correspondence: leopoldo.fazioli@cnr.it; Tel.: +39-0783-229015

Received: 3 December 2018; Accepted: 15 December 2018; Published: 20 December 2018

check for updates

Abstract: Oil extraction platforms are potential sources of oil spills. For this reason, an oil spill forecasting system was set up to support the management of emergencies from the oil fields in the Italian seas. The system provides ready-to-use products to the relevant response agencies and optimizes the anti-pollution resources by assessing hazards and risks related to this issue. The forecasting system covers seven working oil platforms in the Sicily Channel and middle/low Adriatic Sea. It is composed of a numerical chain involving nested ocean models from regional to coastal spatial scales and an oil spill model. The system provides two online services, one automatic and a second dedicated to possible real emergencies or exercises on risk preparedness and responding. The automatic service produces daily short-term simulations of hypothetical oil spill dispersion, transport, and weathering processes from each extraction platform. Products, i.e., risk maps, animations, and a properly called bulletin, are available on a dedicated web-portal. The hazard estimations are computed by performing geo-statistical analysis on the daily forecasts database. The second service is activated in near-real-time producing oil spill simulations for the following 48 h.

Keywords: oil spill; Italian seas; numerical forecasting tool; emergency management

1. Introduction

The success of oil spill mitigation actions is closely dependent on the time necessary to detect the slick and predict its fate (i.e., slick displacement and dispersion), in order to permit governmental response agencies in planning and providing a specific and timely intervention at sea. The oil slick detection can be done in situ or by satellite, while its forecast is usually performed through more or less complex systems of numerical simulation with the application of empirical and semi-empirical algorithms, or the estimate of a surface water masses path starting from the intensity of winds [1–6]. The most advanced systems currently used incorporate both the weather–marine physical forcing and a Lagrangian module that reproduces trajectories, diffusion, and transformation processes of the oil slicks [7,8]. The knowledge of the marine and the weather components is essential to treat any kind of oil pollution, like that coming from maritime accidents with relative spillage of hydrocarbons, or chronic pollution as a continuous release from punctual or diffuse sources. Numerical models are strictly deterministic and based on "primitive equations" for the reproduction of water circulation with a good reliability [9,10]. This approach provides an essential basis for a realistic numerical simulation of the complex processes of transformation and transport of "active" tracers, subject to change as a function of space and time. Therefore, oil spill transport and weathering processes can be provided through the use of numerical models suitably coupled with hydrodynamic ones [7,11–13]. Such a structured system, composed by a three-dimensional hydrodynamic model plus transport simulation and processing of the oil slick, allows users to obtain a space and time evolution of a potential spillage, and then provides important information for the management of a possible emergency. This integrated modelling system permits obtaining maps of potential hazard estimations (i.e., net of the coastal vulnerability) of the stranding/concentration of hydrocarbons in a stretch of coast as well (e.g., References [3,14]).

The Italian seas are subject to oil spill emergencies because there exists hydrocarbons exploration, extraction, and transport within them. In particular, the Sicily Channel and the low/middle Adriatic Sea have seven working oil extraction sites that are potential sources of pollution through accidental oil spills.

Due to an increasing sensibility of the Italian public opinion on marine oil pollution, the Italian Ministry of the Environment and Protection of Land and Sea (MATTM) recently planned to provide an oil spill forecasting system for the Italian seas. Starting from the experience gained in previous national and international projects [15,16], the Italian Research Council (CNR) has developed a forecasting tool regarding the dispersion and transformation of oil at sea and for the estimation of the "riskiness" do to possible oil spillage in the Sicily Channel and in the medium–low Adriatic Sea (Figure 1a,b, respectively).

Figure 1. Spatial distribution of the four active oil platforms (green triangles) in the Sicily Channel (**a**), and the three in the medium–low Adriatic Sea (**b**).

The purpose of the system is two-fold: (a) To provide the responsible authority with possible indications on the spatial/temporal evolution of oil spills at sea in near-real-time in case of emergency events or planned exercises. This is important for optimizing both the resources and the management of the emergency. (b) To determine hazard estimations, based on statistics of the simulated dispersions, of a hydrocarbon slick virtually issued (i.e., simulated) using individual platforms according to different weather and sea conditions from the forecasting system. Such a system wishes to answer to three of the four priorities of the Sendai Framework for Disaster Risk Reduction (2015–2030) [17], adopted by United Nations (UN) Member States in March 2015. These priorities are the first, the second, and the fourth based on oil spill risk assessment, preparedness, and response, but focusing on the management of oil spill events at sea. Specifically, the first priority is on understanding the disaster risk, the second on strengthening disaster risk governance to manage it, and the fourth on enhancing disaster preparedness for an effective response.

In this paper, we describe the forecasting numerical system realized in the framework of the ministerial project "SOS-Piattaforme e Impatti Offshore" and highlight the main features of the service realized at this stage via CNR. Section 2 presents the study area and the oil rigs covered by the system, followed by its components in Section 3. In Section 4 we describe their processes of validation and show resulting products, i.e., the daily bulletin and the hazard assessment. Finally, in Section 5, we give our conclusions and future perspectives.

2. The Oil Fields in the Italian Seas

The oil spill model predicts the fate, i.e., transport and weathering processes, of possible slicks from seven different areas of oil extraction in the Italian seas (see Table 1).

Table 1. The oil fields in the Italian seas covered by the forecasting system (data from the Italian Ministry for the Economic Development—MISE [18], updated on 31 December 2016).

Area	Oil Field	Platforms	Storing Facility	Wells	Distance from Land (Km)	Depth (m)
Sicily Channel	Prezioso	1	Pipeline	9	12	45
	Perla	1	Pipeline	4	13	70
	Gela 1	1	Pipeline	11	2	10
	Vega A	1	FSO	20	20	124
Central Adriatic	Sarago	2	Pipeline	1 + 5	3–4	12
	Rospo	3	FSO	9 + 10 + 12	19–21	80
Southern Adriatic	Aquila	2	FPSO	1 + 1	40–45	820

In these areas we have a variety of extraction facilities, from a single to multiple wells, collecting at a single platform, to wider extraction fields populated by nearby but separated platforms. Once extracted, the oil is stored on FPSO or FSO (floating production storage and offloading) units or moved to the land by underwater pipelines when the platform is near the mainland.

This variety in configurations, described in Table 1, is taken into account in setting up the forecasting system. Therefore, multiple well configurations are simplified by considering a single well with a geographically averaged position, as the distance between platforms in multi-platform fields is comparable with the resolution of the coastal model's grid.

3. The Numerical System

The numerical system developed for the project "SOS-Piattaforme e Impatti Offshore" is based on what was experienced in past projects like Development of TEcnologies for Situational Sea Awareness (TESSA) [19,20], Mediterranean Decision Support System per la Marine Safety (MEDESS-4MS) [8] and SOS-Bonifacio [1,2,15]. It is composed by nested hydrodynamic numerical models at different spatial scales, from the regional to the coastal, through a methodology known as downscaling [21–23]. Daily forecasts from regional scale models are used by sub-regional numerical systems as initial, lateral boundary, and surface boundary conditions for the air–sea interface [24,25]. This repeats from the sub-regional to coastal models then increasing the resolution of the forecasts to a limited area [2,26]. The whole forecast system is presented in Figure 2 and is described in the following paragraphs.

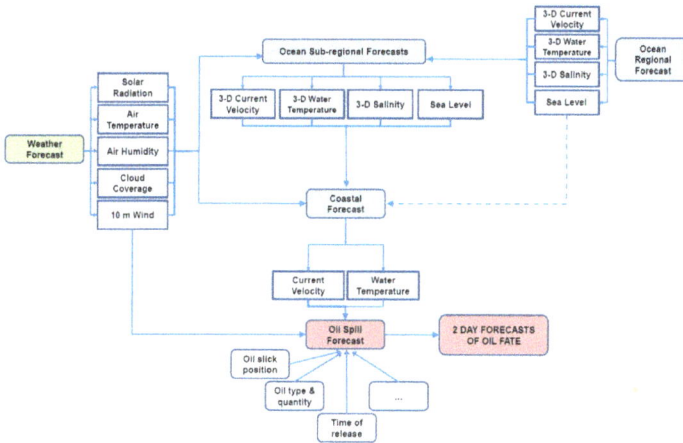

Figure 2. The numerical forecasting chain for coastal circulation and oil dispersion at sea.

3.1. The Oil Slick Model

In our system, the transport and weathering processes of an oil slick at sea are modelled using MEDSLIK-II, widely described in References [12,13]. This is a community model available for free at Reference [27]. MEDSLIK-II is a Lagrangian model that requires meteorological and marine conditions at the air–sea interface, as well as chemical and physical characteristics of the spilled oil to perform a simulation. It simulates the transport of an oil slick with its dispersion due to turbulent fluctuation components, parameterized with a random walk scheme. The transport of the slick in the marine environment is attributed to the advection field, while turbulent fluxes cause its dispersion. The general equation of the model takes into account the variation of the oil concentration in time and space, considering the weathering processes affecting the slick [7,20]. MEDSLIK-II is coupled with the ocean models using hourly oceanographic outputs of current speed and temperature and with the weather model for wind speed, all described in the following paragraphs.

We considered just the early stage weathering processes of evaporation, emulsification, dispersion in the water column, and adhesion to the coast. These processes take their effect in the fate of a typical crude oil for about a week (described in Reference [28]), the usual length of the most oil emergencies at sea. Every day, a simulation of the transport of a specific type (API 33.4) and quantity of oil (at a rate of 15 m$^3 \cdot$h^{-1}) were provided for each active oil platform or groups of Italian platforms.

The realized forecasting system is also used in delayed-mode to calculate statistics and builds hazard estimation maps related to each oil platform or group of platforms and support the eventual emergency containment strategies. In the case of an accident or planned exercise, numerical simulations of the scenario of dispersion and transformation of a real hydrocarbon slick are realized for the following 48 h from the simulation start. In order to perform the numerical simulation, several further inputs were requested like type of oil, areal position and time when the spill started, estimation of the amount of oil spilled, instantaneous or prolonged spill in time, and requested simulation duration.

3.2. The Hydrodynamic Models

The modelization of oil spill trajectories requires operational ocean models. Then, due to the different morphological and oceanographic characteristics of the two basins, different numerical approaches were applied.

3.2.1. The Sicily Channel

In order to describe the circulation in the Sicily Channel, including the processes at the basin scale that influence circulation from mesoscale to coastal [29], we used a sub-regional model called Tyrrhenian-Sicily Channel Sub-Regional Model (TSCRM). It is a free surface three-dimensional primitive equation. It is based on the Princeton ocean model [30] and was implemented between 9–16.50° E and 31.50–43° N with a horizontal resolution of about 2 km (1/48° of degree in latitude) on 30 sigma levels [10]. The initial and lateral conditions were obtained from a one-way asynchronous nesting of the forecasted hourly fields of temperature, salinity, and total velocity [25] from the regional Mediterranean Forecasting System (MFS) provided by the Copernicus Marine Service [31].

Solely for the very shallow Gela platform in the Gulf of Gela, the TSCRM was downscaled to very high resolution using the 3-D hydrodynamic Shallow water HYdrodynamic Finite Element Model (SHYFEM). SHYFEM resolves the system of the primitive equations, vertically integrated over each vertical layer, with Boussinesq approximation used horizontally and hydrostatics used vertically. For vertical diffusivity and viscosity, it used the general ocean turbulence model [32]. It was integrated with a module for the simulation of the transport processes and had a spatial resolution varying from 25 m in very coastal areas or shallow waters to few kilometers off-shore [2,33].

At the surface, both ocean models were forced by using the hourly atmospheric data from the weather limited area model SKIRON. Every day, they produce 3-D 5-day forecasts with an hourly resolution of the main oceanographic parameters as shown in Figure 2.

SKIRON is a numerical weather prediction model developed at the University of Athens [34]. It provides 5-day forecasts of atmospheric parameters at a high frequency (hourly fields) with a horizontal resolution of 10 km. The core of the system is based on the ETA/NCEP model that has been developed at the National Centre for Environmental Prediction of the National Oceanic and Atmospheric Administration (NCEP/NOAA). Initial and boundary conditions are taken from the coarse Global Forecast System model NCEP/GFS. The atmospheric parameters included hourly fields of: mean sea level pressure, air temperature at 2 m, wind speed, and direction at 10 m above sea level (s.l.), convective and accumulated precipitation, cloud cover, sensible and latent heat fluxes, incoming and outgoing shortwave and long-wave radiation fields, and evaporation. These parameters were used in both circulation models to force momentum, turbulent heat, and water fluxes calculated using appropriate bulk formulae [35].

3.2.2. The Adriatic Sea

In the Adriatic Sea, MFS's outputs were used to nest the SHYFEM model to solve for the coastal scales. The numerical computation was performed on a spatial domain that represented the Adriatic basin by means of an unstructured grid. The use of elements of variable sizes was fully exploited to create a seamless transition between different spatial scales. The mesh resolution varied from 4 km in the open sea to 1 km in coastal waters, and up to 300 m around the oil platforms. The sea level and the current velocity boundary conditions at the Otranto Strait were obtained by summing the hourly tidal signal derived from the Finite Element Solution (FES2012) global tidal model [36] (available at Reference [37]), the daily water level, and the baroclinic velocity predicted using MFS. The total water levels were imposed to the boundary nodes, while the total current velocity were nudged using a relaxation time of 3600 s. Water temperature and salinity boundary conditions were computed using the oceanographic fields of MFS. Three-dimensional MFS fields of sea temperature and salinity were nudged during the simulation. Nudging data were given for all nodes of the grid. The value of the relaxation coefficient spatially varied over the model domain (as a function of the grid resolution) from 2 days in the open sea and increasing toward the coast, thus diminishing the restoration contribution.

For the river discharge, where available, daily updated values were derived from automatic hydrometric stations nearest to river mouths, through calibrated stage-discharge relationships (like Isonzo, Piave, Adige, Po, etc.). For the other rivers considered in this study, discharges were prescribed using mean climatological values [38]. Such model implementation resembled the one described in Reference [39].

The meteorological forcing was supplied by the hydrostatic Bologna limited area model (BOLAM), developed and implemented at CNR in Bologna with a daily operational chain [40,41]. The initial and boundary conditions for the BOLAM model were derived from the analyses (00 UTC) and forecasts of the GFS (NOAA/NCEP, Silver Spring, MD, USA) global model [42]. The BOLAM model is implemented over the Euro-Mediterranean region with a horizontal grid spacing of 8.3 km. Forecasts are daily provided at hourly resolution up to 3 days.

At the end of the whole chain, transport, diffusion, and transformation of an oil slick at sea were modelled by MEDSLIK-II.

4. Results

A daily bulletin and hazard estimations are valid and easy products of the system to provide updated information on potential risks at sea and coastal areas and optimize intervention in case of a spill from oil rigs. The quality of the above instruments has been assured through previous validations of the numerical system.

4.1. Numerical System Validation

The forecast current fields from the system for the Sicily Channel and the Adriatic Sea areas have been validated against trajectories of GPS-equipped surface drifters [43] with a small subsurface plastic drogue at a depth of 1 m. In order to obtain forecasts, a process of validation is fundamental to verify the reliability of the implemented integrated system (hydrodynamic data production and oil spill simulation). To carry out this process, several Lagrangian drifters were released in the areas of the platforms with their positions recorded at 10-min intervals for coastal drifters and at 1-h intervals for offshore satellite ones.

In order to evaluate the integrated system capability in reproducing drifters' trajectories, we calculated the root mean square error (RMSE) of the separation distance between observed and simulated trajectories and the skill score [43]. This last parameter is dimensionless based on the cumulative Lagrangian separation distances normalized by the associated cumulative observed trajectory lengths. Its value can vary between 0 and 1, with 1 corresponding to a total overlap between the observed and simulated trajectories, while 0 corresponds to a difference between the path of the drifters and the simulated trajectories on the same path made by the drifters [10,13].

In Figure 3, the 72-h long trajectories of two drifters released in the Sicily Channel are shown. The surface drifters moved south-eastward, following the Atlantic Ionian Stream flow (AIS; [29,44]).

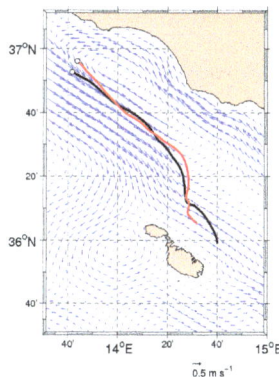

Figure 3. The 72-h long trajectories of two drifters, the red and the black lines starting from the white circle at north to the south, overlapped to the TSCRM current field averaged for the day of the release (29 October 2017). Blue arrows indicate the surface current.

Their trajectories were numerically reproduced by re-initializing the TSCRM simulation every day with updated forecasting fields. Such daily re-initialization is necessary to avoid errors to excessively propagate in calculations that affect the final results. The idea of frequent re-initialization in trajectory simulation was proposed in Reference [43] and was actually used in the Deepwater Horizon oil spill trajectory forecast [45,46].

The simulated drifters' trajectories produced low values of RMSE of less than 10 km (Figure 4a), and a skill score higher than 0.6 (Figure 4b), both in the first 10 h of simulation. Over this time, the RMSE increased exponentially and its average was about 25 km at 24 h. At the same time, the averaged skill score was 0.7. This decrease in efficiency was due to a difficulty in the reproduction of a correct surface current speed but directions were comparable. This deficiency was particularly true for drifters approaching the coast due to the low resolution of the model and/or to lower wind forcing. Stokes drift was demonstrated to be an important factor in bring the oil from the offshore areas to the beaches [47]. Furthermore, although the wave-induced transport was parameterized to simulate the effect of Stoke drift in all numerical experiments, the swell processes were not considered. This factor may be a source of uncertainty in the estimation of the drifters' trajectories.

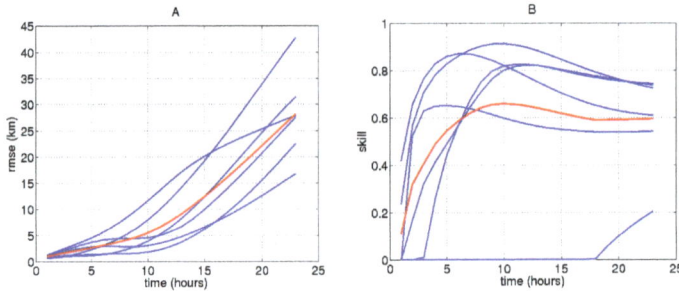

Figure 4. (**A**) The blue lines are the RMSE of separation distances computed from the comparison between TSCRM and both drifters trajectories released on 29 October 2017. The red line is the averaged RMSE. (**B**) The blue lines are the skill score; the red line is the averaged trend.

In the very shallow waters inside the Gulf of Gela, we implemented SHYFEM to solve such a limitation. For its validation, we used the data from two coastal drifters named LCA00128 and LCA00112, deployed on 29 October 2017, for about 30 h. The trajectories were reproduced by groups of particles released at each observed position along the drifters path. For both drifters there is a general correspondence between the simulated trajectories and the observed path (Figure 5).

The path of the simulated particles was both in terms of the distance traveled and the direction that was completely congruent with that followed by the two drifters. In the two experiments, an initial deviation of the path of the particles was observed with respect to the trajectory of the drifters during the first hours of simulation. Here they follow an northwestern (NW) direction, while the numerical particles are directed to the south-western (SW). For the remaining period of the simulation, the simulated and observed trajectories are quite similar.

Furthermore, in both experiments, for each trajectory the accuracy of the results obtained from a series of simulations was quantified. The dimensionless skill score and its average (AV3) obtained for the first 3 h of forecast were calculated at 1, 6, and 12 h of forecast. In Table 2, the skill score for the two drifters was generally low with values at 1 h varying between 0.7 and 0.5, at 6 h between 0.8 and 0.6, and finally at 12 h between 0.8 and 0.7. As far as AV3 was concerned, the most accurately simulated trajectory was the LCA00112 with an AV3 of 0.7.

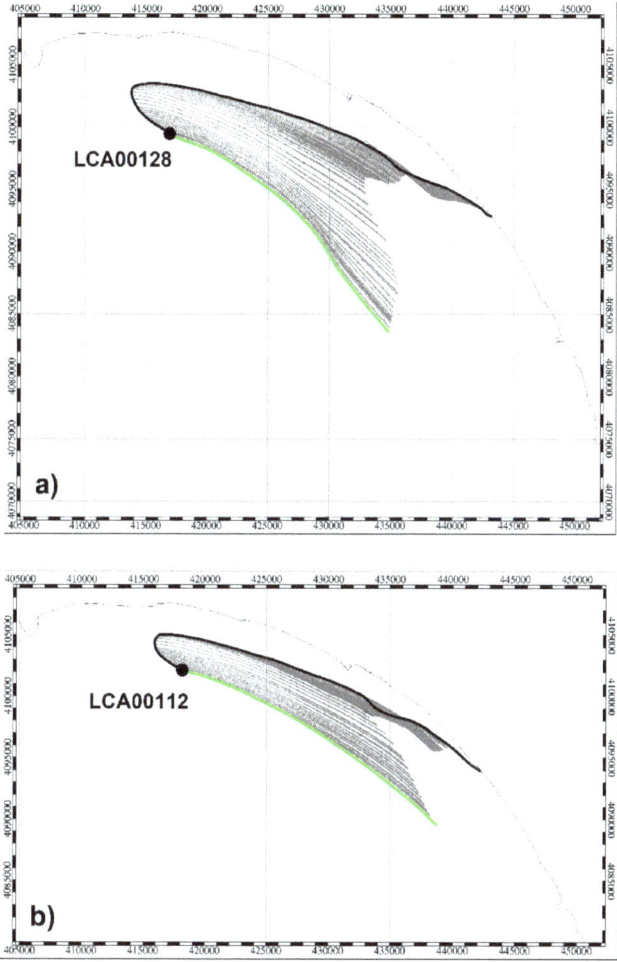

Figure 5. Trajectories simulated by the model for each release made on each position of the drifters LCA00128 (**a**) and LCA00112 (**b**) paths. In black is the trajectory of each drifter, and in green the path of the particles at their first release. In gray is the groups of particles released at each hourly position.

Table 2. Skill score of the results of the numerical simulations performed in the two calibration tests (Test 1 and Test 2) calculated at the forecast intervals of 1, 6, and 12 h from the release and AV3 for each single trajectory (LCA00128 and LCA00112).

	TEST 1			
	1 h	**6 h**	**12 h**	**AV3**
LCA00128	0.6	0.6	0.7	0.6
LCA00112	0.7	0.8	0.8	0.7
	TEST 2			
LCA00128	0.4	0.5	0.7	0.4
LCA00112	0.4	0.7	0.7	0.5

The goodness of the results obtained in simulating the trajectory followed by the two drifters was largely due to the fact that the surface transport was mainly modulated by the wind action on the coastal area. The initial deviation between model results and observations was because the wind was initially of moderate intensity and therefore the transport was probably predominantly led by the large-scale circulation that was not well represented in these simulations.

For the Adriatic Sea, the model results were compared with the trajectories of a satellite global positioning system (GPS)-equipped drifter operating between 14 and 28 May 2018 (14 days, LCE00234-1), and again between 13 and 18 June 2018 (5 days, LCE00234-2) in the central basin. Drifter LCE00234-1 was equipped with a 50 cm long plastic drogue placed at a 20 m depth, thus providing the integral information of the currents in the upper 20 m of the water column. Conversely, during the second drifter release, the plastic drogue was placed at a 1 m depth.

Numerical particles were released every hour along the observed drifters' trajectories in the first 20 m of the water column and at the surface in the two experiments, respectively. The particle-tracking model correctly reproduced the trajectories of the drifters, which moved southward along the coast with a mean speed of 15 cm s^{-1} and 50 cm s^{-1} for drifters LCE00234-1 and LCE00234-2, respectively (Figure 6).

The RMSE and skill score after 24 h of simulation were 5 and 15 km, and 0.57 and 0.67 for the two trajectories, respectively, in the two experiments (Figure 7). The drifter LCE00234-2 was particularly good because it moved along the coast with a high transit velocity (up to 1.2 m s^{-1}) induced by strong southerly winds.

In conclusion, all three models showed good performances that, after 24 h, showed a skill score higher than 0.6 for coastal models and lower for the sub-regional model.

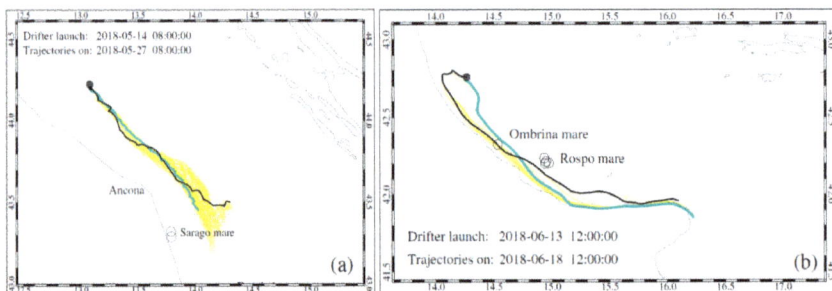

Figure 6. Observed (black thick line) and mean simulated (yellow lines) trajectories for the drifters LCE00234-1 with a drogue at 20 m (**a**), and LCE00234-2 with a drogue at 1 m (**b**), below the sea surface. The green thick line represents the mean trajectory of the particles released at the beginning of simulations. The circles mark the oil platforms.

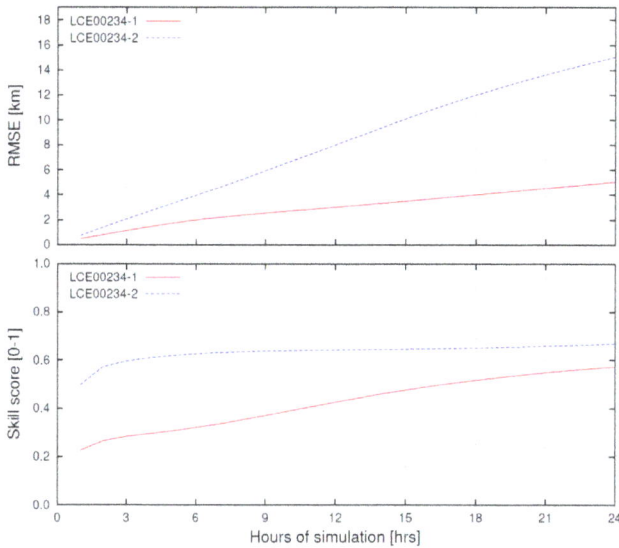

Figure 7. RMSE of separation distance (**top**) and skill score (**bottom**) for the LCE00234-1 drifter with a drogue at 20 m and LCE00234-2 with a drogue at 1 m below the sea surface.

4.2. The Products

4.2.1. The Daily Bulletin

The realized system provides two different operational online services, one automatic and a second in case of emergencies.

The aim of the automatic service is to give a daily forecast of the trajectories of possible oil slicks close to each platform, assuming that the same polluting event repeats daily. It is provided every day at 08:00 UTC through the issue of a web bulletin for each of the seven active oil fields in the Sicily Channel (four fields) and in the Adriatic Sea (three fields) listed in Table 1. Each bulletin contains the scenario of a simulation through 3-hourly maps with the possible distribution of the density in kg m^{-2}. The bulletin is valid for the following 48 h and downloadable from a website with restricted access. The type of oil is the API well number typical of the field, and the time length of the spill is 48 h. Information on transport and weathering processes of the slick are given during the entire simulation. It shows the percentage of oil on the coast along the water column, still on the surface, and evaporated through 3-hourly plots with fields of dispersion at sea and the stranding of the oil released from UTC+1 to UTC+48. The different daily sea conditions produce a mid/long-term numerical dataset to compute hazard and risk statistics.

The service for emergencies is active just in case of real (emergency) or hypothetical (exercise) dispersion in one or more of the seven active oil fields. A user-interface allows a dedicated operator to start any simulation in the areas of the rigs, inputting details about the spill such as coordinates, quantity, and the type of oil released, along with the date and hour of the spill. Also, a near-real-time bulletin is available in case of emergencies.

4.2.2. Hazard Assessment

The hazard estimation is part of the information provided by the system. The hazard assessment is an essential requisite for an attempt and accurate plan for intervention in case of emergencies. It provides probability estimations of the areas that are potentially mostly affected by eventual oil slicks due to spills occurring at the extraction platforms. For the calculation and implementation of hazard maps for each oil platform, we used the forecasting system described above.

The methodology adopted to assess hazard, and further to use hazard as basic informative layer to assess the coastal risk, will be described in a separate work currently in preparation. This system is based on long-term (1 year) statistics performed on forecast outputs. Such hazard maps depict the statistics of the beaching of a hydrocarbon slick after 48 h. The maps represent the statistical distribution of the release for the whole area that is potentially contaminated by hydrocarbons, both at sea and on the land if affected by simulated spills.

Two different geo-statistical descriptive quantities of this probability have been considered:

(a) hazard index or HI, calculated like oil concentration for a defined surface unit normalized at its maximum recorded concentration (at 99th percentile of the distribution of the calculated concentration frequency). It is dimensionless and defined between 0 and 1;

(b) occurrence probability index or PI, indicating the percentage of probability that a given cell is contaminated by an oil spill. Its highest values are recorded on the point of release. It is defined to be between 0 and 100.

The risk assessment can be performed by combining such indices with indicators of vulnerability of the coastal areas like i.e., protection level, shore types, submerged vegetation, beach use, granulometry, etc.

Hazard estimation, for both indexes, was done climatologically on 1-year long outputs and is shown in Figure 8. It will be also done on a seasonal basis with three seasons on the basis of the local circulation and hydrographical characteristics and defined for January–April (mixed conditions), May–August (stratified conditions), and September–December (mixing conditions).

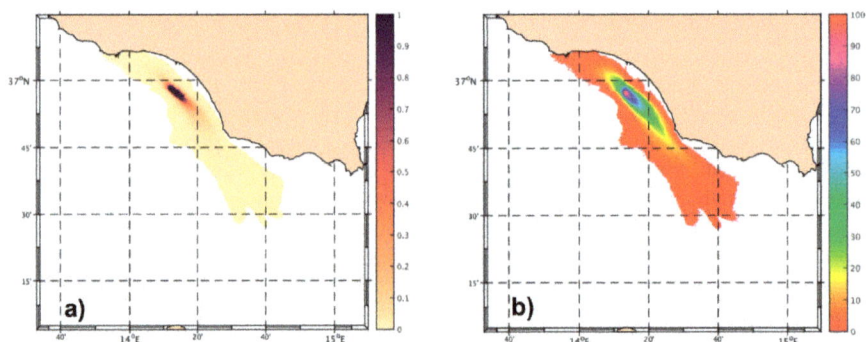

Figure 8. The climatological annual calculation of (**a**) the hazard index (dimensionless) and (**b**) the occurrence probability index (percentage) related to the Perla platform.

5. Summary and Conclusions

In 2017, an oil spill forecasting system for the Italian seas was realized in case of oil spill emergencies on seven active oil extraction areas.

Through a numerical chain composed by nested ocean models, weather models, and Lagrangian models, the system was able to obtain a forecast of the fate of an oil spill for the following 48 h. The integrated system has undergone a positive validation process that has involved the use of drifters' paths collected in the Sicily Channel and in the middle/low Adriatic Sea. It can be used both in operational automatic mode or in case of emergency.

In operational automatic mode, a daily bulletin is provided simulating a virtual oil spill from each of the seven oil platforms using forecasted sea conditions, and then provided a 48 h-scenario of a common quantity of oil at the sea surface. Furthermore, we innovatively adopted an advanced approach combining deterministic calculus with long-term and continuously updating statistics on the trajectories in a database and obtained two hazard indexes, called the hazard and the occurrence probability indexes. They were periodically revised for the areas surrounding platforms on the basis of the continuous updates of the database shaped by the forecast system outputs. These hazard indexes were combined with sensitivity layers to assess risk of the coast and intertidal zone.

In case of emergency, the system produces 48-h forecasts in near-real-time regarding the fate of an oil slick from each of the seven platforms. Currently, a web interface is available to allow an operator to initialize an oil spill scenario.

Author Contributions: All authors contributed to the building of this paper in its different aspects. Conceptualization: A.R. and R.S.; Methodology: A.R., L.F., C.F., G.O., C.T., and R.S.; Software: A.P.; Validation: A.C., C.F., R.S., and C.T.; Formal Analysis: C.F., A.O., G.Q., S.S., and C.T.; Data Acquisition and analysis: A.R., F.A., and A.S.; Writing—Original Draft Preparation: A.R., A.C., M.F.F., L.F., C.F., A.O., G.Q., S.S., C.T., G.U., and R.S.; Writing—Review and Editing: A.R., A.O., and S.R.

Funding: The work is supported by the project *SOS Piattaforme e Impatti Offshore (Servizio Di Previsione Numerica Della Dispersione Di Idrocarburi Dalle Piattaforme Petrolifere Del Canale Di Sicilia E Medio/Basso Adriatico)*, funded by the Italian Ministry of the Environment and Protection of Land and Sea with Executive Agreement prot. m_amte.PNM.REGISTRO UFFICIALE.U.000939.17-01-2017 of 17.01.2017.

Acknowledgments: We thank the colleagues Monica Pinna and Filippo Angotzi for their essential administrative support in Oristano for the management of the project.

Conflicts of Interest: The authors declare no conflict of interest.

References

1. Cucco, A.; Ribotti, A.; Olita, A.; Fazioli, L.; Sorgente, B.; Sinerchia, M.; Satta, A.; Perilli, A.; Borghini, M.; Schroeder, K.; et al. Support to oil spill emergencies in the Bonifacio Strait, western Mediterranean. *Ocean Sci.* **2012**, *8*, 443–454. [CrossRef]

2. Cucco, A.; Sinerchia, M.; Ribotti, A.; Olita, A.; Fazioli, L.; Sorgente, B.; Perilli, A.; Borghini, M.; Schroeder, K.; Sorgente, R. A high resolution real time forecasting system for predicting the fate of oil spills in the Strait of Bonifacio (western Mediterranean). *Mar. Pollut. Bull.* **2012**, *64*, 1186–1200. [CrossRef]

3. Olita, A.; Cucco, A.; Simeone, S.; Ribotti, A.; Fazioli, L.; Sorgente, B.; Sorgente, R. Oil spill hazard and risk assessment for the shorelines of a Mediterranean coastal archipelago. *Ocean Coast. Manag.* **2012**, *57*, 44–52. [CrossRef]

4. Jones, C.E.; Dagestad, K.-F.; Breivik, Ø.; Holt, B.; Röhrs, J.; Christensen, K.H.; Espeseth, M.; Brekke, C.; Skrunes, S. Measurement and modeling of oil slick transport. *J. Geophys. Res. Ocean* **2016**, *121*, 7759–7775. [CrossRef]

5. Alves Tiago, M.; Kokinou, E.; Zodiatis, G.; Lardner, R.; Panagiotakis, C.; Radhakrishnan, H. Modelling of oil spills in confined maritime basins: The case for early response in the Eastern Mediterranean Sea. *Environ. Pollut.* **2015**, *206*, 390–399. [CrossRef]

6. Azevedo, A.; Fortunato, A.B.; Epifânio, B.; den Boer, S.; Oliveira, E.R.; Alves, F.L.; de Jesus, G.; Gomes, J.L.; Oliveira, A. An oil risk management system based on high-resolution hazard and vulnerability calculations. *Ocean Coast Manag.* **2017**, *136*, 1–18. [CrossRef]

7. De Dominicis, M.; Bruciaferri, D.; Gerin, R.; Pinardi, N.; Poulain, P.M.; Garreau, P.; Zodiatis, G.; Perivoliotis, L.; Fazioli, L.; Sorgente, R.; et al. A multi-model assessment of the impact of currents, waves and wind in modelling surface drifters and oil spill. *Top. Stud. Oceanogr.* **2016**, *133*, 21–38. [CrossRef]

8. Zodiatis, G.; De Dominicis, M.; Perivoliotis, L.; Radhakrishnan, H.; Georgoudis, E.; Sotillo, M.; Lardner, R.W.; Krokos, G.; Bruciaferri, D.; Clementi, E.; et al. The Mediterranean Decision Support System for Marine Safety dedicated to oil slicks predictions. *Deep Sea Res.* **2016**, *133*, 4–20. [CrossRef]

9. Coppini, G.; De Dominicis, M.; Zodiatis, G.; Lardner, R.; Pinardi, N.; Santoleri, R.; Colella, S.; Bignami, F.; Hayes, D.R.; Soloviev, D.; et al. Hindcast of Oil Spill Pollution during the Lebanon Crisis, July–August 2006. *Mar. Pollut. Bull.* **2011**, *62*, 140–153. [CrossRef]

10. Sorgente, R.; Tedesco, C.; Pessini, F.; De Dominicis, M.; Gerin, R.; Olita, A.; Fazioli, L.; Di Maio, A.; Ribotti, A. Forecast of drifter trajectories using a Rapid Environmental Assessment based on CTD observations. *Top. Stud. Oceanogr.* **2016**, *133*, 39–53. [CrossRef]

11. De Dominicis, M.; Leuzzi, G.; Monti, P.; Pinardi, N.; Poulain, P.-M. Eddy diffusivity derived from drifter data for dispersion model applications. *Ocean Dyn.* **2012**, *62*, 1381–1398. [CrossRef]

12. De Dominicis, M.; Pinardi, N.; Zodiatis, G.; Lardner, R. MEDSLIK-II, a Lagrangian marine surface oil spill model for short-term forecasting. Part 1: Theory. *Geosci. Model Dev.* **2013**, *6*, 1851–1869. [CrossRef]

13. De Dominicis, M.; Pinardi, N.; Zodiatis, G.; Archetti, R. MEDSLIK-II, a Lagrangian marine surface oil spill model for short-term forecasting. Part 2: Numerical Simulations and Validations. *Geosci. Model Dev.* **2013**, *6*, 1871–1888. [CrossRef]

14. Liubartseva, S.; De Dominicis, M.; Oddo, P.; Coppini, G.; Pinardi, N.; Greggio, N. Oil spill hazard from dispersal of oil along shipping lanes in the Southern Adriatic and Northern Ionian Seas. *Mar. Pollut. Bull.* **2015**, *90*, 259–272. [CrossRef]

15. Ribotti, A.; Cucco, A.; Olita, A.; Sinerchia, M.; Fazioli, L.; Satta, A.; Borghini, M.; Schroeder, K.; Perilli, A.; Sorgente, B.; et al. An integrated operational system for the Coast Guard management of oil spill emergencies in the Strait of Bonifacio. In Proceedings of the Sixth International Conference on EuroGOOS, 4–6 October 2011.

16. Zodiatis, G.; Kirkos, G. Projects on Oil Spill Response in the Mediterranean Sea. In Oil Pollution in the Mediterranean Sea: Part I—The International Context. *Handb. Environ. Chem.* **2017**, *30*. [CrossRef]

17. United Nations Office for Disaster Risk Reduction (UNISDR). The Sendai Framework for Disaster Risk Reduction 2015–2030 Paper. Available online: http://www.unisdr.org/we/inform/publications/43291 (accessed on 30 November 2018).

18. Italian Ministry of Economic Development. Offshore Exploitation Platforms List in the Italian Seas. Available online: http://unmig.sviluppoeconomico.gov.it/unmig/strutturemarine/elenco.asp (accessed on 30 November 2018).

19. Coppini, G.; Marra, P.; Lecci, R.; Pinardi, N.; Creti, S.; Scalas, M.; Tedesco, L.; D'Anca, A.; Fazioli, L.; Olita, A.; et al. SeaConditions: A web and mobile service for safer professional and recreational activities in the Mediterranean Sea. *Nat. Hazards Earth Syst. Sci.* **2017**, *17*, 533–547. [CrossRef]

20. De Dominicis, M.; Falchetti, S.; Trotta, F.; Pinardi, N.; Giacomelli, L.; Napolitano, E.; Fazioli, L.; Sorgente, R.; Haley, P.F.J., Jr.; Lermusiaux, P.; et al. A relocatable ocean model in support of environmental emergencies—The Costa Concordia emergency case. *Ocean Dyn.* **2014**, *64*, 667–688. [CrossRef]

21. Oddo, P.; Pinardi, N. Lateral Open Boundary Conditions for Nested Limited Area Models: Process selective approach. *Ocean Model.* **2008**, *20*, 134–156. [CrossRef]

22. Sorgente, S.; Drago, A.F.; Ribotti, A. Seasonal variability in the Central Mediterranean Sea Circulation. *Ann. Geophys.* **2003**, *20*, 299–322. [CrossRef]

23. Napolitano, E.; Iacono, R.; Sorgente, R.; Fazioli, L.; Olita, A.; Cucco, A.; Oddo, P.; Guarnieri, A. The regional forecasting systems of the Italian seas. *J. Oper. Oceanogr.* **2016**, *9*, s66–s76. [CrossRef]

24. Gabersek, S.; Sorgente, R.; Natale, S.; Ribotti, A.; Olita, A.; Astraldi, M.; Borghini, M. The Sicily Channel Regional Model forecasting system: Initial boundary conditions sensitivity and case study evaluation. *Ocean Sci.* **2007**, *3*, 31–41. [CrossRef]

25. Fazioli, L.; Olita, A.; Cucco, A.; Tedesco, C.; Ribotti, A.; Sorgente, R. Impact of different initialisation methods on the quality of sea forecasts for the Sicily Channel. *J. Oper. Oceanogr.* **2016**, *9* (Suppl. 1), s119–s130. [CrossRef]

26. Drago, A.F.; Sorgente, S.; Ribotti, A. A high resolution hydrodynamical 3D model of the Malta shelf area. *Ann. Geophys.* **2003**, *20*, 323–344. [CrossRef]

27. The MEDSLIK-II Oil Spill Model Community. Available online: http://medslikii.bo.ingv.it (accessed on 30 November 2018).

28. The International Tanker Owners Pollution Federation Limited (ITOPF). Available online: http://www.itopf.com (accessed on 30 November 2018).

29. Sorgente, R.; Olita, A.; Oddo, P.; Fazioli, L.; Ribotti, A. Numerical simulation and decomposition of kinetic energies in the central Mediterranean: Insight on mesoscale circulation and energy conversion. In Special Issue, ECOOP (European Coastal-shelf sea Operational Observing and forecasting system Project). *Ocean Sci.* **2011**, *7*, 503–519. [CrossRef]

30. Blumberg, A.F.; Mellor, G.L. A description of a three-dimensional coastal ocean circulation model. In *Three-Dimensional Coastal Ocean Models*, 2nd ed.; Heaps, N., Ed.; American Geophysical Union: Washington, DC, USA, 1987; Volume 4, pp. 1–39.

31. The Copernicus Marine Environment Monitoring Service. Available online: http://marine.copernicus.eu (accessed on 30 November 2018).

32. Burchard, H.; Petersen, O. Models of turbulence in the marine environment—A comparative study of two equation turbulence models. *J. Mar. Syst.* **1999**, *21*, 29–53. [CrossRef]

33. Umgiesser, G.; Ferrarin, C.; Cucco, A.; De Pascalis, F.; Bellafiore, D.; Ghezzo, M.; Bajo, M. Comparative hydrodynamics of 10 Mediterranean lagoons by means of numerical modeling. *J. Geophys. Res. Ocean* **2014**, *119*, 2212–2226. [CrossRef]

34. Kallos, G. The regional weather forecasting system SKIRON: An overview. In Proceedings of the Symposium on Regional Weather Prediction on Parallel Computer Environments, Athens, Greece, 15–17 October 1997; pp. 109–122.

35. Fairall, C.W.; Bradley, E.F.; Rogers, D.P.; Edson, J.B.; Young, G.S. Bulk parameterization of air-sea fluxes for Tropical Ocean-Global Atmosphere Coupled-Ocean Atmosphere Response Experiment. *J. Geophys. Res.* **1996**, *101*, 3747–3764. [CrossRef]

36. Carrere, L.; Lyard, F.; Cancet, M.; Guillot, A.; Roblou, L. FES2012: A new global tidal model taking advantage of nearly 20 years of altimetry. In Proceedings of the 20 Years of Progress in Radar Altimetry Symposium, Venice, Italy, 24–29 September 2013.

37. The Satellite Altimetry Database of the AVISO+ Mission. Available online: http://www.aviso.altimetry.fr (accessed on 30 November 2018).

38. Ludwig, W.; Dumont, E.; Maybeck, M.; Heussner, S. River discharges of water and nutrients to the Mediterranean and Black Sea: Major drivers for ecosystem changes during past and future decades? *Prog. Oceanogr.* **2009**, *80*, 199–217. [CrossRef]

39. Ferrarin, C.; Davolio, S.; Bellafiore, D.; Ghezzo, M.; Maicu, F.; Mc Kiver, W.; Drofa, O.; Umgiesser, G.; Bajo, M.; De Pascalis, F.; et al. Cross-scale operational oceanography in the Adriatic Sea. *J. Oper. Oceanogr.* **2018**. submitted for publication.

40. Orlandi, E.; Fierli, F.; Davolio, S.; Buzzi, A.; Drofa, O. A nudging scheme to assimilate satellite brightness temperature in a meteorological model: Impact on representation of African mesoscale convective systems. *Q. J. R. Meteorol. Soc.* **2010**, *136*, 462–474. [CrossRef]

41. Italian Research Council—Institute of Atmospheric Sciences and Climate (CNR-ISAC). Daily Numerical Weather Forecasts. Available online: http://www.isac.cnr.it/dinamica/projects/forecasts (accessed on 30 November 2018).

42. NOAA/National Weather Service, National Centers for Environmental Prediction, Environmental Modeling Center. Available online: http://www.emc.ncep.noaa.gov/GFS (accessed on 30 November 2018).

43. Liu, Y.; Weisberg, R. Evaluation of trajectory modeling in different dynamic region using normalized cumulative Lagrangian separation. *J. Geophys. Res. Ocean* **2011**, *116*, C09013. [CrossRef]

44. Lermusiaux, P.F.J.; Robinson, A.R. Features of dominant mesoscale variability, circulation patterns and dynamics in the Strait of Sicily. *Deep Sea Res.* **2001**, *48*, 1953–1997. [CrossRef]

45. Liu, Y.; Weisberg, R.H.; Hu, C.; Zheng, L. Tracking the Deepwater Horizon oil spill: A modeling perspective. *Eos Trans.* **2011**, *92*, 45–46. [CrossRef]

46. Liu, Y.; Weisberg, R.H.; Hu, C.; Zheng, L. Trajectory forecast as a rapid response to the Deepwater Horizon oil spill, in Monitoring and Modeling the Deepwater Horizon Oil Spill: A Record-Breaking Enterprise. *Geophys. Monogr. Ser.* **2011**, *195*, 153–165. [CrossRef]

47. Weisberg, R.H.; Zheng, L.; Liu, Y. On the movement of Deepwater Horizon Oil to northern Gulf beaches. *Ocean Modell.* **2017**, *111*, 81–97. [CrossRef]

Journal of
Marine Science and Engineering

MDPI

Article

Bayesian Statistics of Wide-Band Radar Reflections for Oil Spill Detection on Rough Ocean Surface

Bilal Hammoud [1,2,*], Fabien Ndagijimana [2], Ghaleb Faour [3], Hussam Ayad [1] and Jalal Jomaah [1]

[1]   Doctoral School of Sciences and Technologies, Lebanese University (LU), 1003 Beirut, Lebanon;
      hayad@ul.edu.lb (H.A.); jomaah@enserg.fr (J.J.)
[2]   Grenoble Electrical Engineering Laboratory, Grenoble Alpes University (UGA), 38031 Grenoble, France;
      fabien.ndagijimana@univ-grenoble-alpes.fr
[3]   National Council of Scientific Research (CNRS-L), Remote Sensing Research Center,
      22411 Mansouriyeh, Lebanon; gfaour@cnrs.edu.lb
*   Correspondence: bilal.hmd.14@gmail.com

Received: 26 October 2018; Accepted: 18 December 2018; Published: 10 January 2019

check for
updates

Abstract: In this paper, we present a probabilistic approach which uses nadir-looking wide-band radar to detect oil spills on rough ocean surface. The proposed approach combines a single-layer scattering model with Bayesian statistics to evaluate the probability of detection of oil slicks, within a plausible range of thicknesses, on seawater. The difference between several derived detection algorithms is defined in terms of the number of frequencies used (within C-to-X-band ranges), as well as of the number of radar observations. Performance analysis of all three types of detectors (single-, dual- and tri-frequency) is done under different surface-roughness scenarios. Results show that the probability of detecting an oil slick with a given thickness is sensitive to the radar frequency. Multi-frequency detectors prove their ability to overcome the performance of the single- and dual-frequency detectors. Higher probability of detection is obtained when using multiple observations. The roughness of the ocean surface leads to a loss in the reflectivity values, and therefore decreases the performance of the detectors. A possible way to make use of the drone systems in the contingency planning is also presented.

Keywords: oil spill; remote sensing; reflection coefficient; electromagnetic roughness; multi-frequency detector; multiple observations; probability density function; probability of detection; contingency planning

1. Introduction

Enormous applications and industries use petroleum products worldwide, and thus require the presence of petroleum materials on site. This need stresses the necessity of moving petroleum substances using maritime ships or underwater pipelines internationally between different continents and countries. In addition to the intentional petroleum waste spill in sea water, transportation is vulnerable to involuntary oil spills from tanker collisions with rocky shoals, ship accidents and pipeline ruptures [1]. The European Space Agency (ESA) stated that more than 4.5 million tons of oil is the estimate of the annual spill worldwide, where 45% of the amount is due to operative discharges from ships [2]. Oil spills in sea water are one of the major incidents which adversely cause long-term repercussions for the maritime environment. They are happening on a global scale, and their influence on the ecosystem is extremely severe. Oil spills, including gasoline, fuel, crude, and bulk oil, will affect the ecosystem starting from maritime life and ending in the human life and environmental disasters. Marine oil spills can be highly dangerous since wind, waves and currents can scatter a large oil spill over a wide area within a few hours in the open sea. Environmental rules, regulations and strict operating procedures have been imposed to prevent oil spills, but these measures cannot completely eliminate the risk [3]. Therefore, oil spill detecting and monitoring systems are extremely important in order to react quickly and to limit contamination.

Oil spill detection and monitoring is done with the aid of several techniques and sensors. State-of-the-art sensor technology for oil spill surveillance is listed and described in [4–6]. Infrared sensors are relatively cheap remote-sensing technologies which can be used to detect oil spills. However, thermal radiation from sea weeds appear similar to the radiation arising from the oil which may lead to a false positive result. In addition, infrared sensors require the absence of cloud and heavy fog for good operation [4]. Ultraviolet (UV) sensors cannot detect oil thickness greater than 10 microns. Less UV use is being made for oil spills in today's remote sensing because of the low relevance of thin slicks to oil spill cleanup [6–8]. Microwave radiometer (MWR) is an additional passive sensor that is used for oil spill detection and oil thickness measurements. However, MWR sensors are costly and it is complicated to put them into operation. Microwave radiometer sensors require information about many environmental characteristics and oil properties to accurately detect the oil [4]. There are interferences, and signal differentials may be poor. Currently, microwave is not being used for slick imaging [6].

Radar is a very useful active sensor to detect oil over a large area. Thus, it can be used as a first assessment tool to detect the possible location of an oil spill. SAR (Synthetic Aperture Radar) and SLAR (Side-Looking Airborne Radar) are the two most common types of radar which can be used for oil spill remote sensing [4,9]. Synthetic Aperture Radar is the most widely used sensor on spaceborne platforms for oil spill detection [4]. Imaging SAR systems are off-nadir instruments whose backscattering over the ocean is primarily due to Bragg scattering at relevant incident angles. In [10–13], comparison between different spaceborne and airborne SAR algorithms is done. Synthetic Aperture Radar technique is highly prone to false targets, however, and is limited to a narrow range of wind speeds (approximately 2 to 6 m/s). At winds below this, there are not enough small waves to yield a difference between the oiled area and the sea [3]. At low wind speed it is not possible to distinguish between thick and thin oil slicks. The ocean's slight surface roughness due to very low wind speed (<3 m/s) leads the backscattering to be dominated by the specular component, challenging SAR systems for oil spill detection [12]. Therefore, it would be advantageous to study the radar observations from nadir-looking systems since they cover scenarios that cannot be studied by SAR systems. Being largely independent of surface roughness, the returns from nadir (or near-nadir) systems will benefit from the dominance of the specular scattering and enable detection in very low wind conditions.

Most recent ones are those done remotely using airborne [14,15] and satellite systems [16–21]. Satellites face the limitations of overpass frequency and low spatial resolution [5], whereas airborne systems, despite their high cost due to aircraft dedication, can be used directly when needed for real-time dataset processing [4]. The European Maritime Safety Agency (EMSA) launched in 2016 the need to complement the satellite maritime surveillance systems—that can detect only 25% of pollution accidents—by drones [22]. According to [23], aerial surveillance could be improved significantly by the admission of drones.

According to [7], sensors should provide the following information for oil spill contingency planning:

- *info 1:* the location and spread of an oil spill over a large area,
- *info 2:* the thickness distribution of an oil spill to estimate the quantity of spilled oil,
- *info 3:* a classification of the oil type in order to estimate environmental damage and to take appropriate response activities,
- *info 4:* and timely and valuable information to assist in cleanup operations.

From a system-level perspective, we study the incorporation of both C-band and X-band using remote sensing nadir-looking wide-band radar sensors that can be implemented on drones as oil spill detection systems. The drone-based radar will allow quick assessment of the area where the flag of possible spills is raised by witnesses. Winds and the ocean's currents may spread the spilled oil to a large area within a short period of time [4], hence the proposed systems will be able to provide the most critical information needed for an effective contingency planning by specifying the location of the oil (*info 1*) and its spread over the ocean. Nevertheless, working as nadir-looking systems will definitely decrease the surface of the scanned area viewed by the radar compared to that scanned by "side-looking" radars. However, parallelization in scanning can be used to cover large area in a critical time. This means that instead of using a single drone with several scans to cover a large area, multiple drones can be used at the same time. Using the radar on drone platforms will be a complementary solution to the satellite systems. Once we have a flag of possible oil spill (due to collision of tankers, or to the rupture of pipelines), the drones can be used to start the scans as tactical-response systems. Afterwards, the satellites can be used for strategic planning by providing the synoptic view of the scene. This will assist later in cleanup operations by providing valuable data as needed (*info 4*). Furthermore, scanning with drones provides high spatial resolution compared to satellite and with a principal advantage of relative low cost compared to dedicated airborne detection systems. With respect to (*info 2*), we proposed in [24] using the same drones systems implementing different algorithms to estimate the thickness of the oil. This information is very important to the effectiveness of oil spill contingency planning because it allows the estimation of the total volume spilled.

We developed in [25,26] dual- and multi-frequency algorithms using only single observations for oil spill detection. The model adopted was a planar multi-layer structure. The ocean surface was assumed to be totally smooth corresponding to scenarios where the wind speed is very low. Performance analysis of the proposed algorithms was presented in [27]. In this journal, we extend the previously proposed approach by applying Bayesian statistics on the reflectivity values evaluated at multiple frequencies and collected with multiple observations. A full derivation of the different algorithms is presented. In addition, the multi-layer structure is no more assumed to be planar. The statistical attributes of the ocean surface are presented. The performance of the detectors is tested under different roughness scenarios. Since the oil properties (*info 3*) are not known during the first stages of an oil spill, we study the effect of this missed information on the overall performance of the proposed algorithms.

2. Methods

2.1. Reflection Coefficient Calculation for the Multi-Layer Planar Structure

From a physical point of view, the problem is considered to be a multi-layer structure where we study the reflection of the electromagnetic waves from the sea layer covered by an oil layer. The electrical properties and the physical characteristics are defined for the layers at the boundaries where interaction with electromagnetic waves occurs.

In our model, we assume that there is an oil slick, with d thickness (in mm), on the top of the sea water surface. An oil spill on the sea surface will dampen the waves and hence reduce the surface roughness. Furthermore, at open ocean space, with low wind speeds (2–6 m/s) which are considered to be optimal for oil spill detection [4], the correlation length of the ocean waves is large and the root-mean-square (rms) height of the capillary waves is very small. Hence, we assume that the multi-layer structure is planar. We consider that this assumption is indeed realistic when the ocean's surface is very smooth due to the very low wind conditions. When the roughness increases, it decreases the reflection measurements made or increases the noise level of the environment.

The relative dielectric constants of the air, oil, and sea water are respectively ε_1, ε_2 and ε_3. The different media are assumed to be non-magnetic. The refractive indices n for the different materials are $n_i = \sqrt{\varepsilon_i}$. The electromagnetic waves are assumed to be normally incident on the ocean surface. We assume that the sea water is deep enough so that we can neglect the radar reflections from the sea-floor. The field reflection coefficients for the first interface (between air and oil) and the second interface (between oil and water) are respectively:

$$\rho_{12} = \frac{n_1 - n_2}{n_1 + n_2}, \tag{1}$$

$$\rho_{23} = \frac{n_2 - n_3}{n_2 + n_3}. \tag{2}$$

Across the boundaries, E is conserved. Using continuity property at these interfaces, the reflectivity (power reflection coefficient) for the three-layer structure is derived in [25] as:

$$R = |\rho|^2 = \frac{\rho_{12}^2 + \rho_{23}^2 + 2\rho_{12}\rho_{23}\cos(2\delta)}{1 + \rho_{12}^2 \rho_{23}^2 + 2\rho_{12}\rho_{23}\cos(2\delta)}. \tag{3}$$

The phase shift δ is dependent on the oil-refractive index n_2, the wavelength of the electromagnetic wave λ_0, and the thickness of the oil slick t. It is given by:

$$\delta = \frac{2\pi f}{c} n_2 t = \frac{2\pi}{\lambda_0} n_2 t, \tag{4}$$

where c is the speed of light.

The reflectivity R is a trigonometric function with period T_R that is dependent on the oil-refractive index and the frequency of the electromagnetic wave. The period is expressed as:

$$T_R = \frac{2\pi}{\frac{2\delta}{t}} = \frac{\pi}{\frac{2\pi}{\lambda_0} n_2} = \frac{\lambda_0}{2\sqrt{\varepsilon_2}}. \tag{5}$$

2.2. Smooth or Rough Surface

Several statistical attributes can be calculated for a random surface [28]. The collection of surface height measurements may be described using standard statistical parameters. The surface height measurements are denoted by $z_{(i,j)}$ in two-dimensional array $N \times N$ collected at a horizontal step-size Δx over an area whose length and width are given by L [29]:

$$z_{(x,y)}(\Delta x, L) \tag{6}$$

with $(x,y) \in [1, ..., N][1, ..., N]$.

Collecting big number of measurements, the height probability density function $p(z)$ can be approximated to one of many well-known distributions such as Gaussian, Exponential, or Rayleigh. For most random surfaces, the probability density function (pdf) looks approximately Gaussian in shape, that is [28,29]

$$p(z) = \frac{1}{\sqrt{2\pi s^2}} e^{\frac{-z^2}{2s^2}} \tag{7}$$

where s^2 is the variance of surface heights. With zero-mean distribution, the variance is equal to the standard deviation.

Given $p(z)$, we can calculate several statistical attributes of the random surface, including [28]:

- the height standard deviation s given by

$$s = \sqrt{\int_{-\inf}^{+\inf} z^2 p(z) dz} \tag{8}$$

which is also called the rms-height , and

- the surface correlation function defined by

$$l(\zeta) = \frac{\langle z_{(x,y)} z_{(x',y')} \rangle}{s^2} \tag{9}$$

where (x,y) and (x',y') are two locations on the surface, and ζ is the lateral separation between them.

The correlation function $l(\zeta)$ is a measure of the degree of correlation between the surface at different locations. The value of the correlation function decreases with ζ. Hence, if the spacing between two locations is greater than a certain distance called the correlation length, we can assume that the heights are considered to be statistically uncorrelated [28].

Electromagnetically, the roughness of a surface is measured relatively to the electromagnetic (EM) wavelength λ. According to [28], for a surface with rms height s, its electromagnetic roughness ks is:

$$ks = \frac{2\pi}{\lambda} s \tag{10}$$

For a perfectly smooth surface (flat surface) with rms height $s = 0$, an incident EM wave is reflected along the specular direction, and the reflected power is related to the incident power by the reflectivity formula given in Equation (3).

The component of the scattering pattern of the perfectly smooth surface consists of only a coherent component. If the surface is rough with ks, the scattering pattern will also include a non-coherent component along all other directions. In that case, the reflectivity along the specular direction will be noted as the coherent reflectivity R_{coh}, expressed as

$$R_{coh} = R\, e^{-4\psi^2} \tag{11}$$

where

$$\psi = ks \cos(\theta_i) = \frac{2\pi}{\lambda} s \cos(\theta_i) \tag{12}$$

with θ_i being the incident angle of the EM wave to the interfaces.

With respect to the oil spill problem, according to [11,13], an oil spill on the sea surface dampens the waves and hence reduces the roughness of the surface. Furthermore, at open ocean space, with very low wind speed (<2–3 m/s) which are considered to be optimal for oil spill detection [4], the correlation length of the ocean waves is large and the rms height s of the capillary waves is very small. Hence, the electromagnetic roughness ks is negligible and all interfaces are assumed to be planar.

When the variation of the ocean waves increases due to higher wind speed, the electromagnetic roughness increases respectively with the rms height s. Therefore, the surface is not considered to be totally planar anymore, and its variation will affect its level of roughness. Depending on the electromagnetic roughness factor, the surface can be described to be planar ($ks = 0$), relatively smooth ($ks = 0.2$) or extremely rough ($ks = 2$). The effect of the electromagnetic roughness on the reflectivity value is displayed in Figure 1 [28].

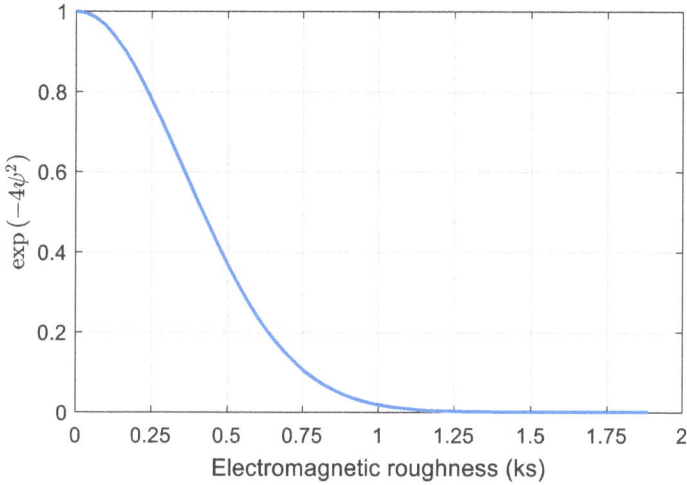

Figure 1. Effect of the electromagnetic roughness on the reflectivity value.

2.3. Detection Algorithms

From the mathematical perspective, the detector algorithms use the statistical characterization of the reflectivity values and its distribution under different oil thicknesses in order to obtain a final decision whether oil exists or not. These reflectivity values are assumed to be independent events. Any previous knowledge about the existence or absence of oil in the surface scanned should be taken into consideration to weight the probability of the decision in the detector block. Nevertheless, without any previous knowledge about the spill situation, the detector decision will be totally based on the statistics of the calculated power reflection ratio.

Let "o, w" be the events indicating the presence of the oil slick and the water, respectively. Let R be the event representing the reflectivity value(s) measured. R could represent one or more reflectivity values repeated at the same frequency or at different frequencies. In all cases, these reflectivity values are assumed to be uncorrelated in time domain (at multiple observations) and in frequency domain (at multiple frequency measurements). The difference between these cases will be studied in the following subsections. With "$\cdot$" being the numerical multiplication, the probability of oil presence and absence given R are respectively

$$Pr(o|R) = \frac{Pr(o \cap R)}{Pr(R)} = \frac{Pr(R|o) \cdot Pr(o)}{Pr(R)} \tag{13}$$

$$Pr(w|R) = \frac{Pr(w \cap R)}{Pr(R)} = \frac{Pr(R|w) \cdot Pr(w)}{Pr(R)} \tag{14}$$

Without any previous knowledge about the spill situation, i.e., with $Pr(o) = Pr(w) = 50\%$, the ratio of probabilities of oil presence to water presence is given by

$$\frac{Pr(o|R)}{Pr(w|R)} = \frac{Pr(R|o) \cdot Pr(o)}{Pr(R|w) \cdot Pr(w)} = \frac{Pr(R|o)}{Pr(R|w)}. \tag{15}$$

The probability of obtaining a measured reflectivity value given that the oil exists is evaluated using the corresponding pdf. Similarly, the probability of obtaining the same reflectivity value given that the water exists is evaluated. If the ratio in (15) gives a result greater than unity, the decision will indicate the oil existence.

2.3.1. Single Observation at Multiple Frequencies

If R in (15) represents single observations of reflectivity measured at different frequencies (up to I total frequencies), then we can express it as

$$R = R_{f_1}, R_{f_2}, ...R_{f_I} \tag{16}$$

Replacing (16) in (15), we obtain

$$
\begin{aligned}
\frac{Pr(o|R)}{Pr(w|R)} &= \frac{Pr(R_{f_1}, R_{f_2}, ...R_{f_I}|o)}{Pr(R_{f_1}, R_{f_2}, ...R_{f_I}|w)} \\
&= \frac{Pr(R_{f_1}|o) \cdot \, ... \, \cdot Pr(R_{f_I}|o)}{Pr(R_{f_1}|w) \cdot \, ... \, \cdot Pr(R_{f_I}|w)} \\
&= \frac{Pr(R_{f_i}, R_{f_j}, ..|o)}{Pr(R_{f_i}, R_{f_j}, ..|w)} \cdot \frac{Pr(o)}{Pr(w)} \\
&= \frac{\prod_i Pr(R_{f_i}|o)}{\prod_i Pr(R_{f_i}|w)} \\
&= \prod_i \frac{Pr(R_{f_i}|o)}{Pr(R_{f_i}|w)}
\end{aligned} \tag{17}
$$

2.3.2. Multiple Observations at Single Frequency

If R in (15) represents multiple observations (up to M total observations) of reflectivity measured at single frequency, then we can express it as

$$R = R_{f_1}^{(1)}, R_{f_1}^{(2)}, ...R_{f_1}^{(M)} \tag{18}$$

Replacing (18) in (15), we obtain

$$
\begin{aligned}
\frac{Pr(o|R)}{Pr(w|R)} &= \frac{Pr(R_{f_1}^{(1)}, R_{f_1}^{(2)}, ...R_{f_1}^{(M)}|o)}{Pr(R_{f_1}^{(1)}, R_{f_1}^{(2)}, ...R_{f_1}^{(M)}|w)} \\
&= \frac{Pr(R_{f_1}^{(1)}|o) \cdot \, ... \, \cdot Pr(R_{f_1}^{(M)}|o)}{Pr(R_{f_1}^{(1)}|w) \cdot \, ... \, \cdot Pr(R_{f_1}^{(M)}|w)} \\
&= \frac{\prod_m Pr(R_{f_1}^{(m)}|o)}{\prod_m Pr(R_{f_1}^{(m)}|w)} \\
&= \prod_m \frac{Pr(R_{f_1}^{(m)}|o)}{Pr(R_{f_1}^{(m)}|w)}
\end{aligned} \tag{19}
$$

2.3.3. Multiple Observations and Multiple Frequencies

If R in (15) represents multiple observations (up to M total observations) of reflectivity measured at multiple frequencies (up to I total frequencies) , then we can express it as

$$R = R_{f_1}^{(1)}, ..., R_{f_1}^{(M)}, R_{f_2}^{(1)}, ..., R_{f_2}^{(M)}, ..., R_{f_I}^{(1)}, ..., R_{f_I}^{(M)} \tag{20}$$

Replacing (20) in (15), we obtain

$$
\begin{aligned}
\frac{Pr(o|R)}{Pr(w|R)} &= \frac{Pr(R_{f_1}^{(1)},...,R_{f_1}^{(M)},R_{f_2}^{(1)},...,R_{f_l}^{(M)}|o)}{Pr(R_{f_1}^{(1)},...,R_{f_1}^{(M)},R_{f_2}^{(1)},...,R_{f_l}^{(M)}|w)} \\[2mm]
&= \frac{Pr(R_{f_1}^{(1)}|o)\cdot ... \cdot Pr(R_{f_1}^{(M)}|o)}{Pr(R_{f_1}^{(1)}|w)\cdot ... \cdot Pr(R_{f_1}^{(M)}|w)} \\[2mm]
&\quad\times \frac{Pr(R_{f_2}^{(1)}|o)\cdot ... \cdot Pr(R_{f_2}^{(M)}|o)}{Pr(R_{f_2}^{(1)}|w)\cdot ... \cdot Pr(R_{f_2}^{(M)}|w)} \\[2mm]
&\quad\times ... \\[2mm]
&\quad\times \frac{Pr(R_{f_l}^{(1)}|o)\cdot ... \cdot Pr(R_{f_l}^{(M)}|o)}{Pr(R_{f_l}^{(1)}|w)\cdot ... \cdot Pr(R_{f_l}^{(M)}|w)} \\[2mm]
&= \frac{\prod_m Pr(R_{f_1}^{(m)}|o)}{\prod_m Pr(R_{f_1}^{(m)}|w)} \times ... \frac{\prod_m Pr(R_{f_l}^{(m)}|o)}{\prod_m Pr(R_{f_l}^{(m)}|w)} \\[2mm]
&= \prod_i\prod_m \frac{Pr(R_{f_i}^{(m)}|o)}{Pr(R_{f_i}^{(m)}|w)}
\end{aligned}
\tag{21}
$$

3. Results and Discussion

3.1. Simulation Setup

Probability of detection calculations are performed using Monte Carlo Simulations in MATLAB. The dielectric constant of the air is $\varepsilon_1 = 1$. The dielectric constant of the thick oil is assumed to be real $\varepsilon_2 = 3$ (the imaginary part of order $0.01\,j$ can be neglected without affecting the results). Sea water dielectric constant, ε_3, is function of the water temperature t_w, water salinity s_w and the frequency of the electromagnetic signal used. For its calculation, we use the model mentioned in [28] with $t_w = 20°$ C and $s_w = 35$ ppt. The noise variance in the system is considered to be additive white Gaussian (AWG) in linear scale, with variance of $\sigma^2 = 0.02$.

3.2. Reflectivity Behavior with Smooth and Rough Surfaces

Figure 2 shows the reflectivity values (coherent component) calculated from the planar multi-layer structure versus the oil thickness, under different electromagnetic roughness ($ks = 0$, 0.2 and 0.5). The plot of $R_{coh,w}$ is simply a copy of the value obtained at $d = 0$ mm at the given frequency. For totally smooth surfaces, the plot of $R_{coh,o}$ at 4 GHz is almost monotonically decreasing in the range (0–10 mm). It has a very small slope at small thickness values (0–3 mm), but this slope increases with the increase of the oil slick thickness (3–7 mm). At some thicknesses, any error in the power reflectivity measurements at 4 GHz would mislead the oil detection due to the very small variation between $R_{coh,o}$ and $R_{coh,w}$.

The variation of the reflectivity values at 12 GHz is quite high for consecutive values of thicknesses. This variation allows the oil detection for small thicknesses (1–3 mm). As discussed in Equation (5), the reflectivity is a trigonometric function and has a cyclic behavior. This is observed clearly at $f = 12$ GHz; the reflectivity repeats every 7.2 mm. Due to the cyclic behavior, many thickness values give the same reflectivity value leading to false interpretations. Therefore, it is important to use more than one frequency to improve the detection. When the electromagnetic roughness of the surfaces increases, the reflectivity values decrease as presented in Equation (11). Increasing the surface roughness ks from 0.2 to 0.5 leads approximately to 4 dB loss in the reflectivity.

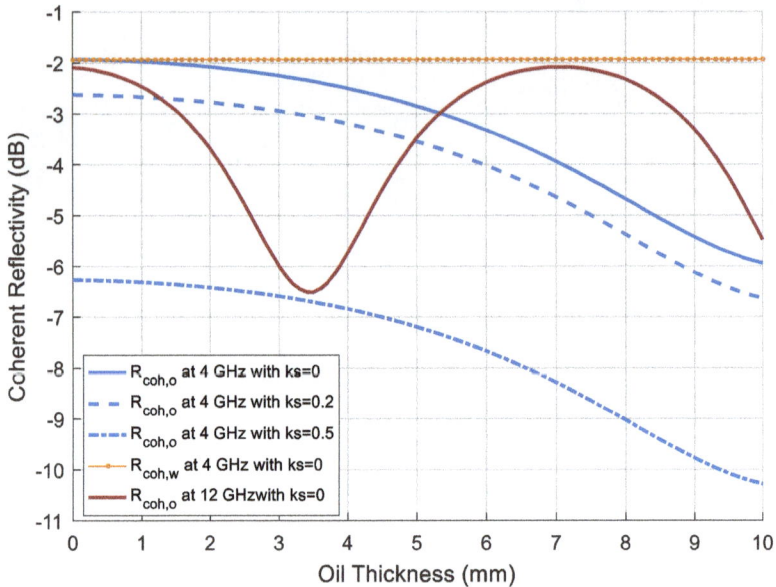

Figure 2. Comparison between the reflectivity values (in dB) versus the oil thickness (in mm) at 4 GHz and 12 GHz for different roughness scenarios. R_o and R_w in each graph represents the oil and water reflectivity values, respectively.

3.3. Effect of the Oil Properties on the Performance of the Single-Frequency Detectors

In real scenarios, when oil spill takes place, it is not always the case that the oil type is well known and defined. This rises the following question from system-monitoring aspect: Does the absence of the exact value of the oil property affects the reflectivity values? To answer this, Figure 3 presents the reflectivity values for different relative dielectric constants of the oil ($\varepsilon_2 = 2.9, 3$ and 3.3) at two frequencies (4 and 12) GHz. At 4 GHz, the difference in the relative dielectric constant values does not modify the reflectivity values at (0–3) mm. However, for higher thickness values, the reflectivities start to change for different dielectric constants. At 4 GHz, the difference between the reflectivities when $\varepsilon_2 = 2.9$ and $\varepsilon_2 = 3.3$ reaches approximately 1 dB at around 7.8 mm. At 12 GHz, the difference is high at around 3.5 mm, decreases to null at 7.8 mm and increases again to 2 dB at 10 mm. Now, we know that the oil type slightly affects the reflectivity value, but more importantly, is it the case for the performance of the detectors?

Figure 4 compares between the probability of detection versus the oil thickness (in mm) for different single-frequency detectors with single scan ($M = 1$) and variant oil properties. Using the detector with single observation at 4 GHz, the probability of oil detection increases with the oil thickness from 51% to 70%. At 8 GHz, the detector records highest detection of 69% at 5 mm but it fails for thickness ranges (1–2 mm) and (8.5–10 mm) recording a value smaller than 55%. At 12 GHz, the detector fails at 1 mm and between (6–8 mm), where the higher detection occurs at 3.5 mm. The results for all these single-frequency detectors (with single observation) validate the reflectivity behavior explained previously in Figure 2.

The effect of the variation in the reflectivity values, due to the variation of the oil properties, in the performance of the single-frequency detectors is also presented in Figure 4. From the obtained results, we notice that although at different thickness values there exists some noticeable difference in the reflectivity values at the same frequency for different dielectric constants (as shown in Figure 3), but these variations do not affect the performance of the detectors more than 2%. Therefore, even when monitoring an oil spill, the proposed algorithms can be used with an approximate value of the oil permittivity without affecting the performance of the technique on the detection. This analysis is highly useful for the monitoring system because it defines the vulnerability of the drone systems against the absence of some information about the oil properties that are not be present during tactical response.

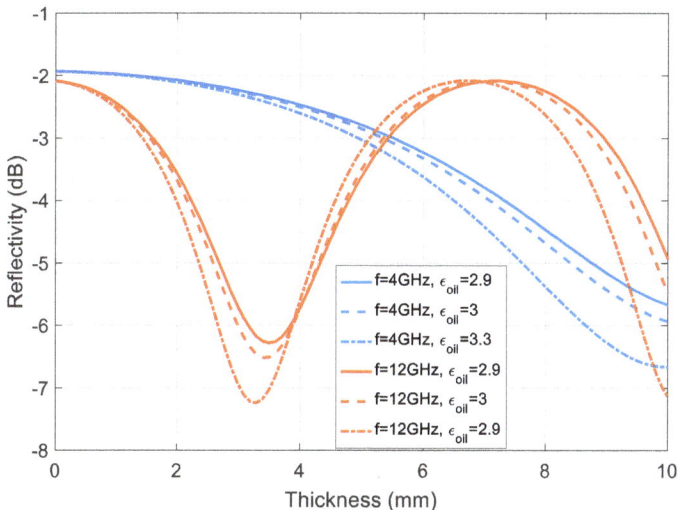

Figure 3. Reflectivity R (in dB) versus oil thickness (in mm) at different frequencies (4 GHz, and 12 GHz) and different oil dielectric constants. The electromagnetic roughness is $ks = 0.5$.

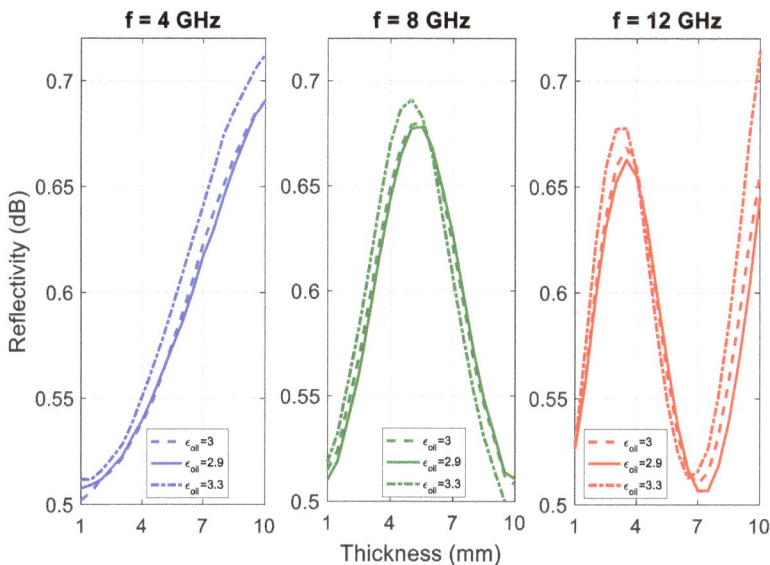

Figure 4. Comparison between the probability of detection versus the oil thickness (in mm) for single-frequency detectors at 4 GHz, 8 GHz and 12 GHz, using single observation ($M = 1$) and different oil properties. The electromagnetic roughness is $ks = 0.5$.

3.4. Performance Analysis of the Multi-Frequency Detectors

Figure 5 compares between the probability of detection versus the oil thickness (in mm) for different single-frequency detectors when the number of observations used by the detectors is varied up to 5. The electromagnetic roughness of the surface is 0.5. The performance of all detectors improves when the number of observations increases from 1 to 5. Using many observations for the detection reduces the AWG noise. However, no improvement is recorded for thickness values equal to multiple of wavelengths. Therefore, even when increasing the number of observations to reduce the effect of the noise, there are still some thickness ranges where the decision is totally wrong. With the use of one frequency, it is not possible to cover all the possible range of thickness values because of the periodicity of the reflectivity. The reflectivity's cyclic behavior highlights the need to use multiple frequencies to achieve accurate decision about the oil spill situation.

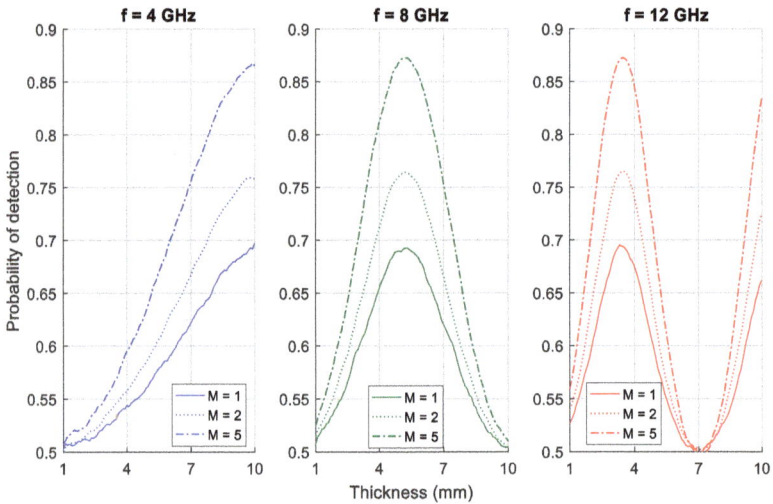

Figure 5. Comparison between the probability of detection versus the oil thickness (in mm) for single-frequency detectors at 4 GHz, 8 GHz and 12 GHz, using single observation ($M = 1$) and multiple observations ($M = 2$ and $M = 5$), with electromagnetic roughness $ks = 0.5$.

Figure 6 compares the probability of detection for different single- and dual-frequency detectors under different surface-roughness scenarios. Using more than one frequency in the detector increases the range of thicknesses over which the detection is correct and omits to a certain extent the drawbacks of the reflectivity cyclic behavior. When combining 4 GHz with 8 GHz, the probability of detection increases to more than 70% for any thickness value exceeding 4.5 mm. Using more observations ($M = 3$ and 5), the dual-frequency detectors performs respectively around 10% and 15% better than using single observation only. The performance of the detection is much efficient when the surface is planar. It is evaluated to be higher than 75% for thickness values greater than 2 mm when $M = 5$. The probability of detection generally improves for a larger number of scans M because the noise will be averaged out. However, with dual-frequency detectors, the probability of detection in the low frequency range is still low. What if we increase the number of frequencies used when scanning?

Figure 7 shows the effect of increasing the number of frequencies on the performance of the detectors. For planar surfaces, the tri-frequency detector using the combination of (4 GHz, 6 GHz and 8 GHz) shows better performance than the combination of (4 GHz, 8 GHz and 12 GHz) in high thickness ranges. The cyclic behavior of the reflectivity at 12 GHz leads to the 8% drop witnessed at 8.5 mm. However, its slope is steepest than the slope of the reflectivity at 6 GHz. This explains the 5–15% less efficiency at low thickness values 1–4 mm.

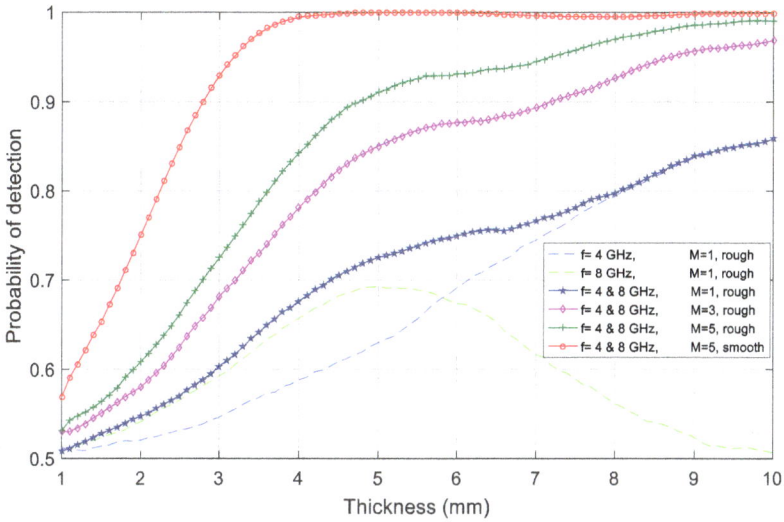

Figure 6. Comparison between the probability of detection versus the oil thickness (in mm) for different detector algorithms: single-frequency detectors at 4 GHz and 8 GHz, and dual-frequency detectors using combinations of these frequencies, for different surface-roughness scenarios.

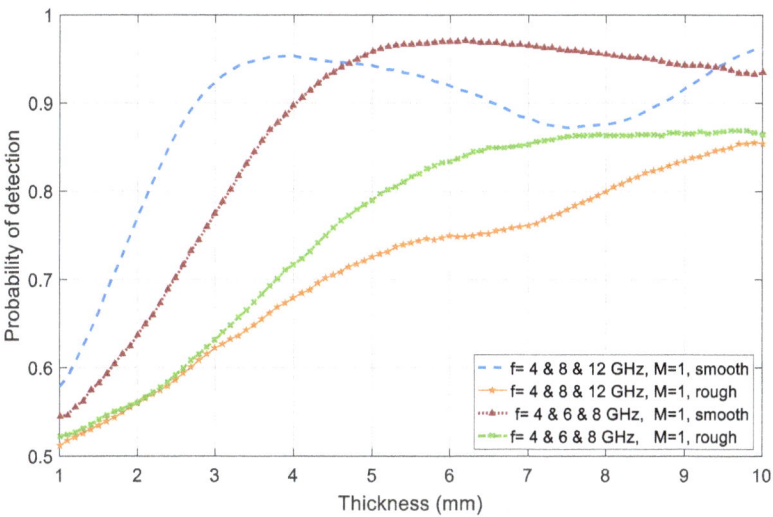

Figure 7. Comparison between the probability of detection versus the oil thickness (in mm) using different multi-frequency detectors for smooth and rough surfaces.

Higher frequencies witness more electromagnetic loss at the same rms height (s) of the surface. Therefore, it would be more advantageous to use lower frequencies when the ocean surface is rough. For electromagnetic roughness $ks = 0.5$, the loss of the reflectivity at 12 GHz is much higher than the loss at 6 GHz. For that reason, the tri-frequency detector that uses 6 GHz instead of 12 GHz gives better performance over all the possible thickness values. By increasing the number of scans, more benefit can be achieved at low thickness values.

Overall, based on the previous obtained results, we can propose the following plan using drone systems for oil spill detection:

1. Receiving alarm for possible oil spill (witnesses, underwater pipelines ruptures, collisions of tankers, etc.)

    - Send multiple drones to test the candidate scene.
    - The drones should use wide-band radar to collect several measurements at different frequencies in each scan.
    - Depending on the weather conditions and the ocean's waves, scan M times the scene to reduce the effect of the noise on the collected measurements as needed.
    - Data collected should be transmitted directly to base stations for post-processing.

2. Apply the detection algorithms on the collected data and analyze it. If the probability of detection exceeds the threshold set by the persons in charge, launch the oil spill alarm and the need to start the contingency plan.

3. Apply the oil thickness estimation algorithms [24] on the collected data to estimate the severity of the spill. High values (>1 mm) indicate the need for quick intervention because the oil slick will be considered to be "thick" and "heavy", and it will persist for a long period of time [4].

4. Use the satellite systems to provide a synoptic view of the scene during the upcoming days to track the spill.

5. Once the contingency plan is started, make use of other sensors to get all other needed information more accurately (for example, the large-size and weight fluoro-sensor can be used to identify the oil type [4].)

4. Conclusions

In this paper, we present a new probabilistic approach using wide-band radar for drone-based oil spill detection applications. We derived multi-frequency algorithms that use the statistical characterization of the power reflectivity and its distribution under various oil thicknesses and electromagnetic wave frequencies. We first introduce multi-frequency single-observation detector that uses single measurement of power reflection coefficient at different frequencies. Then, we present the single-frequency multiple-observations detector that uses multiple measurements of power reflection coefficients over several scanning for the sea area under study. We finally derive the multi-frequency multiple-observations detector that uses different frequencies at the same time and repetitively to provide a final decision about oil spill presence or absence. Performance analysis of all three types of detectors is done. Results show the inability of the single-frequency detectors to effectively distinguish between oil slicks and water for the total range of possible thicknesses. Nevertheless, increasing the number of observations leads to an increase in the effectiveness of the detectors. Dual-frequency and tri-frequency detectors prove their ability to overcome the drawbacks of the single-frequency detector by providing accurate detection especially for multiple observations. The performance of these detectors is reduced when the roughness of the ocean surface due to winds increases. The proposed algorithms can be implemented on nadir-looking systems such as the drones to be complementary systems for oil spill detection. Using multiple drones at the same time allow for quick intervention and real-time data collection for post-processing. During the early stages of a possible oil spill, the drone systems act as tactical-response systems complementing the large-scale view obtained by satellite systems. Once the spill is confirmed, the drones can track the spill using the high spatial resolution feature provided by the wide-band radars.

Author Contributions: Conceptualization, methodology, investigation, and writing of the paper, B.H.; results discussion, F.N.; funding acquisition, F.N. and Gh.F.; supervision, F.N., H.A. and J.J.

Funding: This work was supported by the National Council of Research at Lebanon (CNRS-L).

Conflicts of Interest: The authors declare no conflict of interest.

References

1. Alpers, W. Remote sensing of oil spills. In Proceedings of the Maritime Disaster Management Symposium, Dhahran, Saudi Arabia, 19–23 January 2002; pp. 19–23.
2. Oil Pollution Monitoring. Remote Sensing Exploitation Division. ESRIN—European Space Agency (ESA). p. 2. Available online: http://www.esa.int/esapub/br/br128/br128_1.pdf (accessed on 18 December 2018).
3. Fingas, M. *The Basics of Oil Spill Cleanup*; CRC Press: Boca Raton, FL, USA, 2012.
4. Jha, M.N.; Levy, J.; Gao, Y. Advances in remote sensing for oil spill disaster management: State-of-the-art sensors technology for oil spill surveillance. *Sensors* **2008**, *8*, 236–255. [CrossRef] [PubMed]
5. Leifer, I.; Lehr, W.J.; Simecek-Beatty, D.; Bradley, E.; Clark, R.; Dennison, P.; Hu, Y.; Matheson, S.; Jones, C.E.; Holt, B.; et al. State of the art satellite and airborne marine oil spill remote sensing: Application to the BP Deepwater Horizon oil spill. *Remote Sens. Environ.* **2012**, *124*, 185–209. [CrossRef]
6. Fingas, M.; Brown, C.E. A review of oil spill remote sensing. *Sensors* **2017**, *18*, 91. [CrossRef] [PubMed]
7. Grüner, K.; Reuter, R.; Smid, H. A new sensor system for airborne measurements of maritime pollution and of hydrographic parameters. *GeoJournal* **1991**, *24*, 103–117. [CrossRef]
8. Yin, D.; Huang, X.; Qian, W.; Huang, X.; Li, Y.; Feng, Q. Airborne validation of a new-style ultraviolet push-broom camera for ocean oil spill pollution surveillance. Remote Sensing of the Ocean, Sea Ice, and Large Water Regions 2010. *Int. Soc. Opt. Photonics* **2010**, *7825*, 78250I.
9. Fingas, M.F.; Brown, C.E. Review of oil spill remote sensing. *Spill Sci. Technol. Bull.* **1997**, *4*, 199–208. [CrossRef]
10. Yang, C.S.; Kim, Y.S.; Ouchi, K.; Na, J.H. Comparison with L-, C-, and X-band real SAR images and simulation SAR images of spilled oil on sea surface. In Proceedings of the 2009 IEEE International Geoscience and Remote Sensing Symposium, IGARSS 2009, Cape Town, South Africa, 12–17 July 2009; Volume 4, pp. IV–673.
11. Skrunes, S.; Brekke, C.; Eltoft, T. Oil spill characterization with multi-polarization C-and X-band SAR. In Proceedings of the 2012 IEEE International Geoscience and Remote Sensing Symposium (IGARSS), Munich, Germany, 22–27 July 2012; pp. 5117–5120.
12. Skrunes, S.; Brekke, C.; Eltoft, T.; Kudryavtsev, V. Comparing near-coincident C-and X-band SAR acquisitions of marine oil spills. *IEEE Trans. Geosci. Remote Sens.* **2015**, *53*, 1958–1975. [CrossRef]
13. Marzialetti, P.; Laneve, G. Oil spill monitoring on water surfaces by radar L, C and X band SAR imagery: A comparison of relevant characteristics. In Proceedings of the 2016 IEEE International Geoscience and Remote Sensing Symposium (IGARSS), Beijing, China, 10–15 July 2016; pp. 7715–7717.
14. Collins, M.J.; Denbina, M.; Minchew, B.; Jones, C.E.; Holt, B. On the use of simulated airborne compact polarimetric SAR for characterizing oil–water mixing of the deepwater horizon oil spill. *IEEE J. Sel. Top. Appl. Earth Obs. Remote Sens.* **2015**, *8*, 1062–1077. [CrossRef]
15. Hensley, S.; Jones, C.; Lou, Y. Prospects for operational use of airborne polarimetric SAR for disaster response and management. In Proceedings of the 2012 IEEE International Geoscience and Remote Sensing Symposium (IGARSS), Munich, Germany, 22–27 July 2012; pp. 103–106.
16. Laneve, G.; Luciani, R. Developing a satellite optical sensor based automatic system for detecting and monitoring oil spills. In Proceedings of the 2015 IEEE 15th International Conference on Environment and Electrical Engineering (EEEIC), Rome, Italy, 10–13 July 2015; pp. 1653–1658.
17. Dan, W.; Jifeng, S.; Yongzhi, Z.; Pu, Z. Application of the marine oil spill surveillance by satellite remote sensing. In Proceedings of the 2009 International Conference on Environmental Science and Information Application Technology, Wuhan, China, 4–5 July 2009; Volume 1, pp. 505–508.
18. Rocca, F. Remote sensing from space for oil exploration. In Proceedings of the 2015 IEEE International Geoscience and Remote Sensing Symposium (IGARSS), Milan, Italy, 26–31 July 2015; pp. 2876–2879.
19. Minchew, B.; Jones, C.E.; Holt, B. Polarimetric analysis of backscatter from the Deepwater Horizon oil spill using L-band synthetic aperture radar. *IEEE Trans. Geosci. Remote Sens.* **2012**, *50*, 3812–3830. [CrossRef]
20. Bayındır, C.; Frost, J.D.; Barnes, C.F. Assessment and Enhancement of SAR Noncoherent Change Detection of Sea-Surface Oil Spills. *IEEE J. Ocean. Eng.* **2018**, *43*, 211–220. [CrossRef]
21. Xu, L.; Wong, A.; Clausi, D.A. An Enhanced Probabilistic Posterior Sampling Approach for Synthesizing SAR Imagery With Sea Ice and Oil Spills. *IEEE Geosci. Remote Sens. Lett.* **2017**, *14*, 188–192. [CrossRef]

22. Lecomte, E. En Fevrier 2017, des Drones vont Traquer la Pollution Maritime. 2017. Available online: https://www.sciencesetavenir.fr/high-tech/drones/en-fevrier-2017-des-drones-vont-traquer-la-pollution-maritime_109732 (accessed on 1 January 2017).

23. Kirkos, G.; Zodiatis, G.; Loizides, L.; Ioannou, M. Oil Pollution in the Waters of Cyprus. In *The Handbook of Environmental Chemistry*; Springer: Berlin/Heidelberg, Germang, 2017; doi:10.1007/698_2017_49.

24. Hammoud, B.; Ayad, H.; Fadlallah, M.; Jomaah, J.; Ndagijimana, F.; Faour, G. Oil Thickness Estimation Using Single-and Dual-Frequency Maximum-Likelihood Approach. In Proceedings of the 2018 International Conference on High Performance Computing & Simulation (HPCS), Orleans, France, 16–20 July 2018; pp. 65–68.

25. Hammoud, B.; Mazeh, F.; Jomaa, K.; Ayad, H.; Ndadijimana, F.; Faour, G.; Fadlallah, M.; Jomaah, J. Multi-Frequency Approach for Oil Spill Remote Sensing Detection. In Proceedings of the 2017 International Conference on High Performance Computing & Simulation (HPCS), Genoa, Italy, 17–21 July 2017; pp. 295–299.

26. Hammoud, B.; Mazeh, F.; Jomaa, K.; Ayad, H.; Ndagijimana, F.; Faour, G.; Jomaah, J. Dual-frequency oil spill detection algorithm. In Proceedings of the 2017 Computing and Electromagnetics International Workshop (CEM), Barcelona, Spain, 21–24 June 2017; pp. 27–28.

27. Hammoud, B.; Faour, G.; Ayad, H.; Ndagijimana, F.; Jomaah, J. Performance Analysis of Detector Algorithms Using Drone-Based Radar Systems for Oil Spill Detection. *Multidiscip. Digit. Publ. Inst. Proc.* **2018**, *2*, 370. [CrossRef]

28. Ulaby, F.T.; Long, D.G.; Blackwell, W.J.; Elachi, C.; Fung, A.K.; Ruf, C.; Sarabandi, K.; Zebker, H.A.; Van Zyl, J. *Microwave Radar and Radiometric Remote Sensing*; University of Michigan Press: Ann Arbor, MI, USA, 2014; Volume 4

29. Campbell, B.A. *Radar Remote Sensing of Planetary Surfaces*; Cambridge University Press: Cambridge, UK, 2002.

Journal of
Marine Science and Engineering

MDPI

Review

Environmental Decision Support Systems for Monitoring Small Scale Oil Spills: Existing Solutions, Best Practices and Current Challenges

Davide Moroni, Gabriele Pieri * and Marco Tampucci

Institute of Information Science and Technologies-National Research Council, Via Moruzzi, 1–56124 Pisa (IT), Italy; Davide.Moroni@isti.cnr.it (D.M.); Marco.Tampucci@isti.cnr.it (M.T.)
* Correspondence: Gabriele.Pieri@isti.cnr.it; Tel.: +39-050-621-3120

Received: 29 October 2018; Accepted: 17 January 2019; Published: 21 January 2019

check for updates

Abstract: In recent years, large oil spills have received widespread media attention, while small and micro oil spills are usually only acknowledged by the authorities and local citizens who are directly or indirectly affected by these pollution events. However, small oil spills represent the vast majority of oil pollution events. In this paper, multiple oil spill typologies are introduced, and existing frameworks and methods used as best practices for facing them are reviewed and discussed. Specific tools based on information and communication technologies are then presented, considering in particular those which can be used as integrated frameworks for the specific challenges of the environmental monitoring of smaller oil spills. Finally, a prototype case study actually designed and implemented for the management of existing monitoring resources is reported. This case study helps improve the discussion over the actual challenges of early detection and support to the responsible parties and stakeholders in charge of intervention and remediation operations.

Keywords: marine information systems; environmental monitoring; proactive systems; decision support systems; signal integration; oil spills

1. Introduction

It is well-known that large spills of oil and related petroleum products in the marine environment can have serious biological and economic impacts. According to [1], the 2010 Deepwater Horizon disaster is still exacting an ongoing and largely unknown toll. Public and media scrutiny is usually intense after a spill, demanding that the location and extent of the oil spill be properly identified and quantified. Remote sensing is playing an increasingly important role in oil spill response efforts [2]. Through the use of modern remote sensing instrumentation, oil can be continuously monitored on the open ocean. With knowledge of slick locations and movement, response teams can more effectively plan countermeasures in an effort to curtail the effects of the induced pollution.

Pollution sources in the sea are disparate in size, origin, and nature of the pollutants, and are not limited to major accidents. It is possible to distinguish several classes of pollution sources and to evaluate the impact of each class. Particular classes are: (i) pollution sources caused by oil exploration and production; (ii) pollution sources caused through transporting oil by sea; (iii) natural oil pollution sources; (iv) pollution sources generated by general maritime traffic and shipping operations; and (v) pollution sources caused by coastal activities. While events deriving from (i) and (ii) might produce accidents of great impact to coastal populations, they are relatively rare compared to pollution events, due to the general shipping traffic and coastal activities.

Although operational discharges may be considered small when compared to spills caused by shipping accidents, they tend to be repetitive and even chronic, being concentrated in ports and along shipping routes. Therefore, these spills will have an impact on local marine habitats, including physical disturbances, toxic inputs to sensitive species, and organic sediments enrichment [3]. Ships of all kinds discharge oily residues into the sea during routine operations. Further, ships periodically clean their ballast and bilge water tanks, comprising a considerable source of pollution. It has been estimated that most oil spills are the result of daily operations, most often occurring in oil or port terminals [4]. Indeed, the International Tanker Owners Pollution Federation Limited (ITOPF) reports that small and medium-sized spills account for 95% of the total number of all the incidents recorded [5]. Furthermore, the impact of coastal activities as a source of oil spills does not yet seem to be too well-understood [6–8].

Operational oil spills pose a serious threat to the environment, especially because attention and mitigation measures tend to be focused on large accidental spills. Most of the existing frameworks based on remote sensing and systems for environmental decision support are mainly focused on large catastrophic events, while small-scale oil spills have received somewhat less attention. For instance, the European Maritime Safety Agency (EMSA) provides the CleanSeaNet service [9], covering all European sea areas, which are analyzed in order to detect and track possible oil spills on the sea surface. Besides operational services, the interest of the research community is witnessed by the numerous projects and prototypical systems for marine pollution monitoring [10,11]. Recent works include cloud-based solutions [12], in which a cloud-based image processing the facility for oil spill detection is integrated with a web-based geographical information system, and the framework is introduced in [13] in which a high-resolution hydrodynamic model is used for accurately forecasting oil spill evolution and weathering.

Addressing small and micro oil spills exhibits some challenging differences with respect to large ones. For example, they are on a small spatial scale, and most of them are difficult to detect solely on the basis of remote sensing. For instance, aerial surveys of the North Sea have shown that between 500 and 1200 oil spills have been observed each year, with 73–88% of oil spills having a volume less than 1 cubic meter [14], which make them difficult to be accurately detected and analyzed by satellite-borne sensors alone.

One possible approach to the problem of small-scale oil pollution monitoring is to fuse satellite images with other data sources. For instance, integrating data collected in situ by a suitable network of sensors may improve the pervasiveness of monitoring in a marine area of particular environmental value, while simultaneously helping to resolve ambiguities and filtering false positives deriving from the analysis of data coming from a specific and single modality, i.e., the use of data acquired and processed from only one source. Multisource, i.e., more sensors that can be from different devices that can offer the same typology of data (e.g., SAR with different resolutions or from different satellites), and multimodal, i.e., with reference to the physical features recorded by devices using different typologies for acquisition (e.g., buoys, in situ, airborne, AIS, satellite, SAR/optical), surveillance of the sea thus has good capabilities in addressing small-scale oil spills, but it demands for additional problems to be solved. Collection, cross-correlation, and comparison of multiple data sources cannot be routinely performed manually by authorities and stakeholders in charge of the intervention and remediation operations. For instance, it is difficult to establish possible correspondences between vessels and oil slick positions sampled at different times by: (a) satellite-borne sensors, (b) Automatic Identification System (AIS), and (c) in situ devices, without including and integrating the data into a single information system endowed with models where all encompassing forecasting and retrodiction of slick and ship positions are available. Furthermore, while major pollution events are managed by special contractors for carrying out intervention and remedy actions, small ones are addressed—at least in the first stages—primarily through the use of local monitoring and remedy resources. The orchestration and optimization in the use of such resources also poses some problems in the routine management of small pollution events.

From the above considerations, it may be evinced that while multimodal data integration has strong potential in dealing with small and micro oil spills, suitable algorithms and models are needed to properly exploit such data and optimize the use of both a monitoring network and a local intervention chain.

The main contribution of this paper is to review the relevant literature concerning decision support in environmental monitoring, especially for the marine and maritime domain, and then to establish a rationale for the design and integration of an Environmental Decision Support System (EDSS) devoted to oil spill management. Advanced data gathering functionalities and coordinated management of available models emerge as key components for the design of a successful and useful Information and Communication Technologies (ICT) system. On the basis of the proposed analysis, guidelines are suggested for steering the design of an EDSS, as well as for defining its functional requirements. Such guidelines are then put into practice and properly demonstrated in a case study. EDSS is designed and integrated into the Marine Information System (MIS) presented in [15]. Such integration shows the advantages of decision support services for more efficient management of small and micro oil spills.

2. Related Works

An environmental system is complex, dynamic, spatially distributed, and highly non-linear. Its processes operate on a multitude of interdependent scales in time and space [16,17]. In addition, many of the governing processes are not directly observable, and therefore are not easily understood. Along with such inherent difficulties, every decision related to environmental planning and management is characterized by multiple and usually conflicting objectives, as well as multiple criteria; thus, it is important to be aware of the problem of uncertainty and also of issues arising as a consequence of the increasingly wide public participation in decision-making processes. In this framework, EDSSs are emerging as fundamental tools to aid analysis and planning of all the decisional processes that are pertinent to environmental management. The advantage of using EDSS, in particular for small-scale oil spill, is at least twofold: it aids decision-makers in their activity by facilitating the use of data, models, and structures; and it favors reproducibility and transparency of decision-making. In the following, a general treatment of EDSS is provided, while simultaneously focusing on the requirements and functionalities needed to properly address small-scale events, to which the rest of the present work is devoted.

The three primary axes of intervention of EDSS might be identified as (see also [18]): (i) integrating information into a coherent framework for analysis and decision-making, discerning key information that impacts decision-making from more basic information; (ii) identifying realistic management choices; and (iii) providing a framework for transparency (i.e., all parameters, assumptions, and data used to reach the decision should be clearly documented) and ensuring that the decision-making process itself is documented.

In most applied contexts, environmental monitoring processes imply a continuous intelligent monitoring system, an increasing volume of data, and, in many instances, decreasing time for making decisions. This is particularly true in the case of marine and maritime monitoring for taking care of pollution events, where the so-called *near-real-time* is the amount of time between the occurrence of the event and its notification to the appointed authorities, who take charge of the notification and start possible remediation operations. This time span represents the interval that can be used by automated tools to perform more or less autonomous tasks which lead to a better and more precise description of the polluting event, or to disregard the event if it is a false alarm.

The provision of support services is generally based on Artificial Intelligence (AI) paradigms. In [19], an overview of the impact of AI techniques on the definition and development of the first EDSS during the last fifteen years is reported. Cortés et al. highlights the desirable features that an EDSS must show, and their paper concludes with a selection of successful applications to a wide range of environmental problems. By contrast, in [20], the authors understand that these tools often fail to be truly adopted by the intended end-users, and try to identify and assess key challenges in EDSS development and offer recommendations to resolve them. In particular, to tackle the described challenges, the authors provide a set of best-practice recommendations to improve ease of use, establish trust and credibility, and promote EDSS acceptance.

More general integrated environmental modeling, which fairly comprises Environmental Information Systems (ENVISs), is presented in [21] where the problem is faced from a socio-economic environmental point of view and the decision-making approach follows an integration of resources and analyses to address the problems as they occur in the real-world, including input from appropriate stakeholders [22]. In [23], the authors perform a review of five common modeling approaches for an integrated environmental assessment and management. Integration is defined by the purpose the specific model wants to achieve, from prediction to decision-making, in the context of different environmental assessments. In particular, regarding the approach to modeling complex situations, knowledge-based models offer advice to the users, based both on their own knowledge and on the user's response to a number of *if–then* questions [24]. In parallel, [25] introduced another probabilistic approach producing a framework and claiming to have a holistic view of the global risk assessment. In particular, they operate for what concerns the risk factors on a single species endangered by a polluting event, and providing a model for a quantitative risk estimate. Mokhtari et al. [26] developed a spatial predictive model for estimating the probability of oil spills' occurrence. Their model estimates the probability of oil spills at a pixel level as a function of four specific variables: ship routes, coastlines, oil facilities, and oil wells. It uses a Generalized Linear Model (GLM) with a polynomial function. The number of variables taken into account is very limited, but the spatial mapping approach produces informative raster maps for aggregating and presenting risk estimation, similar to the dynamic risk maps presented by us in [15]. Finally, [27] developed another system for the evaluation of spatially distributed ecological risk related to oil spills. Their system features a tanker accident model based on Bayesian networks which has been placed upstream to an existing oil spill simulation model for evaluating the impact on a set of threatened species.

Oil spill prediction services have also been proposed in the literature and tested in conjunction with operational oil spill detection and monitoring frameworks. In the Mediterranean sea, an oil spill prediction service has been set up, known as Mediterranean Decision Support System for Marine Safety (MEDESS-4MS), whose underlying concept is the integration of existing regional models and national ocean forecasting systems with the Copernicus Marine Environmental Monitoring Service (CMEMS) and their interconnection, through a dedicated network data repository, facilitating access to all these data and to the data from the oil spill monitoring platforms, including satellite data [28]. MEDESS-4MS offers a range of service scenarios, access to multiple models tuned for specific Mediterranean areas, and interactive capabilities to suit the needs of Regional Marine Pollution Emergency Response Centre for the Mediterranean Sea (REMPEC) and EMSA. Such variegated prediction services are based on the comparison of oil spill simulation exercises carried out during EU projects, such as ECOOP [29], MERSEA-IP [30], MyOcean [31], and NEREIDs [8,32], which include well-established oil spill models of the Mediterranean region, as well as new oil plume models to simulate the oil from spills located at any given depth below the sea surface [33]. Although the approach of MEDESS-4MS is comprehensive, it is focused on prediction services and, thus, its direct impact on the routine workflow of regional stakeholders is limited.

A particular mention is for the NEREIDs project ended in 2014 [34]. Its goal has been to foster effective cross-border co-operation, while setting best practices for other members of the European Civil Protection Mechanism in order to use innovative ideas and tools as a base to build on training, preparedness, and research.

Again in the Mediterranean, but more specifically in the East, other exercises were conducted and precise models were derived based on novel and high-resolution bathymetric, meteorological, oceanographic, and geomorphological data. Seabed morphology has been correlated to the direction of the oil slick expansion, since it is able to alter the movement of sea currents [35]. The work derived precise a priori information for the management of oil spills, and while it provided aid for civil protection authorities and mitigation teams, it is not yet a real-time decision support method. Specifically, the above work suggests that oil spills in the Eastern Mediterranean Sea should be mitigated within a few hours of their onset, and before wind and currents disperse them, thus prompting the need for an EDSS for prompt mitigation actions. Protocols should be prioritized between neighboring countries to mitigate any oil spills. Similarly, the work in [36,37] shows shoreline susceptibility varies significantly depending on differences in morphology, degree of exposure to wave action, as well as the existence of uplifted wave-cut platforms, coastal lagoons, and pools. The added presence of tourists and environmentally sensitive zones suggests that mitigation work should take into account the high shoreline susceptibility of parts of the Eastern Mediterranean Sea. A significant suggestion arising from experiences like the abovementioned projects is to increase the monitoring of oil-spills.

Another example of an integrated monitoring system is presented in a project regarding the Venice Lagoon, named *Atlas of the lagoon* [38], where heterogeneous dynamic data and tools are published and focused on various thematic maps using a large amount and typology of environmental data.

Other examples of susceptibility analysis include small basins and gulfs, such as the Gulf of Finland. In [39], the presented results characterize the role of surface currents in the transport of contaminants located in the uppermost layer in the Gulf of Finland on a time-scale of the first few weeks. Their work contributes to the understanding of the potential use of the dynamics of currents for environmental management of offshore activities. Again in this case, the proposed services may support the stakeholders in designing and optimizing fairway, but no relevant real-time support can be offered in the case of crises.

Finally, another important and current topic regards the raising of environmental consciousness and awareness, and several projects, even some which have already been mentioned in this section include this aspect as its importance is well-recognized, as shown by an Erasmus+ project called Sea4ALL [40], which specifically addresses this problem through school games and an educational portal.

3. Rationale for an EDSS Devoted to Oil Spill Management

In this section, we survey the main decisional points in which an EDSS can be beneficial, highlighting the importance of both data integration and service orchestration, meant as the coordinated deployment and arrangement of multiple services according to a precise logic. Below, details are provided starting from a general oil spill management system, then looking more closely at what EDSS tasks might be, and ending up with specific details about its design and overall operation.

3.1. Main Tasks to Address in Oil Spill Management

The management of problems related to oil spill detection in a certain site includes a number of tasks that can benefit from the intervention of automatic systems and computational models for a more efficient treatment. Among them, it is important to mention:

(i)   Collection of information about the site: In order to be as accurate as possible, the number, frequency, and location of the site-specific data to be collected should be decided on;

(ii)   Assessment of the risk: Based on the initial site-characterization data, models for interpolation, extrapolation, and prediction should be applied for evaluating the hazard and guiding the decisions on recovery strategies;

(iii) Projections of contamination levels: Decisions should regard which strategy should be followed for an effective recovery and, to this end, whether more data are needed to better define the region that requires recovery, or to improve the remedy selection or remedy design;

(iv)   Monitoring and evaluation of the interventions made: Further decisions should be made on what and where to monitor, the duration of monitoring, and, of course, the effective monitoring of the selected areas.

There is a number of basic decisions that should be made before the actual decision process is developed, such as design choices (what to sample, when to sample, what technologies should be used), as well as policies for determining which risk levels might be considered acceptable.

It is unlikely that any single person will have the knowledge to perform every analysis required to support all of the abovementioned decisions. Typically, a team of people with different areas of expertise are involved in interpreting basic information and providing it in a form useful for other people in the decision process chain. EDSS can be employed to offer support to the team as a whole, by providing more thorough monitoring or strengthening the skills and technical expertise of its members using computational models and automatic reasoning methods.

3.2. Main EDSS Points of Intervention and Reasoning Paradigms

On the basis of the tasks identified above in the management of oil spills, an EDSS is useful for addressing different activities [19], also shown in Figure 1:

(i)   Hazard identification, by filtering and screening criteria and reasoning about the activity being considered: This phase may be characterized as a continuous monitoring activity of the system looking for possible adverse outcomes, and includes the search for further data to enhance its own performance.

(ii) Risk assessment, by quantitative and qualitative measurements of the hazard: The heterogeneity of data coming from various sources and with many different levels of precision may be faced by using a number of approaches, including model-based, rule-based and case-based reasoning (see, e.g., [41] for a review of such approaches).

(iii) Risk evaluation: Once potential risks have been assessed, it is possible to introduce judgments regarding the degree of concern about a certain hypothesis. This is possible if the system has accumulated experience solving similar situations using, for instance, a case-based reasoning approach or inferential modeling, where previous experience of risk evaluation is used to assist future judgments.

(iv) Intervention decision-making: The system needs appropriate methods for controlling or reducing risks. The system also requires knowledge about the context where the activity takes place and must be able to interpret its results and knowledge about the risk/benefit balancing methods.

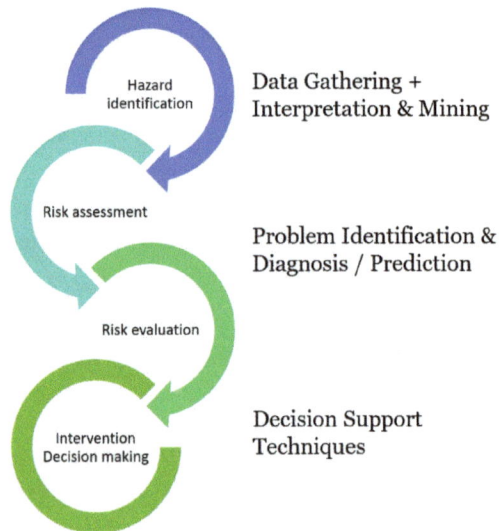

Figure 1. Main and subsequent steps of an EDSS during its activity.

In order to be effective and useful, an EDSS should be able to collect all the relevant data and interpret these data according to prediction models for understanding the situation and assessing the risk. This entails managing the monitoring resources for acquiring more useful data or planning a remedy intervention. Such requirements lead to the definition of a particular design of EDSS, as explained in Section 3.3 below.

3.3. EDSS Design

Design of an EDSS requires an understanding of the environmental problem domain and identifying the experts and authorities to cooperate with. The identification of the problem to be solved by exploiting the EDSS aid has been particularly important, as well as the functionalities by which the system can intervene and improve the current oil spill detection and management procedures. Coherently with the outcome of interviews carried out with experts and authorities, the following three main functionalities have been included: (i) *Data Gathering*, (ii) *Diagnosis and/or Prediction*, and (iii) *Decision Support*.

For (i), the EDSS has to cope with very different types of data, which can be produced and received even in real time from a variety of sensors. Indeed, data can be gathered from various monitoring resources, including Synthetic Aperture Radar (SAR) images, hyperspectral images collected during flight campaigns, data collected by sensorized buoys [42] and Autonomous Underwater Vehicles (AUVs), forecast data obtained by applying simulation models, data about ship traffic through AIS systems, and other miscellaneous reports, possibly including Volunteered Geographical Information (VGI) [43]. Heterogeneity of these data suggests the necessity of distributing interpretation tasks among different subsystems of an EDSS. Identified requirements on data gathering are discussed below.

Resource Management Services (RMSs) should be provided for diagnosis and prediction (ii). In particular, the environmental data acquired by various monitoring resources should be fused by applying simulation and optimization models for site characterization and observation, in order to detect possible marine pollution events.

Finally, for (iii), assistance in decision-making should be supplied by drawing an optimized plan for the exploitation of available monitoring resources and of the modules for data analysis, so as to confirm the detection of an event and raise an alert if required. Suitable presentation and documentation of events are to be supplied, along with feasible suggestions aimed at supporting sustainable event management and recovery interventions.

Once EDSS general requirements have been explained, the EDSS shall have strict interaction with the various components of the information system. Such an interaction will be needed in order to guarantee fulfillment of the following features:

- Ability to acquire, represent and structure the knowledge in the specific domain under investigation;
- Ability to separate data from models, in order to be re-usable;
- Ability to deal with geo-referenced data;
- Ability to provide expert knowledge related to the specific domain;
- Ability to give the end-users (both on the manager/experts side, and the external users) assistance for interfacing with the system and selecting resolution methods.

A prerequisite for such skills is represented by the possibility to transfer data seamlessly among different elements and actors involved in the decision chain (data sources, EDSS modules, and stakeholders), so as to bring different data together easily and in a consistent form, and to facilitate dynamic links between different models and analytical processes. To this end, a network of distributed subjects ensuring integration of every single dataset in the collection, storage, retrieval, and dissemination of environmental information is highly desirable. In brief, an EDSS is required to achieve interoperability at several levels, such as the *measurement and monitoring level*, where the issues principally concern the consistency of observational methods and monitoring network design; the *models and data analysis level* where key issues relate to the consistency and suitability of input data, and to the validity and robustness of the models and algorithms used; the *metadata level*, where agreement needs to be reached on the data attributes described and the form of these descriptions, and the *service level* needed to facilitate data exchange and information retrieval and dissemination.

Despite the recent, great advances in standards and specifications, there is still a real need to test and demonstrate their deployment in large, integrated systems, and to realize the vision of Infrastructure for Spatial Information in the European Community (INSPIRE) [44] for a paneuropean spatial data platform.

4. EDSS Integration into a MIS Platform

In order to have a complete understanding of EDSSs, it is necessary to also briefly review the general platform in which it is integrated and consolidated, i.e., a special kind of ENVIS dedicated to the marine environment, which we will call the Marine Information System (MIS) in the following. It must be kept in mind that all the data, information, and models that are acquired, processed, and applied are all part of the MIS and, to this end, a brief recall of a general MIS architecture and of its main components is provided in this section.

The MIS aims at an effective and feasible detection and management of marine pollution events, by integrating a number of monitoring resources that are exploited to get useful and relevant information about the controlled sites. Each resource collects a specific type of data which are processed by a dedicated module that can be nominally considered a subsystem of the MIS.

The main task of the MIS is, then, to serve as a catalyst for integrating data, information, and knowledge from various sources in the environmental sector by means of adequate ICT tools. The MIS has been conceived as a connected group of subsystems for performing data storage, data mining, and analysis over data warehouses, decision support, as well as a web portal for the access and usage of products and services released to system managers and end-users.

Architecturally, six main units have been identified when designing the MIS—that is, the *Service Unit*, the *Notification Unit*, the *Operational Storage Unit*, the *Graphical User Interface (GUI) Unit*, the *Knowledge Discovery Unit*, and, of course, the *EDSS Unit*, which is the core component surveyed in this paper. A scheme of this composition is shown in Figure 2.

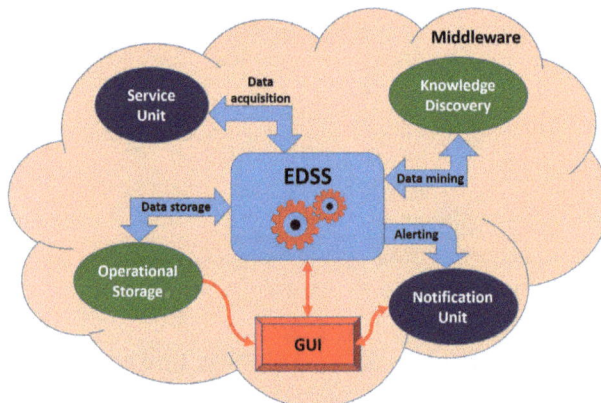

Figure 2. Architecture of a prototypical Marine Information System with its component units.

The Service Unit and the Notification Unit, having a direct interface with the external data sources (i.e., the different technologies and sensors used for data acquisition and processing), provide and allow data access and data exchange from and to the MIS for each external data source. In particular, the Service Unit is in charge of acting as a data manager for integrating information from all available data sources, applications (such as mathematical simulation models and image analysis methods), and repositories (like AIS data). The Notification Unit dispatches messages, such as alerts and suggestions, to personnel enrolled in the system. The Operational Storage Unit constitutes an internal storage unit of the MIS useful for guaranteeing timely access to operational data. In particular, a geo-enabled Data Base (DB) and a multimedia repository constitute the core of this unit. The GUI Unit represents the graphical front-end of the MIS, encompassing the interface for end-users and system manager. The Knowledge Discovery Unit and the EDSS are the most advanced services of the MIS. The first is oriented to off-line trend analysis and to the discovery of hidden patterns in the data in order to learn suitable data models, while the latter aims at providing real-time suggestions to system users, as discussed in detail in Section 5. The management of the data flow and of the main communications among these units is in charge of a central orchestrator, i.e., *Middleware*, detailed in the following Section 4.1.

4.1. The Middleware

The Middleware provides interfaces and methods to allow components to cooperate, and exchange information, products, or results among them in a reliable and efficient way, and with an optimized approach. The middleware allows general access both to the data and software units available in the MIS. From a Service Oriented Architecture (SOA) perspective [45], a middleware layer allows seamless consumption of data into models, making transparent to both users and software agents the transactions with the actual service providers, fully supporting paradigms such as Data-as-a-Service (DaaS) [46] and Software-as-a-Service (SaaS) [47]. Thus, the middleware is not a mere communication bus to transfer data, but consists of an *interface engine* that guarantees the functioning of the entire MIS system. This is the reason why the middleware is actually composed of two modules: the *workflow manager* and the *communication infrastructure*. The former orchestrates business processes in the MIS. An internal business logic engine has to be included for managing complex sequences of process executions and for coping with branching in case of connection failures. The workflow manager has to incorporate a scheduler of the event-driven stream of information/requests. The latter, i.e., the communication infrastructure, covers the connectivity logic part of the MIS and manages message-based communications between the single units and services, routing and transforming the needed data and requests. Communications between MIS units and services is based on generating proper messages (e.g., as XML structured documents or JSON) containing the data to be exchanged. Each unit has to be endowed with a dedicated listener able to retrieve incoming messages, as well as to parse and understand them. The reception of a message will start the process required to manage the contained data. The workflow manager might also be in charge of acting as a logging facility, by keeping track of the platform workflow. When an operation is performed, the relative identifying code is saved into the log, along with the involved units and the operation outcome. In this way, reproducibility, auditing, and transparency of every process, including decision-making, are met.

5. A Prototype Case Study of an EDSS

In MIS architecture, the EDSS Unit has a central role because it is responsible for the combination of all the multi-source data entering the system through the various units introduced in order to detect and monitor oil spills, issue alarms, and support their operational management.

In this section, we discuss how a prototypical case study of an EDSS can be related to other MIS components; then, we detail its functionalities and discuss its main modules.

Whenever the likelihood of a polluting event is determined, either by the risk analysis or reported by the processing results of one of the other MIS subsystems, the EDSS is responsible for developing an optimized exploitation plan of monitoring resources and models, in order to confirm the detection of the event and issue an alarm.

The presentation and documentation of suitable alarms should be provided, together with possible EDSS suggestions aimed at supporting event management and recovery interventions.

5.1. EDSS Main Components and Their Interconnection with the MIS Platform

According to the design of an EDSS introduced in Section 3.3, a prototype EDSS should be logically organized according to a three-level structure that consists in: (i) data-gathering, (ii) analysis and/or prediction, and (iii) decision support. With reference to the MIS platform:

(i)   Data need to be gathered through the Service Unit and stored into the Operational Storage Unit. Planning of their collection and retrieval needs to be performed through requests that are orchestrated by the Middleware.

(ii) Analysis and prediction are realized through the risk assessment models within the EDSS that can be applied to the collected data previously retrieved from the Operational Storage Unit through the Middleware.

(iii) Decisions are supported, first of all, by the definition within the EDSS of an optimized exploitation plan of available resources in order to confirm the detection of the event and issue an alarm through the Notification Unit. In addition, suggestions should be provided to support the implementation of event management and recovery interventions.

This logical structure corresponds architecturally to the two main components of the EDSS, namely the Risk Analysis Model (RAM) and the Resource Management Service (RMS). The RAM implements the analysis and prediction function, while the RMS is responsible for organizing the monitoring resources. Their combination results in the decision-support function of the system. In particular, since much of the work of the EDSS regards the planning of various activities, such as the acquisition of monitoring data, its general functioning can be structured according to a workflow model. This approach is also particularly convenient because the entire organization of the MIS operation is organized in a workflow, and therefore, part of the EDSS work can be delegated to the Workflow Manager in the Middleware (see Section 4.1), thus increasing the possibility to control and orchestrate all activities more effectively.

5.2. Functionalities of a Prototypical EDSS

The EDSS Unit consists of a multi-criterion decision support system aimed at helping decision-makers by providing them with criteria for assessing the most suitable way for the prevention, control, and recovery of oil spill pollution events. The system can be defined as the main intelligence of the MIS, and can be imagined as a consultancy and supervision service that implements predictive activities and planning of environmental monitoring. This is achieved by providing the following functional characteristics:

(i)   Detection and characterization of possible oil slick events and consequent alerting;
(ii)   Organization and management of the different monitoring resources;
(iii) Orchestration and combination of the results of the different data acquisition and processing subsystems;
(iv)   Harmonization and issue of alerts;
(v)   Suggestions on possible intervention protocols;
(vi)   Provision of specific and well-documented alarms to the competent authorities.

In more detail, the system should act according to two different modalities—it could be (a) **reactive**, or (b) **proactive**. In the reactive mode, the EDSS is triggered when new data or new reports coming from the monitoring resources are uploaded into the system and analyzes them in order to detect possible pollution events. In case some anomalies are detected, the system can try to better clarify the situation by collecting all the other related information, such as previous, current, or future data that can be acquired from other monitoring resources. By analyzing all the available monitoring data, the system decides when to issue an alarm or a warning and supply all related information. This could be done automatically, or simply by assisting an operator in the analysis workflow of the different resources. In the case of a proactive mode, the EDSS should apply a dynamic risk analysis derived from a model available for the identification of areas that could be under-monitored, i.e., where there is a high risk of a pollution event, but the resources deployed in situ are not enough to supervise the area. When this happens, the system prioritizes the resources, for example by increasing the rate of data acquisition from a specific device, or by organizing another specific in situ mission in those areas. The analysis of data obtained in this way is then periodically scheduled, and the system could move to the reactive functioning mode.

Once an alarm is issued, the EDSS can provide a protocol that can be followed by the authorities in charge for the intervention activities. This might be selected among a number of possible procedures that are properly represented and stored. Suitable reasoning mechanisms for this selection might be employed, such as simple case-based processing. To supply the above-described functionalities, the EDSS needs to be aware of the monitoring resources available in the whole platform and be able to optimize their employment by suitably handling the different events that can occur. This might be achieved by developing proper optimization and models for risk analysis, and by implicitly encoding the relevant knowledge into orchestration and organization procedures. In particular, as already introduced in the previous chapter, the EDSS Unit comprises:

- a RAM, which concerns the identification of areas with a high risk of oil spills, and needs to be developed following a model-based approach;
- an RMS, which is dedicated to prioritization of resources to detect possible oil spill events, and needs to be developed according to an optimization approach (see e.g., [48]) .

In the following Figure 3, a schematic view is given of the composition of the central Environmental Decision Support System with its described components and the interaction with the rest of the Marine Information System.

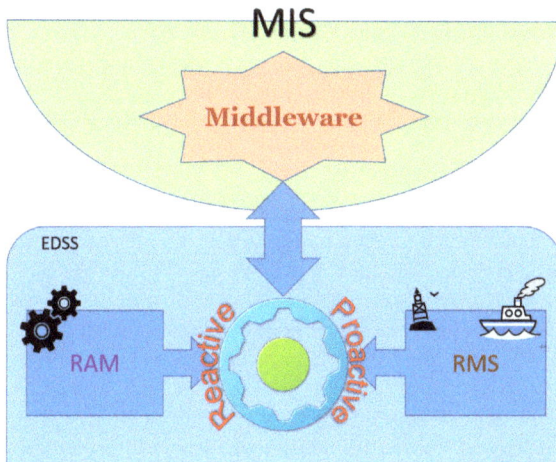

Figure 3. Structure of the Environmental Decision Support System and its components and behavior.

The application of the models, the provision of services, and the integration of the results coming from different resources and subsystems of the MIS might be orchestrated according to a business logics approach based on event handling. This means that the flow of data and actions of the system might be codified according to a workflow-based representation.

5.3. Risk Analysis Model (RAM)

The main task of risk assessment lies in the complex environmental problem of evaluating the likelihood of the occurrence of a hazard balanced with the severity of its consequences [49]. A RAM calculates the probability of spill occurrence, as well as the likely paths or trajectories of spills in relation to the locations of recreational and biological resources which may be vulnerable. The analytical methodology can easily incorporate estimates of weathering rates, slick dispersion, and possible mitigating effects of cleanup. The developed method for providing a risk assessment in *near-real-time*, defined as the actual time lapse passing between the occurrence of the hazard (i.e., oil spill) and the first official notification to the deputed authorities, has been developed with the goal to produce risk-related information in a geographic area of interest which can be both automatically analyzed by proactive services, and visually analyzed by the users involved within the intervention chain in order to reduce the risks and possibly improve the efficiency of the remediation operations.

For example, an implicit codification of the risk is represented by the amount of monitoring resources actually existing in an area, where an augmented number of them decreases the risk level (heavy monitoring and quicker intervention), whereas a smaller number of them, as well as distant location (sparse monitoring and delay in intervention), increase the risk level.

A dynamic risk map can be defined for assessing the hazard of oil slicks by evaluating several risk factors through the combination of the data collected by the MIS. This map can be used for planning and monitoring a prioritization of the resources in order to improve the degree of control of a high-risk area. Aiming at increasing the precision of the risk map, and at the same time lowering the amount of data to be transferred, the map cells with higher risk have smaller dimensions (i.e., thus a higher granularity) with respect to the ones with lower risk.

Risk calculation takes account of data gathered by the MIS, combining and correlating them in order to better estimate the risk of oil slick occurrences. Gathered data contribute to risk calculation by increasing or lowering it: for instance, the presence of vessels in a small area increases the risk at a different degree depending on the typology of the vessel (e.g., a tanker will bring a much increased risk level); meanwhile, a negative analysis from a remote device (e.g., a sensor-equipped buoy) will result in the lowering of the risk in the covered zone. All these variables, and many others which can bring their input data into the MIS, represent factors with specific weights, which, combined, altogether yield a final risk value for each specific location (i.e., cell) of the dynamic risk map.

5.4. Models for the Resource Management Service (RMS)

The RMS is devoted to the optimization of the monitoring resources that compose the available devices in the model in order to cover the monitoring areas and get the most valuable information possible about polluting events. The idea is that when an oil spill event is detected by one of the monitoring resources, the EDSS, with its RMS, organizes the use of the other resources to get more information about the site of interest and to provide suggestions about the recovery strategy to be followed. Methods for the development of RMS are based on optimization models that try to drive the effective and efficient use of resources, according to the tasks to be performed. In particular, the aim is to define strategies for assigning resources to different activities focusing on both process and resource use to optimize task operations. Different options should usually be explored for resource allocation, availability, relevance, and data. The development of descriptive and analytical models is required to accurately represent and simulate the processes that involve resource deployment. These models should be dynamic and provide a new resource deployment plan each time it is required. From a methodological point of view, defining the resource optimization models requires outlining objectives, decision variables, and constraints. How these are involved in the optimization process is illustrated in Figure 4, and described according to the following steps that are being performed to develop the resource optimization model:

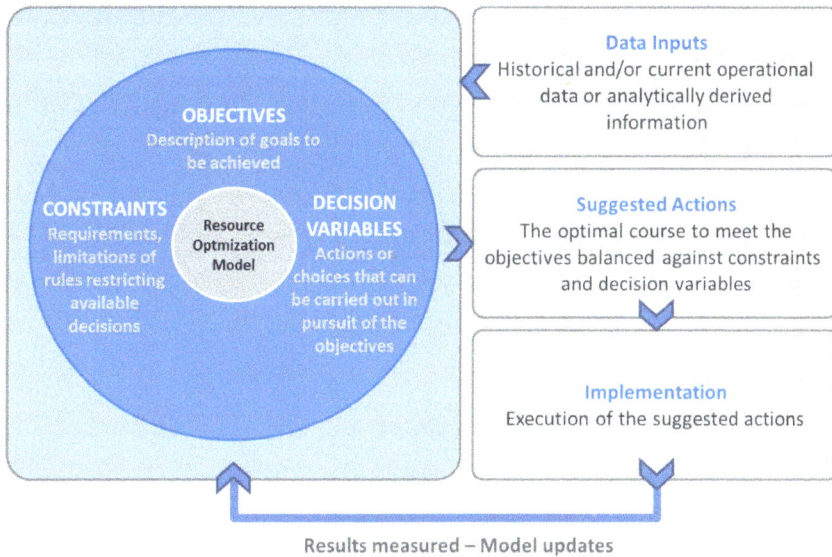

Figure 4. The components of optimization. The optimization models analyzes all possible decisions or actions based on given objectives and constraints.

Step 1. Define the objective to reflect the model mission and strategy.

The objectives to be pursued need to be determined—what the resources are meant to do, and how they are characterized, described, and cataloged for assessing their relevance and availability to the tasks to be performed. Moreover, the activities to reach these objectives should be outlined, as well as how success or failure will be measured.

Step 2. Establish the context.

The requirements, rules, and constraints need to be established to precisely define the action scene of the optimization model and the decisions that will be made.

Step 3. Define the conceptual model.

All the elements of the model should be inserted into a conceptual framework. First of all, input data should be defined, and then decision variables and actions listed in accordance to objectives and constraints.

Step 4. Formulate the resource optimization model.

The conceptual model is then translated into an analytic model with more rigor and detail, represented in mathematical terms. The key elements of the optimization model—the objective, constraints, and decision variables—are initially coded. There is no single "correct" way to use mathematical expressions to represent the elements of a decision problem. Every formulation represents a compromise because no mathematical representation can reflect every detail of a real-world scenario. Good modeling balances realism and workability.

Step 5. Implement and update the model.

The model should finally be implemented, and analytical software can be useful for this task. The real application of the implemented model can then supply some hints about necessary changes to the model for improving performance.

To be more clear, we report as an example within the case study EDSS, the use of an actual device implemented and integrated—in particular, the device is a static floating buoy equipped with several sensors for water quality control, such as hydrocarbons detection, tide measurement, wind and waves measures, and several other environmental variables. Specific details of the sensorized buoy can be found in [42], but for this specific case-study we report it as one of the monitoring resources which could be used for prioritization and active monitoring of a marine area. For example, reception of an oil spill report has been simulated close to a sensorized buoy. Once this information was gathered by the MIS and understood by the EDSS, the latter could proactively perform further investigations autonomously by checking eventual resources in the surrounding area and querying them. In this case, one of the resources was the sensorized buoy, which was questioned and asked to perform new sampling on the water, without any need of user intervention. Once the results from the buoy were received and collected by the MIS, the dynamic risk map was recomputed and a new updated risk value issued for the area being monitored. Once this action was taken, the new risk map could be analyzed, and decisions could be made regarding whether a further proactive action should be made (i.e., maybe querying some different available resources) or whether the end users in charge of the monitoring should be notified about the general situation, in order to take into consideration all available variables [50].

6. Conclusions

In this paper, we addressed the problem of small-scale oil spills at sea, focusing on the scientific and technological advances in Marine Information Systems.

As a first step, existing solutions and approaches have been analyzed, surveying the ever-increasing number of initiatives and funded projects addressing complementary and different aspects—from the assessment of the environmental risks, to the preparedness with respect to polluting events, and from the spread of knowledge and the raising of environmental consciousness and awareness, to the skill development for the training of civil protection, marine pollution professionals, volunteers, and other related stakeholders.

Afterwards, a rationale for an integrated framework was proposed, which we assert to be an important tool as support for the deputed authorities and stakeholders, particularly in view of prompt remediation operations.

We are optimistic that the impact of these advances will grow and improve quality of life and the sea environment. The dedication of research scientists and technological advances in such areas as remote sensing, modeling, and electronic communications have taught us far more about the seas, the surrounding environment, and their resources.

We shared our experiences with regard to actions that it may have undertaken to reduce sea and coastal pollution. Our survey shows that we have learned to make better predictions about how the marine environment resources are responding at both the individual species and ecosystem levels. Improvements and advances in monitoring programs have also been considered—this allows for a more accurate assessment of changes and for a more effective dissemination of such information to policy-makers who would implement science-based management actions [51]. Tools have been developed, along with the knowledge, to design an instrument for the support of policy-making that increases the ability of tomorrow's generation to understand its position in the local, global, coastal, and marine environment, and to sustain that position. The produced tools and instruments proved to be effective and of actual use within the intervention chain of an oil-spill event, when these tools can act as a valuable Environmental Decision Support System for the deputed authorities and stakeholders involved.

Author Contributions: All the authors contributed equally to the paper's preparation.

Funding: This research has been partially supported by the EU FP7-Transport Project ARGOMARINE (Automatic oil-spill recognition and geopositioning integrated in a marine monitoring network, Grant No. 234096).

Conflicts of Interest: The authors declare no conflict of interest.

References

1. Fingas, M. *Oil Spill Science and Technology*; Gulf Professional Publishing: Houston, TX, USA, 2016.
2. Fingas, M.; Brown, C. Review of oil spill remote sensing. *Mar. Pollut. Bull.* **2014**, *83*, 9–23. [CrossRef] [PubMed]
3. Boteler, B.; Coastal, M.W. European Maritime Transport and Port Activities: Identifying Policy Gaps towards Reducing Environmental Impacts of Socio-Economic Activities. Available online: http://www.ecologic.eu/sites/files/presentation/2014/european-maritime-transport-and-port-activities_0.pdf (accessed on 21 January 2019).
4. Abdulla, A. *Maritime Traffic Effects on Biodiversity in the Mediterranean Sea: Legal Mechanisms to Address Maritime Impacts on Mediterranean Biodiversity*; IUCN: Gland, Switzerland, 2008.
5. The International Tanker Owners Pollution Federation Limited. Oil Tanker Spill Statistics 2017. Available online: https://www.itopf.org/fileadmin/data/Photos/Statistics/Oil_Spill_Stats_2017_web.pdf (accessed on 9 January 2019).
6. Showstack, R. Research urged on impacts of chronic oil releases to marine environment. *Eos Trans. Am. Geophys. Union* **2002**, *83*, 254. [CrossRef]
7. Hyder, K.; Wright, S.; Kirby, M.; Brant, J. The role of citizen science in monitoring small-scale pollution events. *Mar. Pollut. Bull.* **2017**, *120*, 51–57. [CrossRef] [PubMed]
8. Margarit, G. Integrated maritime picture for surveillance and monitoring applications. In Proceedings of the 2013 IEEE International Geoscience and Remote Sensing Symposium-IGARSS, Melbourne, Australia, 21–26 July 2013; pp. 1517–1520.
9. CleanSeaNet. Available online: http://www.emsa.europa.eu/csn-menu.html (accessed on 18 January 2019).
10. Jordi, A.; Ferrer, M.; Vizoso, G.; Orfila, A.; Basterretxea, G.; Casas, B.; Álvarez, A.; Roig, D.; Garau, B.; Martínez, M.; et al. Scientific management of Mediterranean coastal zone: A hybrid ocean forecasting system for oil spill and search and rescue operations. *Mar. Pollut. Bull.* **2006**, *53*, 361–368. [CrossRef] [PubMed]
11. Ferraro, G.; Bernardini, A.; David, M.; Meyer-Roux, S.; Muellenhoff, O.; Perkovic, M.; Tarchi, D.; Topouzelis, K. Towards an operational use of space imagery for oil pollution monitoring in the Mediterranean basin: A demonstration in the Adriatic Sea. *Mar. Pollut. Bull.* **2007**, *54*, 403–422. [CrossRef]
12. Fustes, D.; Cantorna, D.; Dafonte, C.; Arcay, B.; Iglesias, A.; Manteiga, M. A cloud-integrated web platform for marine monitoring using GIS and remote sensing. Application to oil spill detection through SAR images. *Future Gener. Comput. Syst.* **2014**, *34*, 155–160. [CrossRef]
13. Janeiro, J.; Zacharioudaki, A.; Sarhadi, E.; Neves, A.; Martins, F. Enhancing the management response to oil spills in the Tuscany Archipelago through operational modelling. *Mar. Pollut. Bull.* **2014**, *85*, 574–589. [CrossRef]
14. Carpenter, A. The Bonn agreement aerial surveillance programme: Trends in North Sea oil pollution 1986–2004. *Mar. Pollut. Bull.* **2007**, *54*, 149–163. [CrossRef]
15. Moroni, D.; Pieri, G.; Tampucci, M.; Salvetti, O. A proactive system for maritime environment monitoring. *Mar. Pollut. Bull.* **2016**, *102*, 316–322. [CrossRef]
16. Fedra, K. Integrated risk assessment and management: Overview and state of the art. *J. Hazard. Mater.* **1998**, *61*, 5–22. [CrossRef]
17. Fedra, K. Environmental Decision Support Systems: A Conceptual Framework and Application Examples. Ph.D. Thesis, Université de Genéve, Geneva, Switzerland, 2000.
18. Mansfield, R.; Moohan, J. The evaluation of land remediation methods. *Land Contam. Reclam.* **2002**, *10*, 25–31. [CrossRef]
19. Cortés, U.; Sànchez-Marrè, M.; Ceccaroni, L.; R-Roda, I.; Poch, M. Artificial intelligence and environmental decision support systems. *Appl. Intell.* **2000**, *13*, 77–91. [CrossRef]
20. McIntosh, B.S.; Ascough, J.; Twery, M.; Chew, J.; Elmahdi, A.; Haase, D.; Harou, J.J.; Hepting, D.; Cuddy, S.; Jakeman, A.J.; et al. Environmental decision support systems (EDSS) development—Challenges and best practices. *Environ. Model. Softw.* **2011**, *26*, 1389–1402. [CrossRef]
21. Laniak, G.F.; Olchin, G.; Goodall, J.; Voinov, A.; Hill, M.; Glynn, P.; Whelan, G.; Geller, G.; Quinn, N.; Blind, M.; et al. Integrated environmental modeling: A vision and roadmap for the future. *Environ. Model. Softw.* **2013**, *39*, 3–23. [CrossRef]

22. EPA (US Environmental Protection Agency). *Toward Integrated Environmental Decision-Making*; Science Advisory Board: Washington, DC, USA, 2000.

23. Kelly, R.A.; Jakeman, A.J.; Barreteau, O.; Borsuk, M.E.; El Sawah, S.; Hamilton, S.H.; Henriksen, H.J.; Kuikka, S.; Maier, H.R.; Rizzoli, A.E.; et al. Selecting among five common modelling approaches for integrated environmental assessment and management. *Environ. Model. Softw.* **2013**, *47*, 159–181. [CrossRef]

24. Forsyth, R. The expert systems phenomenon. In *Expert Systems Principles and Case Studies*; Chapman & Hall: Upper Saddle River, NJ, USA, 1989; pp. 3–21.

25. Nevalainen, M.; Helle, I.; Vanhatalo, J. Preparing for the unprecedented—Towards quantitative oil risk assessment in the Arctic marine areas. *Mar. Pollut. Bull.* **2017**, *114*, 90–101. [CrossRef] [PubMed]

26. Mokhtari, S.; Hosseini, S.M.; Danehkar, A.; Azad, M.T.; Kadlec, J.; Jolma, A.; Naimi, B. Inferring spatial distribution of oil spill risks from proxies: Case study in the north of the Persian Gulf. *Ocean Coast. Manag.* **2015**, *116*, 504–511. [CrossRef]

27. Jolma, A.; Lehikoinen, A.; Helle, I.; Venesjärvi, R. A software system for assessing the spatially distributed ecological risk posed by oil shipping. *Environ. Model. Softw.* **2014**, *61*, 1–11. [CrossRef]

28. Zodiatis, G.; De Dominicis, M.; Perivoliotis, L.; Radhakrishnan, H.; Georgoudis, E.; Sotillo, M.; Lardner, R.; Krokos, G.; Bruciaferri, D.; Clementi, E.; et al. The mediterranean decision support system for marine safety dedicated to oil slicks predictions. *Deep Sea Res. Part II Top. Stud. Ocean.* **2016**, *133*, 4–20. [CrossRef]

29. Perivoliotis, L.; Krokos, G.; Nittis, K.; Korres, G. The Aegean sea marine security decision support system. *Ocean Sci.* **2011**, *7*, 671–683. [CrossRef]

30. Zodiatis, G.; Lardner, R.; Solovyov, D.; Panayidou, X.; De Dominicis, M. Predictions for oil slicks detected from satellite images using MyOcean forecasting data. *Ocean Sci.* **2012**. [CrossRef]

31. Bahurel, P.; Adragna, F.; Bell, M.J.; Jacq, F.; Johannessen, J.A.; Le Traon, P.Y.; Pinardi, N.; She, J. Ocean monitoring and forecasting core services: The European MyOcean Example. *Proc. Ocean.* **2009**, *9*, 2.

32. Pallotta, G.; Horn, S.; Braca, P.; Bryan, K. Context-enhanced vessel prediction based on Ornstein-Uhlenbeck processes using historical AIS traffic patterns: Real-world experimental results. In Proceedings of the 17th International Conference on Information Fusion (FUSION), Salamanca, Spain, 7–10 July 2014; pp. 1–7.

33. Lardner, R.; Zodiatis, G. Modelling oil plumes from subsurface spills. *Mar. Pollut. Bull.* **2017**, *124*, 94–101. [CrossRef] [PubMed]

34. NEREIDs Project. Available online: http://www.nereids.eu (accessed on 21 January 2019).

35. Alves, T.M.; Kokinou, E.; Zodiatis, G.; Radhakrishnan, H.; Panagiotakis, C.; Lardner, R. Multidisciplinary oil spill modeling to protect coastal communities and the environment of the Eastern Mediterranean Sea. *Sci. Rep.* **2016**, *6*, 36882. [CrossRef] [PubMed]

36. Alves, T.M.; Kokinou, E.; Zodiatis, G. A three-step model to assess shoreline and offshore susceptibility to oil spills: The South Aegean (Crete) as an analogue for confined marine basins. *Mar. Pollut. Bull.* **2014**, *86*, 443–457. [CrossRef] [PubMed]

37. Alves, T.M.; Kokinou, E.; Zodiatis, G.; Lardner, R.; Panagiotakis, C.; Radhakrishnan, H. Modelling of oil spills in confined maritime basins: The case for early response in the Eastern Mediterranean Sea. *Environ. Pollut.* **2015**, *206*, 390–399. [CrossRef] [PubMed]

38. Tosi, L.; Lio, C.D.; Teatini, P.; Menghini, A.; Viezzoli, A. Continental and marine surficial water–groundwater interactions: The case of the southern coastland of Venice (Italy). *Proc. Int. Assoc. Hydrol. Sci.* **2018**, *379*, 387–392. [CrossRef]

39. Delpeche-Ellmann, N.C.; Soomere, T. Investigating the Marine Protected Areas most at risk of current-driven pollution in the Gulf of Finland, the Baltic Sea, using a Lagrangian transport model. *Mar. Pollut. Bull.* **2013**, *67*, 121–129. [CrossRef]

40. Miliou, A.; Quintana, B.; Kokinou, E.; Alves, T.; Nikolaidis, A.; Georgiou, G. Enhancing Students Critical Thinking About Marine Pollution Using Scientifically-Based Scenarios. In Proceedings of the CRETE 2018—Sixth International Conference on Industrial & Hazardous Waste Management, Tinos, Greece, 13–17 July 2018.

41. Liao, S.H. Expert system methodologies and applications—A decade review from 1995 to 2004. *Exp. Syst. Appl.* **2005**, *28*, 93–103. [CrossRef]

42. Moroni, D.; Pieri, G.; Salvetti, O.; Tampucci, M.; Domenici, C.; Tonacci, A. Sensorized buoy for oil spill early detection. *Methods Ocean.* **2016**, *17*, 221–231. [CrossRef]

43. Martinelli, M.; Moroni, D. Volunteered Geographic Information for Enhanced Marine Environment Monitoring. *Appl. Sci.* **2018**, *8*, 1743. [CrossRef]
44. Infrastructure for Spatial Information in the European Community (EU INSPIRE). Directive: Directive 2007/2/EC of the European Parliament and of the Council of 14 March 2007 establishing an Infrastructure for Spatial Information in the European Community (INSPIRE). *Off. J. Eur. Union L* **2007**, *108*, 50.
45. Laskey, K.B.; Laskey, K. Service oriented architecture. *Wiley Int. Rev. Comput. Stat.* **2009**, *1*, 101–105. [CrossRef]
46. Wang, L.; Von Laszewski, G.; Younge, A.; He, X.; Kunze, M.; Tao, J.; Fu, C. Cloud computing: A perspective study. *New Gener. Comput.* **2010**, *28*, 137–146. [CrossRef]
47. Brunelière, H.; Cabot, J.; Jouault, F. Combining Model-Driven Engineering and Cloud Computing. In Proceedings of the Sixth European Conference on Modelling Foundations and Applications (ECMFA 2010), Paris, France, 15–18 June 2010.
48. Snyman, J.A.; Wilke, D.N. *Practical Mathematical Optimization*; Springer: Berlin/Heidelberg, Germany, 2018.
49. Gasparotti, C.; Rusu, E. Methods for the risk assessment in maritime transportation in the Black Sea basin. *J. Environ. Prot. Ecol.* **2012**, *13*, 1751–1759.
50. Pieri, G.; Cocco, M.; Salvetti, O. A marine information system for environmental monitoring: ARGO-MIS. *J. Mar. Sci. Eng.* **2018**, *6*, 15. [CrossRef]
51. Global Congress on Integrated Coastal Management EMECS 10—MEDCOAST 2013 Joint Conference. Marmaris Declaration. Available online: https://www.medcoast.net/uploads/documents/Marmaris_Declaration.pdf (accessed on 2 November 2013).

Journal of
Marine Science and Engineering

MDPI

Article

Oil Slick Characterization Using a Statistical Region-Based Classifier Applied to UAVSAR Data

Patrícia C. Genovez [1,*], Cathleen E. Jones [2], Sidnei J. S. Sant'Anna [1] and Corina C. Freitas [1]

[1] Image Processing Division (DPI), Earth Observation Coordination, Brazilian Institute for Space Research (INPE): Avenida dos Astronautas 1758, São José dos Campos 12227-010, Brazil; sidnei.santanna@inpe.br (S.J.S.S.); corina.freitas@gmail.com (C.C.F.)

[2] Jet Propulsion Laboratory (JPL), California Institute of Technology, 4800 Oak Grove Dr., Pasadena, CA 91109, USA; cathleen.e.jones@jpl.nasa.gov

* Correspondence: genovez.oilspill@gmail.com; Tel.: +55-12-3208-6444

Received: 31 October 2018; Accepted: 25 December 2018; Published: 6 February 2019

check for updates

Abstract: During emergency responses to oil spills on the sea surface, quick detection and characterization of an oil slick is essential. The use of Synthetic Aperture Radar (SAR) in general and polarimetric SAR (PolSAR) in particular to detect and discriminate mineral oils from look-alikes is known. However, research exploring its potential to detect oil slick characteristics, e.g., thickness variations, is relatively new. Here a Multi-Source Image Processing System capable of processing optical, SAR and PolSAR data with proper statistical models was tested for the first time for oil slick characterization. An oil seep detected by NASA's Uninhabited Aerial Vehicle Synthetic Aperture Radar (UAVSAR) in the Gulf of Mexico was used as a study case. This classifier uses a supervised approach to compare stochastic distances between different statistical distributions (fx) and hypothesis tests to associate confidence levels to the classification results. The classifier was able to detect zoning regions within the slick with high global accuracies and low uncertainties. Two different classes, likely associated with the thicker and thinner oil layers, were recognized. The best results, statistically equivalent, were obtained using different data formats: polarimetric, intensity pair and intensity single-channel. The presence of oceanic features in the form of oceanic fronts and internal waves created convergence zones that defined the shape, spreading and concentration of the thickest layers of oil. The statistical classifier was able to detect the thicker oil layers accumulated along these features. Identification of the relative thickness of spilled oils can increase the oil recovery efficiency, allowing better positioning of barriers and skimmers over the thickest layers. Decision makers can use this information to guide aerial surveillance, in situ oil samples collection and clean-up operations in order to minimize environmental impacts.

Keywords: oil slicks characterization; oil thickness; polarized SAR data; polarimetric SAR data (PolSAR); statistical region-based classification; uncertainty maps; UAVSAR

---

1. Introduction

Petrogenic oil slicks in offshore areas can occur naturally through oil seeps or be caused by anthropogenic activities related to oil exploration, production and transportation. Depending on the amount and characteristics of the oil, as well as the sea state and drift direction, oil can reach coastal regions, increasing environmental damages.

During emergency response, the oil containment and recovery are the main cleanup operations with potential to minimize these impacts. The thickness of the oil slick has a significant effect on the recovery efficiency rates, being higher over the thicker layers [1,2].

Synthetic Aperture Radars (SAR) data are most frequently used by operational oil spill surveillance service providers to detect and monitor oil slicks on the sea surface. Within the microwave region of the electromagnetic spectrum, oil slicks and look-alikes dampen the sea surface roughness and are detected as low backscatter regions [1–6].

Previous research [2,5,7] indicated a relationship between the oil slick thickness and the damping effect on the sea surface roughness. Thicker layers cause more damping of the capillary and gravity-capillary waves, and hence appear darker than thinner layers of oil in SAR imagery.

The potential of the polarimetric SAR (PolSAR) data in full, dual, and compact polarization modes was indicated to distinguish petrogenic from biogenic oil slicks and other look-alikes in some circumstances [8–10]. However, methodologies able to characterize oil slicks by extracting additional information regarding the thickness variations within the slicks have been less explored until recently [2,5,11–14].

PolSAR data has the potential to detect a wide range of scattering mechanisms that may be related to oil slick thickness, weathering, as well as different concentrations of water in oil-mixtures and emulsions [2,5,8,12,13]. However, there is evidence that low noise airborne SAR instruments are also able to characterize oil slicks with single polarization SAR data. In these cases, the most sensitivity was obtained transmitting and receiving the electromagnetic pulse in the vertical direction (VV). The VV damping ratio, a contrast measure, or the VV intensity [2,11,13] can also be used.

The potential of SAR to detect and characterize an oil slick varies depending on several factors [4,6,8,15] such as: (*i*) physical characteristics of oil and oil layer: the denser, more viscous and thicker the oil, the higher the damping effect; (*ii*) Wind intensity: greater contrast within the limits considered ideal for oil detection, between 3 and 10 m/s; (*iii*) Currents intensity and wave height: the larger, the less the contrast of the slicks with the ocean; (*iv*) Radar frequency: the higher, the greater the interference from adverse atmospheric conditions, but also the greater the sensitivity to ocean capillary; (*v*) Polarization: the greater the number of polarimetric channels available, the greater potential to detect different scattering mechanisms; (*vi*) Incidence angle: greater contrast of the slicks in the near range within Bragg scattering limits, and; (*vii*) Signal to Noise Ratio (SNR): a higher SNR increases the potential of each polarization channel to detect oil slicks at sea surface.

The numerous aerial and orbital platforms available, acquiring PolSAR and SAR data with different configurations and formats (single look complex, intensity or amplitude) increase this challenge, and there is a need to understand which is the better format and statistical model to improve oil slick detection and characterization.

Considering all aforementioned factors, the better configuration to detect and characterize an oil slick may be different according to different acquisition scenarios. From the operational point of view, the use of only one polarization channel simplifies data acquisition, the statistical modeling needed, and reduces the complexity and the time required for processing. However, in some cases the polarimetric data may extract key information to characterize the oil slicks.

In this context, testing a Multi-Source Image Processing System developed to integrate SAR & PolSAR data of different formats and with different statistical properties, aiming to discriminate and characterize oil slicks, represents strategical research. This Multi-Source system, based on information theory and using stochastic distances to perform the region-based classification process, was previously developed [16–18]. The supervised classifier uses stochastic distances between different statistical distributions (*fx*) and hypothesis tests to classify the regions and associate confidence levels to the classification results. The multi-source approach permits the integration of SAR (intensity and amplitude), PolSAR (single look complex) and optical data considering proper statistical modelling for each type of data, and processing single or multi-source data in a customized computational system.

Previous research using satellite acquired PolSAR data [5,19] has shown the potential of this system, not to characterize, but rather to discriminate oil slicks considering different oil types. In that research, it is shown that the polarimetric information provided improvements that are statistically significant in terms of accuracy. It also shows that the classification performance is dependent on the input data format and, consequently, on the statistical modeling used to represent the diversity of formats properly.

The objective of the work reported here is to evaluate the potential of the Multi-Source Image Processing System to characterize thickness variations within the oil slick, indicating the better format and statistical modeling for this application, considering as a study case an oil seep detected by NASA's UAVSAR, with components ranging from thin sheen through thick emulsions [2]. As the system is being tested for the first time with this focus, a complete scientific investigation using the amount of available data and combining all formats is recommended.

To accomplish this investigation, the polarized information in intensity format as well as all polarimetric complex information contained in the PolSAR data was considered by using the full and dual-polarization (dual-pol) covariance matrices. The integration of the uncertainty levels in the interpretation process provided additional information to more reliably indicate the regions of likely thicker oil within the slick.

The fast processing of multiple products, all of them integrated to generate operational maps to inform the position, area and likely thicker oil layers, constitutes an important method to be used by the contingency team during the clean-up operations. This specific demand for spatial intelligence during oil spill emergencies emphasizes the importance of developing and testing robust image processing systems, as proposed in this research. As indicated by reference [2], once validated, a system like that could be implemented on an aircraft and incorporated into an on-board processor (OBP) to be used operationally, transmitting all information in near real time (NRT) to the incident command system (ICS).

2. Statistical Modeling Classification Based on Stochastic Distances

The statistical classifiers require proper models to represent the statistical nature of the pixels (or image regions). Depending on the sensor used (optical, microwave, etc.), the acquired data format (polarimetric, intensity, amplitude, etc.) and the scene backscattering characteristics, different probability density functions (fx) are needed to represent and process the data properly [20].

Research to develop and test computational systems able to process this diversity of data is essential, especially considering the wide range of remote sensors available operating in multi-frequency and multi-resolution. Optimized systems to integrate different data sources, as well as to deliver operational reports in near real time, are needed to plan and implement response actions during emergencies involving oil spills.

In this way, a Multi-Source Image Processing System able to utilize this data diversity and complexity has been developed and tested by the Brazilian Institute for Space Research (INPE) [16–18] in different real applications, including oil slick detection. The Multi-Source statistical region-based classifier performs a supervised classification using stochastic distances (d) and statistical tests (S), considering proper statistical modeling for different data formats [16–18]. Its architecture (Figure 1) is designed in four processing modules according to the statistical distribution (fx) of the input data:

- Polarimetric: consider as input full or dual-pol covariance matrices in a complex format, assuming that the data comes from a Scaled Complex Wishart distribution (fx = SCW);
- Intensity Pair: consider as input a pair of SAR images in intensity format, assuming as statistical distribution the multi-look Intensity Pair (fx = IP);
- Intensity Single-channel: consider as input each channel individually in intensity format, assuming the Gamma distribution for the data (fx = G);
- Multivariate Amplitude: consider as input optical and/or SAR data in amplitude format, assuming the Multivariate Gaussian (fx = MG) as statistical distribution.

This is an innovative approach that permits us to process and integrate different data types in a single computational system, considering two approaches: (*i*) Mono-Source: process each data type independently, and; (*ii*) Multi-Source: integrate the mono-source classifications searching for the best classification performance.

Figure 1 illustrates the architecture of the Multi-Source Image Processing System, indicating the proper statistical modeling for each data type and the stochastic distances available for each module. The type of mathematical solution used—analytical or numerical—was also indicated.

In order to understand the trade-off between each data type in capability to characterize an oil slick, a mono-source approach, with proper statistical modeling, was used to process three different types of SAR & PolSAR data, those being: (*i*) Polarimetric full & dual-pol; (*ii*) intensity pairs (Intensity Pair); and (*iii*) single-channel intensity (Intensity single-channel).

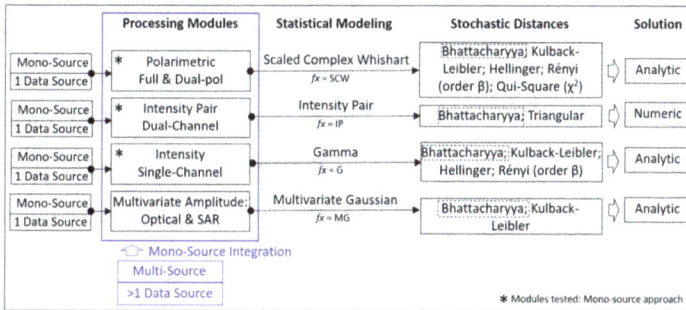

Figure 1. The Multi-Source image processing architecture indicating the four module configurations.

A region-based classification provides K disjoint segments, $R_1, \ldots R_k$; and a set of training samples defined for each class. The data (pixels) in the segment k are denoted by Z_{ik}, with $k = 1, \ldots, K$ and $i = 1, \ldots, N_k$, where N_k is the number of pixels within the segment k.

It is important to highlight that Z_{ik} is a matrix whereby the SAR data can be represented in complex, intensity or amplitude format assuming different statistical distributions (*fx*). In this way, assuming that Z_{ik} ($k = 1, \ldots, K$) follows a *fx* distribution with parameters Σ_k and L (number of looks), the maximum likelihood (ML) estimator of Σ_k is:

$$\hat{\Sigma}_k = N_k^{-1} \sum_{i=1}^{N_k} Z_{ik},\tag{1}$$

The purpose of the classifier is to classify each of the k segments into one of the C classes, assuming that (*i*) the training samples follows a *fx* distribution with parameter L and Σ_{C_j}, with $j = 1, \ldots, C$; and (*ii*) the ML estimator of Σ_{C_j}, based on samples of size M_j, is denoted by $\hat{\Sigma}_{C_j}$. This classification is performed calculating stochastic distances (*d*) between different statistical distributions (*fx*) associated to each region R_k ($k = 1, \ldots, K$) and to each training sample of class C_j. Then, using these distances and the results given in references [21,22], a statistic S_{kj} is used to perform a statistical test to verify the hypothesis that $\Sigma_k = \Sigma_{C_j}$ for all k, j ($k = 1, \ldots, K$ and $j = 1, \ldots, C$).

At the end of the process, each segment R_k is assigned to a class which presents: (*i*) the lower distance (*d*) between $\hat{\Sigma}_k$ and $\hat{\Sigma}_{C_j}$, (*ii*) the lower or equivalently statistical test (S_{kj}), and a higher associated *p*-value p_{kj}. The Bhattacharyya stochastic distance (d_B) was used because it is available in all modules (Figure 1), being the only option to compare the classifier potential combining all data formats and polarimetric channels.

For the Polarimetric module, the d_B between two SCW distributions (d_{WB}), one associated to segment k and the other associated to the class j, is given by [21]:

$$d_{WB}\left(\hat{\Sigma}_K, \hat{\Sigma}_j\right) = L \left[\frac{\log|\hat{\Sigma}_K| + \log|\hat{\Sigma}_{C_j}|}{2} - \log\left| \left(\frac{\hat{\Sigma}_K^{-1} + \hat{\Sigma}_{C_j}^{-1}}{2} \right)^{-1} \right| \right],\tag{2}$$

For the Intensity Single-channel module, the d_B distance was derived between Gamma distributions (d_{GB}), being [18]:

$$d_{GB}\left(\hat{\Sigma}_K, \hat{\Sigma}_j\right) = \log\left(\frac{\left(\hat{\Sigma}_K + \hat{\Sigma}_{C_j}\right)^L}{2^L \left(\hat{\Sigma}_K \hat{\Sigma}_{C_j}\right)^{(L/2)}}\right),$$ (3)

Under the conditions stated by [11,12], the hypothesis test $H_0 : \Sigma_k = \Sigma_{C_j}$ can be performed using the test statistics S_{fx} defined by each stochastic distance. The equations developed for the statistical tests between SCW distributions (S_{WB}) and between the Gamma distributions (S_{GB}) are given in the Equations (4) and (5), following [16,18]:

$$S_{WB}\left(\hat{\Sigma}_K, \hat{\Sigma}_{C_j}\right) = \frac{8N_K M_j}{N_K + M_j} d_{WB}\left(\hat{\Sigma}_K, \hat{\Sigma}_{C_j}\right),$$ (4)

$$S_{GB}\left(\hat{\Sigma}_K, \hat{\Sigma}_{C_j}\right) = \frac{4N_K M_j}{N_K + M_j} d_{GB}\left(\hat{\Sigma}_K, \hat{\Sigma}_{C_j}\right),$$ (5)

The equations for the Intensity Pair module were defined by reference [23], being the distance (d_{IPB}) and statistical test (S_{IPB}) derived from a Bivariate Gamma distribution extracted from the SCW distribution. However, the distance and the statistical test need an extensive and complex numerical solution, available in reference [16].

The null hypothesis is rejected at α significance level if the probability $\Pr\left(\chi_\nu^2 > S_{fx}\left(\hat{\Sigma}_K, \hat{\Sigma}_{C_j}\right)\right) \leq \alpha$, where χ_ν^2 represents a chi-square distribution with ν degrees of freedom, where ν is the number of parameters of the distribution. The classification based on a minimum test statistic consists in assigning the segment (R_k) to the class C_l if:

$$S_{fx}\left(\hat{\Sigma}_K, \hat{\Sigma}_{C_l}\right) < S_{fx}\left(\hat{\Sigma}_K, \hat{\Sigma}_{C_j}\right), \quad \forall j \neq l,$$ (6)

When a segment R_k is assigned to the class l, the p-value ($p_{k,l}$) is calculated as:

$$p_{Kl} = \Pr\left(\chi_\nu^2 > S_{fx}\left(\hat{\Sigma}_K, \hat{\Sigma}_{C_l}\right)\right),$$ (7)

It is a measure of certainty that the segment k belongs to the class l. At the end of the process, the classification and the uncertainty ($1 - p_{kl}$) maps are provided. To illustrate the region-based classification, Figure 2 indicates the class assignment process considering only one segment as an example.

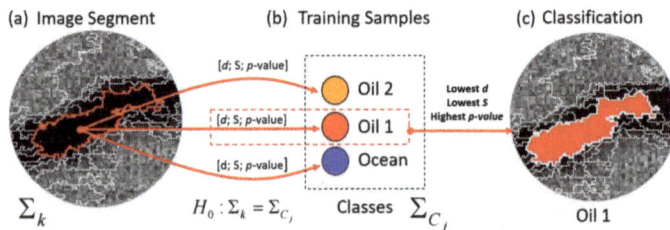

Figure 2. Demonstration of the region-based classification method: The algorithm calculates the distance (d), the statistical test (S) and p-value between the segment analyzed (**a**) and the training samples collected for each class (**b**). The class of the training sample, which presented the lowest d, the lowest S and the highest p-value in relation to the segment will be assigned to this segment (**c**). In this example, the segment used (**a**) was classified as Oil 1 (**c**), presenting the lowest d, the lowest S and the highest p-value related with this class.

3. SAR & PolSAR Data Description and Methodology

During Hurricane Ivan in 2004 a persistent seep source developed following damage caused to production infrastructure in the Mississippi Canyon 20 area of the Gulf of Mexico. This study case used L-Band PolSAR data acquired over this region on 17 November 2016 by the NASA's UAVSAR sensor. In this occasion, a large oil slick with thickness ranging from thin sheen through thick emulsion was detected [2].

In this study, a Multi-source Image Processing System—able to process different data formats, with different statistical properties—was tested to characterize and extract regions with different oil thicknesses using both UAVSAR SAR & PolSAR data. The goal was to differentiate sheen from thicker layers, as well as indicate the better format and statistical modeling to do this.

The PolSAR data provides the amplitude and phase information, while the polarized SAR data only provides the amplitude of the backscattered signal. The electromagnetic pulse can be transmitted and received by the antenna in different directions, Vertical (V) or Horizontal (H), being: (*i*) co-polarized—transmitting and receiving in one single direction (VV or HH), or; (*ii*) cross-polarized – transmitted and received in orthogonal directions (HV) or (VH). Different combinations are possible, defining the SAR systems as: (*i*) single: VV, HH or HV; (*ii*) dual: HH-HV, HH-VV or VV-HV, and; (*iii*) full (quad): HH-HV-VH-VV. The term ´polarimetric´ is applied only when the amplitude and phase information are available. For monostatic antennas, HV is considered equivalent to VH, configuring a full polarimetric system with 3 bands HH-HV-VV. Details about different mathematical forms to represent SAR and PolSAR data including the scattering matrix, covariance matrix (C) and others can be found in reference [24].

Standard full-polarimetric UAVSAR products in ground projected format (grd) contain the calibrated complex cross products HHHH, HVHV, VVVV, HHHV, HVHV and HVVV used to calculate the elements of the multi-looked covariance matrix (C). The PolSAR image was acquired with 20 km swath width, incidence angle 22° (near range) to 67° (far range), and 7 m spatial resolution after multilooking, and the intensity products were extracted from the main diagonal of the covariance matrix.

To perform the case study, three types of data were used as input. The Polarimetric module used as input the covariance matrices in a full-pol (C3: HH-HV-VV) and dual-pol (C2: HH-HV or HH-VV or VV-HV) format. The Intensity Pair module uses as input pairs of intensity images (IP) combining the polarized channels as indicated: HH-HV or HH-VV or VV-HV. The Intensity Single-channel module processes each channel individually using the intensity (I) format as input. The segmented image, the training and test samples are also needed as input to process the classification in all modules. Figure 3 depicts the classification methodology, indicating the input and output data.

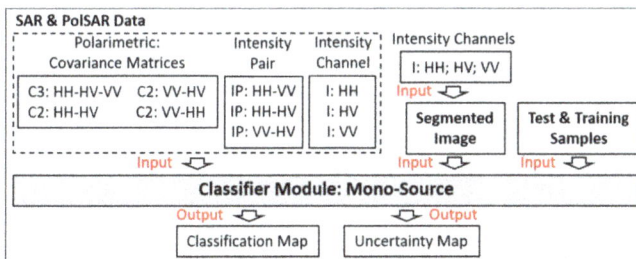

Figure 3. The statistical region-based classification methodology, indicating the input and output data.

The segmented image (Figure 4) was obtained applying the multi-level region-growing algorithm, MultiSeg [25]. MultiSeg is a hierarchical segmentor which uses the information contained at the top level to segment the subsequent levels through a pyramidal compression, integrating region growing and clustering techniques, edge detection, minimum area threshold and homogeneity tests to split

& merge pixels in segments. The cartoon model, used in this work, considers that the image is formed by homogeneous regions, being themselves clustered according to predefined parameters. The larger the similarity index, the greater the area of the generated segments. In this study case the following parameters were considered: (*i*) 5 levels of compression; (*ii*) minimum area of 20 pixels; (*iii*) 1 dB of similarity degree, and (*iv*) 9.17 equivalent number of looks.

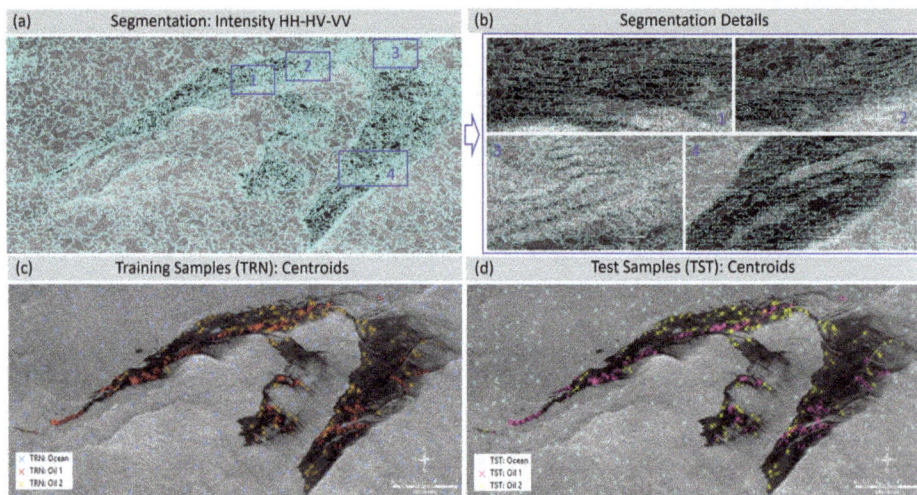

Figure 4. (**a**) Multiresolution Segmentation, (**b**) details about dark spots detection, and (**c**,**d**) Centroids position of the training (**c**) and test (**d**) samples.

The segmentation process was completely unsupervised and considered as input the three polarized channels HH-HV-VV in intensity format. The algorithm automatically recognized 11,772 representative segments (Figure 4a), which delineated even the smaller dark regions (Figure 4b) originated by the influence of the internal waves and currents. This result is relevant considering that without representative segments it is impossible reach a good classification accuracy.

Considering the aforementioned oil slick properties, three classes were defined, namely, (*i*) Ocean, (*ii*) Oil 1: representing the thicker layers (crude oil or emulsion), and (*iii*) Oil 2: representing the thinner layers (sheen). The classes definition was reinforced by previous research which detected thickness variations within the same oil slick using the VV damping ratio [2]. As described by reference [2], the VV damping ratio is the contrast between the clean sea water vs. slicked water, using the VV channel in intensity format (VVclean/VV).

The samples collection considered that thicker layers cause more damping, appearing darker than thinner layers, as observed previously [2,7,11–13]. Therefore, the training and test samples were collected manually, based on the interpreter's experience, searching for darker areas to represent the thicker oil (class Oil 1) and the less dark areas to represent the thinner oil (class Oil 2). The ocean samples were collected over the brighter areas in the background. Figure 4c,d illustrates the centroids position of the training (TRN) and test (TST) samples collected per class, being the total number of pixels collected per class: 1887 for Ocean-TRN, 1890 for Oil 1-TRN, 1886 for Oil 2-TRN, 1887 for Ocean-TST, 1890 for Oil 1-TST and 1886 for Oil 2-TST.

The class assignment process associates the class of the training sample to the analyzed segment, considering the lowest statistic and the highest *p*-value, all computed for each segment k ($k = 1, \ldots, 11{,}772$) and each class C_j ($j = 1, \ldots, 3$). The statistical reports, as well as the classification and uncertainty maps, are the classifier output. The validity of the mineral oil slick characterization was evaluated considering visual interpretation, statistical accuracy indexes and the statistical uncertainty levels.

It is important to note that the proposed methodology is designed to be applied when mineral oil slicks are confirmed in the field, determining a real emergency situation. In this context, the goal is to characterize the oil slicks and extract additional information such as the relative thickness variation within the slicks. The spatial location, the area of the slicks, as well as identifying the thickest layers are all extremely important data to support decision making during clean-up operations.

4. Oil Slick Characterization

The classification results obtained for all formats tested are available in Table 1 and Figure 5 providing (*i*) the overall accuracies, (*ii*) the variances of the overall accuracies, (*iii*) the Kappa coefficient of agreement, and (*iv*) the Kappa Variance. The Kappa coefficient is another index to do the accuracy assessment in remote sensing data classification. An explanation can be found in references [26–28].

The oil slick characterization using a statistical approach and applying stochastic distances achieved global accuracies above 99% for all data types. However, the best result was obtained by the IP: HH-VV (99.84%). This result is statically equivalent to C3: HH-HV-VV (99.70%), C2: HH-VV (99.68%) and I: VV (99.68%) (Table 1: results highlighted in blue) and statistically superior to the remaining results, at the 95% confidence level.

Despite these differences, the results obtained by the best data formats were very similar and statistically equivalent. Therefore, for this study case, the oil slick characterization can be done using a fully polarized data, as well as using only the VV channel in intensity, as these results are statistically equivalent.

Table 1. Statistical evaluation of the classification results.

		Overall Accuracy	Variance	Kappa	Kappa Variance
Full-Pol	HH-HV-VV	0.9970	5.28×10^{-7}	0.9955	1.19×10^{-6}
Dual-Pol	HV-VV	0.9938	1.09×10^{-6}	0.9907	2.44×10^{-6}
	HH-HV	0.9935	1.14×10^{-6}	0.9902	2.58×10^{-6}
	HH-VV	0.9968	5.64×10^{-7}	0.9952	1.26×10^{-6}
Intensity Pair	HV-VV	0.9954	8.09×10^{-7}	0.9931	1.82×10^{-6}
	HH-HV	0.9935	1.14×10^{-6}	0.9902	2.58×10^{-6}
	HH-VV	0.9984	2.82×10^{-7}	0.9976	6.31×10^{-7}
Intensity Single-Channel	HH	0.9811	3.28×10^{-6}	0.9716	7.37×10^{-6}
	HV	0.9793	3.58×10^{-6}	0.9690	8.05×10^{-6}
	VV	0.9968	5.64×10^{-7}	0.9952	1.26×10^{-6}

The classification maps and the detected area (km²) per class are available at Figure 5a–d, only the best overall accuracies are considered statistically equivalent. It is interesting to note that the potential to recognize different patterns within the oil slick was more stable for the thicker layers of oil, represented by the class Oil 1 (red regions). For this class, the detected area was very similar between these different data types, ranging between 13 and 14 km².

A higher confusion visible in the background of all classification maps was observed between the classes Oil 2 (orange regions) and Ocean (blue regions). However, a higher instability was observed for the I: VV and C2: HH-VV, which presented higher classification noise in the background generated by the confusion between these classes, returning the smaller oceanic areas ($\approx$156 km²).

To conduct a detailed analysis regarding the confidence levels applied to oil slick characterization, only the C3 classification result was considered (Figure 6).

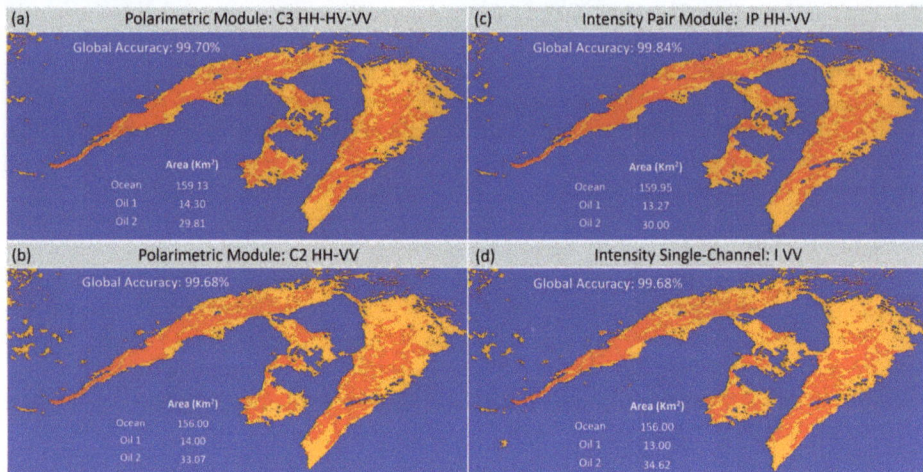

Figure 5. Classification results obtained by the better overall accuracies, all of then indicated as statistically equivalent: (**a**) covariance matrix full-pol (C3: HH-HV-VV): (**b**) covariance matrix dual-pol (C2: HH-VV), (**c**) Intensity Pair (IP: HH-VV), and (**d**) Intensity Single-channel (I: VV).

The criteria used for this choice were: (*i*) the best results are very similar so it is feasible to choose any one of them to evaluate details, (*ii*) the C3 classification had lower confusion between the classes Ocean and Oil 2, (*iii*) using covariance matrix results, it is possible to discuss the results in terms of scattering mechanisms if the returns are significantly above the noise floor.

Figure 6a illustrates the oil slick detected by the UAVSAR using the VV channel. The classification map (Figure 6b) and uncertainty map (Figure 6c) are available only for the result C3. In Figure 6c low uncertainties are represented in black, while high uncertainties are represented in white. Figure 6d–g provides details of the oil slick classification along selected ocean features, allowing us to compare the dark spots backscattering, the classification results and the uncertainty levels.

The results show the potential of the Multi-source Image Processing System for characterizing the oil slick. The two oil classes show general differences with respect to their location within the slick, likely related to oil thickness variations, with Oil 1 class (red regions) likely related with the thicker layers, as discussed below, and the thinner layers, represented by the Oil 2 class (orange regions), spreading around the thicker layers.

Usually the thickest layers are concentrated at the center of the slicks, becoming gradually thinner towards the edges, where the spreading mechanisms are stronger, as shown in Figure 6d. However, the presence of the intense oceanic fronts and currents may create convergence zones (Figure 6e,g), which influence the dispersion, shape and concentration of the thickest layers within the oil slicks. A similar effect occurs when internal waves or fronts are observed (Figure 6f). In the UAVSAR data used for this study, these two patterns are present, influencing the concentration of the thickest oil layers within the slick.

The first pattern is visible in the Figure 6d and the second in Figure 6e,g, where the thickest layers are concentrated near the borders in the convergence zones, being influenced by the currents, as well as following the geometry of the oceanic fronts (Figure 6 e,g) and the internal waves (Figure 6f). This is consistent with the oil being trapped by the internal waves and concentrated along the oceanic fronts. The detection of zones within slicks was also demonstrated in other published research using only the VV channel [2,11,13] or including the polarimetric information [5,12]. The authors of reference [2], evaluating the same slick as the one studied here, showed that the VV-intensity contrast between clean and slicked water (damping ratio) could be used to identify likely concentrated oil along convergence features in the scene.

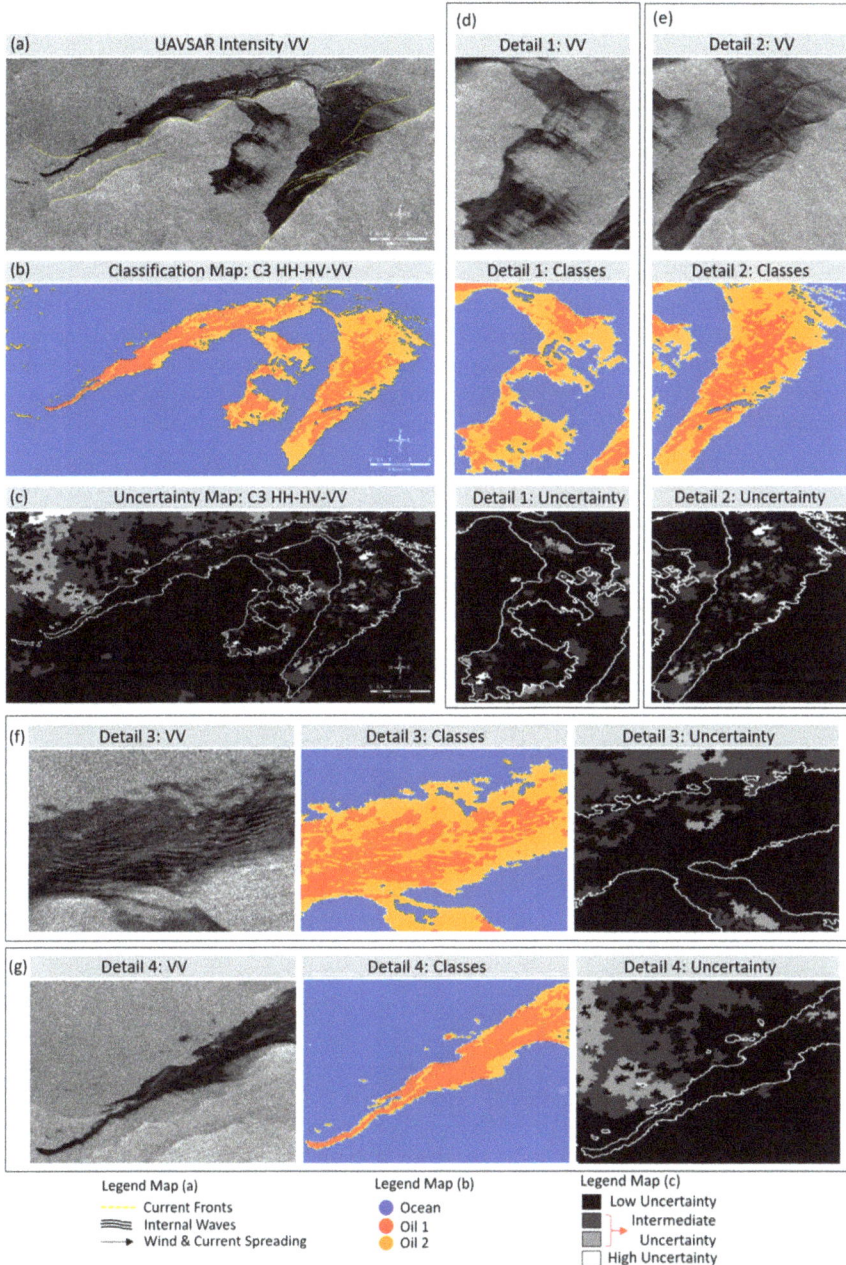

Figure 6. Classification results obtained by the covariance matrix (C3: HH-HV-VV): (**a**) Oil Slick study case; (**b**) Classification map; (**c**) Uncertainty map; (**d–g**) three image regions in detail.

Regarding the classification uncertainty levels, for all classes, most of the regions show low uncertainties in the class assignment process. These regions are seen in dark in the uncertainty map (Figure 6c) and within detailed regions shown in Figure 6d–g. The ocean class has the greater uncertainty variability, with the higher incidence of regions with intermediate and high uncertainties.

Almost all regions classified as Oil 1 presented low uncertainty levels, as seen in Figure 6d–g. Oil 2 regions returned mainly low uncertainty levels but with some regions of moderate and high uncertainties, albeit lower than that observed for the ocean class. This makes sense considering that the thickest oil layers likely evolve in thin layers through spreading and some weathering processes, increasing the uncertainties during the class assignment process.

It is important to note that the multilevel segmentation algorithm was essential to provide the high level of accuracy achieved by the statistical region-based classifier. The satisfactory delineation of smaller dark paths was obtained through a hierarchical process, splitting and merging pixels from segments, according to statistical tests applied along the five compression levels.

Over the slick-free ocean, where the homogeneity is higher, the segmentor was able to merge more pixels in larger segments and to reduce the classification noise in the background (Figure 7a). However, the lower homogeneity within the oil slick generated a higher number of smaller segments, making it feasible to delineate and to detect small thicker layers (in red, Figure 7b). Figure 7b–f exemplify in details the potential of the tested method to characterize the oil slick, indicating the contour of the segments and the regions classified as Oil 1 (thicker layers: red) and Oil 2 (thinner layers: orange).

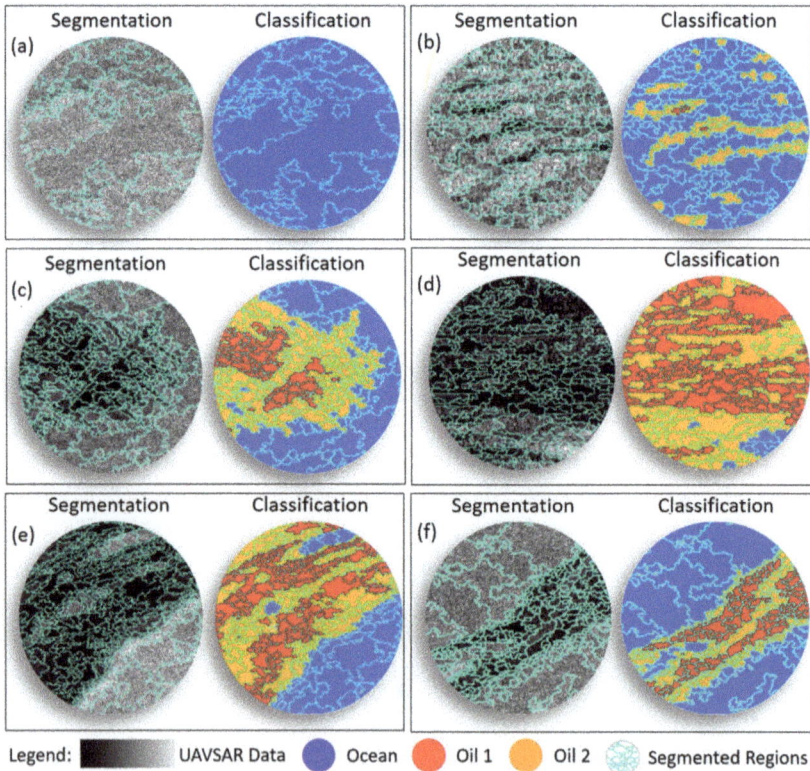

Figure 7. Oil slick characterization illustrating details of the segmented and classified regions over: (a) slick-free ocean regions, and; (b–f) oil slick covered regions with different thicknesses, sizes and geometries.

This type of information, organized in thematic maps and integrating in situ measurements about the sea state and meteorological conditions, is a valuable instrument to guide the response actions in the field. During emergencies involving rapid spillage of large volumes of oil or continuous long-term spills, the action of wind and currents can form kilometer-long slicks, which can be fragmented, contain weathered oil, and drift in multiple directions depending upon the time on surface and changing winds and currents. Especially in these situations, using remote sensing instruments to rapidly identify the location of the thicker oil layers is highly valuable to efficiently and effectively guide the aircraft flight planning and oil dispersion or recovery activities.

5. Conclusions

The Multi-source Image Processing System provided promising results in separating the slick into two classes of oil and differentiating sheen from thicker oil, a distinction that is important to responders. The classifier was able to detect zoning regions within the slick and to identify specific locations - convergence zones - where the oil was concentrated by the influence of the oceanic waves and currents. These areas are likely sites of the thicker oil layers. The statistical analysis showed that this method could classify most of the regions within oil slicks containing sheen and thicker layers with high global accuracy and low uncertainty levels. Further study in controlled releases of known amounts of oil or with in situ validation data would provide more stringent validation, but those studies are costly and their data are not often available.

In this study case, oil slick characterization is possible using fully polarized data, as well as only the VV channel in intensity format, with both being statistically equivalent. Other research using UAVSAR data characterized oil slicks using only the VV through the damping ratio [2,11,13]. One reason for this is the higher signal-to-noise ratio (SNR) provided by airborne SAR sensors compared with satellite SAR sensors [2]. Because polarimetric data contains all possible information about the sea surface backscattering, it is an important data source for understanding different scattering mechanisms. Thus, given that UAVSAR provides a complete polarimetric dataset, it is interesting to explore the possible combinations of multi-polarization and polarimetric data.

The equivalence of the classification using the three types of input data tested does not invalidate the results and importance of the Multi-Source classifier. Different classification accuracies may be reached by oil slicks detected in different wind and current conditions, acquired by airborne or satellite sensors, in multi-frequency, multi-resolutions and with different acquisition geometries. Within this broader trade space, polarimetric data may contribute to better discriminate and characterize oil slicks detected under diverse and unknown conditions, which is the situation faced by operational surveillance agencies, including differentiation of different types of surface slicks, such as from biogenic and petrogenic oils.

In particular during environmental emergencies when any and all remote sensing data are used for responses, the possibility to use and integrate all data available, including polarimetric and optical data, within the same processing system has both tactical and strategic advantages. The system tested was shown capable of extracting essential information on the location of relatively thicker oil from each SAR data set both individually and in combination, offering the capability to customize an operational tool to deliver to the incident command system (ICS) in near real time thematic maps and accuracy reports.

For the operational activities, the use of only one polarization channel simplifies the data acquisition, the statistical modeling needed, as well as the complexity and the time required for processing. In some cases, the polarimetric data may extract key information to characterize the oil slicks, but at the expense of increasing the time needed to process and evaluate the results. This reinforces the relevance and importance of continuing this avenue of research into SAR and PolSAR-based oil classification, and suggests extending it to include study cases acquired with multi-frequency, polarimetry, or different resolution, and including optical imagery. Therefore, consolidating a database with several examples of mineral oils and look-alikes validated in the field using a multi-sensor approach would be of high value to evaluating the full potential of the proposed system.

Considering that oil thickness has a significant effect on recovery efficiency, the possibility to identify the thicker layers of the spilled oil using SAR and PolSAR data is a significant contribution to ongoing efforts to improve emergency responses. Methods that work to locate the thickest oil layers irrespective of the particular type of available remote sensing data will aid in directing responders to the best regions for barriers and skimmer deployment, thereby increasing the oil recovery efficiency and ultimately minimizing environmental impacts.

Author Contributions: This publication was conceptualized, validated and supervised by all authors: P.C.G., C.E.J., S.J.S.S. and C.C.F. P.C.G. was in charge of writing the original draft, including and consolidating the contributions and revision provided by the authors. The data curation and acquisition was provided by C.E.J. from JPL-NASA. All software used was developed by INPE with support provided by the researchers S.J.S.S. and C.C.F. The methodology was defined by all authors and conducted by P.C.G.

Funding: This research was funded by the Brazilian Institute for Space Research (INPE) and CNPQ under Grants for Projects #303752/2013-0 and #314248/2014-5. The research of CEJ was funded by the National Aeronautics and Space Administration (NASA) task number NNN13D788T.

Acknowledgments: We are grateful to the Image Processing Division (DPI), from Brazilian Institute for Space Research (INPE), for providing all in-house software used to process the SAR & PolSAR data and by the support dedicated by the researchers Evlyn M. Novo and José C. Mura during this research development. UAVSAR data are courtesy of NASA/JPL-Caltech. The research was carried out in part at the Jet Propulsion Laboratory, California Institute of Technology, under contract with the National Aeronautics and Space Administration.

Conflicts of Interest: The authors declare no conflict of interest.

References

1. International Petroleum Industry Environmental Conservation Association (IPIECA). *International Association of Oil & Gas Producers (IOGP) Report 522: At-Sea Containment and Recovery Good Practice Guidelines for Incident Management and Emergency Response Personnel*; IPIECA: London, UK, 2015.
2. Jones, C.E.; Holt, B. Experimental L-band airborne SAR for oil spill response at sea and in coastal waters. *Sensors* **2018**, *18*, 641. [CrossRef] [PubMed]
3. Fingas, M.; Brown, C.E. A Review of Oil Spill Remote Sensing. *Sensors* **2018**, *18*, 91. [CrossRef] [PubMed]
4. Brekke, C.; Solberg, A.H.S. Oil spill detection by satellite remote sensing—Review. *Remote Sens. Environ.* **2005**, *95*, 1–13. [CrossRef]
5. Genovez, P.C.; Freitas, C.C.; Sant'Anna, S.; Bentz, C.M.; Lorenzzetti, J.A. Oil slicks detection from polarimetric data using stochastic distances between Complex Wishart distributions. *IEEE J. Sel. Top. Appl. Earth Obs. Remote Sens.* **2017**, *10*, 463–477. [CrossRef]
6. Genovez, P.C.; Ebecken, N.F.; Freitas, C.C.; Bentz, C.M.; Freitas, M.R. Intelligent hybrid system for dark spot detection using SAR data. *Expert Syst. Appl.* **2017**, *81*, 384–397. [CrossRef]
7. Angelliaume, S.; Dubois-Fernandez, P.C.; Jones, C.E.; Holt, B.; Minchew, B.; Amri, E.; Miegebielle, V. SAR Imagery for Detecting Sea Surface Slicks: Performance Assessment of Polarization-Dependent Parameters. *IEEE Trans. Geosci. Remote Sens.* **2018**, *56*, 4237–4257. [CrossRef]
8. Migliaccio, M.; Nunziata, F.; Buono, A. SAR polarimetry for sea oil slick observation. *Int. J. Remote Sens.* **2015**, *36*, 3243–3273. [CrossRef]
9. Espeseth, M.; Skrunes, S.; Jones, C.E.; Brekke, C.; Holt, B.; Doulgeris, A. Analysis of Evolving Oil Spills in Full-Polarimetric and Hybrid-Polarity SAR. *IEEE Trans. Geosci. Remote Sens.* **2017**, *55*, 4190–4210. [CrossRef]
10. Skrunes, S.; Brekke, C.; Eltoft, T. Characterization of marine surface slicks by Radarsat-2 multipolarization features. *IEEE Trans. Geosci. Remote Sens.* **2014**, *52*, 5302–5319. [CrossRef]
11. Minchew, B. Determining the mixing of oil and sea water using polarimetric synthetic aperture radar. *Geophys. Res. Lett.* **2012**, *39*, L16607. [CrossRef]
12. Minchew, B.; Jones, C.E.; Holt, B. Polarimetric Analysis of Backscatter from the Deepwater Horizon Oil Spill using L-Band Synthetic Aperture Radar. *IEEE Trans. Geosci. Remote Sens.* **2012**, *50*, 3812–3830. [CrossRef]
13. Jones, C.E.; Espeseth, M.M.; Holt, B.; Brekke, C.; Skrunes, S. Characterization and discrimination of evolving mineral and plant oil slicks based on L-band synthetic aperture radar (SAR). In Proceeding of the SPIE 10003, SAR Image Analysis, Modeling, and Techniques XVI, 100030K, Bellingham, WA, USA, 18 October 2016. [CrossRef]
14. Fingas, M. The Challenges of Remotely Measuring Oil Slick Thickness. *Remote Sens.* **2018**, *10*, 319. [CrossRef]

15. Skrunes, S.; Brekke, C.; Jones, C.E.; Holt, B. A Multisensor Comparison of Experimental Oil Spills in Polarimetric SAR for High Wind Conditions. *IEEE J. Sel. Top. Appl. Earth Obs. Remote Sens.* **2016**, *9*, 4948–4961. [CrossRef]

16. Silva, W.B.; Freitas, C.C.; Sant'Anna, S.J.S.; Frery, A.C. Classification of segments in PolSAR imagery by minimum stochastic distances between Wishart distributions. *J. Sel. Top. Appl. Earth Obs. Remote Sens.* **2013**, *6*, 1263–1273. [CrossRef]

17. Braga, B.C. Stochastic Distances and Associated Hypothesis Tests Applied to the Images Classification from Multiple Independent Sensors. Mater's Thesis, Department Remote Sensing, INPE, São José dos Campos, Brazil, 2016.

18. Santos, M.D.L. Mono and Multisource Image Classification System Based on Stochastic Distances and Associated Hypotheses Tests. Mater's Thesis, Applied Computing Division, Brazilian Institute for Space Research, São José dos Campos, São Paulo, Brazil, 2018.

19. Genovez, P.C.; Freitas, C.C.; Sant'Anna, S.; Bentz, C.M.; Lorenzzetti, J.A. *Oil Slicks Classification Using Multivariate Statistical Modeling Applied to SAR and PolSAR Data*; Anais Simpósio Brasileiro de Sensoriamento Remoto (XVIII SBSR), Mendes Convention Center: Santos, Brazil, 2017; ISBN 978-85-17-00088-1.

20. Nascimento, A.D.C. Statistical Information Theory Applied for Univariate and Polarimetric Synthetic Aperture Radar Data. Ph.D. Thesis, Federal University of Pernambuco (UFPE), Recife, Brazil, 2012; 280p.

21. Frery, A.C.; Nascimento, A.D.C.; Cintra, R.J. Analytic expressions for stochastic distances between relaxed complex Wishart distributions. *IEEE Trans. Geosci. Remote Sens.* **2014**, *52*, 1213–1226. [CrossRef]

22. Nascimento, A.D.C.; Cintra, R.J.; Frery, A.C. Hypothesis testing in speckled data with stochastic distances. *IEEE Trans. Geosci. Remote Sens.* **2010**, *48*, 373–385. [CrossRef]

23. Lee, J.S.; Hoppel, K.W.; Mango, S.A.; Miller, A.R. Intensity and Phase Statistics of Multilook Polarimetric and Interferometric SAR Imagery. *IEEE Trans. Geosci. Remote Sens.* **1994**, *32*, 1017–1028.

24. Richards, J.A. Remote sensing with imaging radar. In *Signals and Communication Technology*; Springer: New York, NY, USA, 2009.

25. Sousa, M.A., Jr. Multilevel Segmentation and Multi-Models for Radar and Optical Images. Ph.D. Thesis, Applied Computing Division, Brazilian Institute for Space Research, São José dos Campos, São Paulo, Brazil, 2005.

26. Lillesand, T.M. *Remote Sensing and Image Interpretation*, 5th ed.; Chapter 7; Digital Image Processing; John Wiley and Sons: New York, NY, USA, 2007; pp. 550–610.

27. Hudson, W.D.; Ramm, C.W. Correct formulation of the kappa coefficient of agreement. *Photogramm. Eng. Remote Sens.* **1987**, *4*, 421–422.

28. Congalton, R.G.; Green, K. *Assessing the Accuracy of Remotely Sensed Data: Principles and Practices*; CRC Press: Boca Raton, FL, USA, 2009.

Journal of
Marine Science and Engineering

MDPI

Article

An Improved Method to Estimate the Probability of Oil Spill Contact to Environmental Resources in the Gulf of Mexico

Zhen Li * and Walter Johnson

Bureau of Ocean Energy Management, Office of Environmental Programs, Sterling, VA 20166, USA;
Walter.Johnson@boem.gov
* Correspondence: Zhen.Li@boem.gov; Tel.: +1-703-787-1721

Received: 30 November 2018; Accepted: 3 February 2019; Published: 8 February 2019

check for
updates

Abstract: The oil spill risk analysis (OSRA) model is a tool used by the Bureau of Ocean Energy Management (BOEM) to evaluate oil spill risks to biological, physical, and socioeconomic resources that could be exposed to oil spill contact from oil and gas leasing, exploration, or development on the U.S. Outer Continental Shelf (OCS). Using long-term hindcast winds and ocean currents, the OSRA model generates hundreds of thousands of trajectories from hypothetical oil spill locations and derives the probability of contact to these environmental resources in the U.S. OCS. This study generates probability of oil spill contact maps by initiating trajectories from hypothetical oil spill points over the entire planning areas in the U.S. Gulf of Mexico (GOM) OCS and tabulating the contacts over the entire waters in the GOM. Therefore, a probability of oil spill contact database that stores information of the spill points and contacts can be created for a given set of wind and current data such that the probability of oil spill contact to any environmental resources from future leasing areas can be estimated without a rerun of the OSRA model. The method can be applied to other OCS regions and help improve BOEM's decision-making process.

Keywords: trajectory model; oil spill model; oil spill response; oil spill risk analysis; Gulf of Mexico; Outer Continental Shelf; environmental resources; risk modelling; Princeton Ocean Model; trajectory analysis

1. Introduction

The Gulf of Mexico (GOM), a semi-enclosed sea bordering the western Atlantic Ocean in the east and connected with the Caribbean Sea to the south, remains an important ecosystem that provides the Gulf Coast communities and nations with abundant fisheries and energy resources. Offshore oil production in the U.S. GOM Outer Continental Shelf (OCS) generally has increased over the past several decades, partly due to advancing technology. Today, the GOM OCS remains a significant source of oil and gas for the nation's energy needs. As of December 2017, OCS leases in the GOM produce 17 percent of domestic oil and 5 percent of domestic gas, and oil and gas production in the GOM OCS is forecasted to increase through 2024 [1].

According to the Outer Continental Shelf Lands Act (OCS Lands Act), established in 1953, the U.S. Department of the Interior (USDOI) has jurisdiction over OCS lands—submerged lands located generally 3 miles from state coastlines. Under the OCS Lands Act, the Bureau of Ocean Energy Management (BOEM) within the USDOI is responsible for managing the oil and gas resources in the OCS, with a goal of balancing the benefits derived from development of these resources with environmental protection. Prior to any offshore oil and gas leasing or approval of exploration and development plans, BOEM is required to prepare environmental analyses such as Environmental

Impact Statements (EISs) under the National Environmental Policy Act (NEPA). One of the key components in BOEM's EIS documents is the estimation of the likelihood of oil spill contact with biological, physical, social, and economic resources in the OCS. These resources are referred to as 'environmental resources' herein, with details discussed in Section 2.4.

The oil spill risk analysis (OSRA) model was developed by the USDOI in 1975 to evaluate the oil spill risks associated with the offshore oil and gas leasing and related activities to the environmental resources. A variety of environmental resources are considered in BOEM's OSRA model, including coastlines; water quality; archaeological and culture resources; recreation, tourism, and visual resources; environmental sensitive areas that represent concentrations of wildlife, habitat, or subsurface habitat; and national and state parks, refuges, and protected areas. The first application of OSRA was conducted in 1976 for the North Atlantic OCS Lease Area [2], and the first detailed documentation of the OSRA model was written by Smith et al. in 1982 [3]. The OSRA model has been verified with several oil spill incidents including the Argo Merchant incident off Nantucket Island in 1976 [4] and the Santa Barbara Channel blowout in 1969 [5], and the spill trajectories simulated by the OSRA model closely resembled the observed movements of the spill oils during these incidents.

As shown in Figure 1, the OSRA model delivers three products: conditional probability, oil spill occurrence, and combined probability. The calculation of the conditional probability begins with the construction of the oil spill trajectory. The trajectories are initiated every day and calculated every hour using long-term (decades) hindcast wind and current data. The conditional probability of contact to an environmental resource is calculated by dividing the number of contacts (i.e., number of times a trajectory reaches a location occupied by the environmental resource) in a given time by the total number of trajectories initiated within each hypothetical oil spill area. Only spills greater than or equal to 1000 barrels (referred to as 'large oil spills') undergo trajectory simulation in the OSRA model because smaller spills would not persist on water long enough for such analysis. The term 'conditional' is used to reflect the assumption (condition) that an accidental large oil spill occurs at hypothetical oil spill location. The OSRA model estimates the probability of large oil spills occurring from the prospective production sites and transportation routes of a specific volume over the lifetime of the scenario. The estimate of large spill occurrence at the production sites or transportation routes is based on the projected oil production volume, transportation scenarios, and historical spill occurrence in the U.S. OCS [6–8]. Finally, the OSRA model uses the conditional probability and estimated oil spill occurrence relative to the production volume and transport scenarios to derive the combined probability. The 'combined' probability is the overall probability of one or more large oil spill occurring and contacting the environmental resources over the lifetime of the scenario.

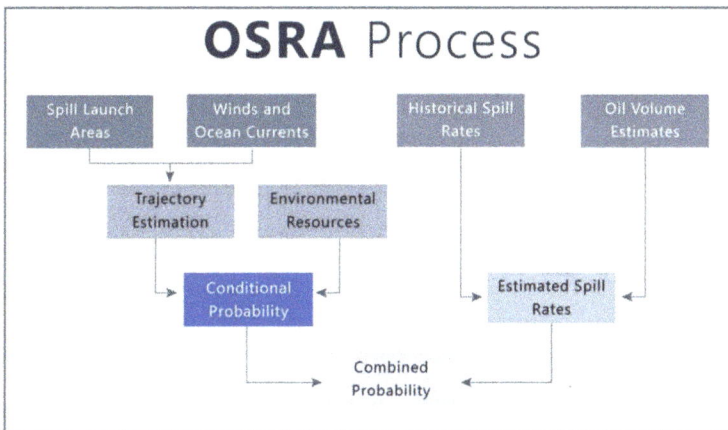

Figure 1. Schematic diagram of the OSRA model process.

BOEM has committed to continuous improvement of the OSRA model over years [9–16]. A more recent overview of the OSRA model is given by Price et al. (2003) [9]. The sensitivity analysis in Price et al. (2004) [10] showed that a time integration step of 1 h with a fourth-order Runge–Kutta scheme and a daily release of the hypothetical oil spills are sufficient to apply the OSRA model to the GOM using the selected wind and current data sets. Guillen et al. (2004) [11] suggested utilizing a 'multivariate statistical method called cluster analysis' to group areas that 'pose similar risk to specific targets or groups of targets'. This method was used in the spill risk analysis for the recent lease sales in the GOM OCS [12]. Johnson et al. (2005) [13] examined the statistics of length of coastline in the GOM contacted by the hypothetical oil spills by varying the number of spillets (adding a random component to both components of velocity at each OSRA model integration time step to represent the spreading of oil spills), the number of the trajectories, and level of concern.

The OSRA model results have been compared with the surface drifter data in the GOM to evaluate the accuracy of the model. Comparison of the estimated landing probabilities on the GOM coastline from the historical (1955–1987) surface drifter data (mostly cards and bottles) with the OSRA model results by Lugo-Fernández et al. (2001) [14] demonstrated that the probabilities were within an order of magnitude. The correlation coefficients were from 0.44 to 0.49 for the total, winter, and nonwinter seasons in their spatial distributions. The OSRA model trajectories were compared statistically against 97 trajectories of 'oil-spill-simulating' drifters (freely moving, satellite-tracked, surface floats) deployed over the northeastern GOM continental shelf during five hydrographic surveys from 1997 through 1999 in Price et al. (2006) [15]. The discrepancies found were largely due to the integration of the imperfect wind and ocean current fields, the empirically derived wind-drift factor, and inability of the ocean model in resolving the smaller-scale processes.

Though the OSRA model was designed to study the surface oil spills, it was shown to statistically capture the pattern of surface oiling from the Deepwater Horizon oil spill of 2010, as detailed in Ji et al. (2011) [16]. For a deepwater oil spill trajectory model, BOEM uses the Clarkson Deepwater Oil and Gas Blowout Model [17–19], which was funded by BOEM in collaboration with 11 industry partners [20,21] and simulates the transport of oil and natural gas from a blowout or a pipeline rupture in deepwater.

Although most of the oil spill models are designed for use in real-time forecast mode, such as the National Oceanic and Atmospheric Administration's (NOAA's) General NOAA Operational Modeling Environment (GNOME) [22,23], the OSRA model was specifically developed to inform the decision-making process for OCS oil and gas lease sales. It was designed to estimate the long-term (decades) risk associated with the OCS lease sales. The model characterizes an entire lease sale area by simulating hundreds of thousands of trajectories under decade-long, historical wind and current conditions to derive the climatology of spill contact probabilities, without having to make assumptions on the exact locations of the leases, numbers of wells drilled, and the oil properties. As such, the OSRA model adopts a conservative approach without considering the oil weathering process. The specifics of one or more appropriate oils for weathering estimates are described in the EIS using the oil weathering model from SINTEF (*Stiftelsen for INdustriell of TEknisk Forskning ved NTH*—Foundation for Industrial and Technical Research) [24]. The use of a stand-alone weathering modelling allows BOEM the flexibility of examining the weathering characteristics of different types of crude oils rather than a single oil type for multiple different reservoirs.

BOEM's NEPA documents for lease sales are governed by a number of environment laws, regulations, and executive orders, including Clean Air Act, Clean Water Act, Coastal Zone Management, Endangered Species Act, Magnuson–Stevens Fishery Conservation and Management Act, Marine Mammal Protection Act, Migratory Bird Treaty Act, Tribal Consultation and Environmental Justice. To comply with these laws and regulations, the OSRA model compiles a large list of environmental resources that include broad categories of onshore and offshore resources [12] and estimates the likelihood (probability) of oil spill contact to the resources. The impact of oil spill on all considered resources is analyzed separately in the EIS that are prepared prior to conducting any leasing sales, and therefore factors in measures such as weathering and the effects of cleanup activities.

The existing OSRA model estimates the conditional probability of contact from a specific launch area to an environmental resource mapped to OSRA model grid at three time intervals—3, 10, and 30 days. The environmental resources, typically on the order of hundreds of resources, are treated as inputs to the existing OSRA model; as such, a rerun of the OSRA model is needed if additional environmental resources are considered later. In this study, a method is developed to calculate the conditional probability by treating entire waters in the GOM as a multitude of environmental resources consisting of ocean grid cells. The number of contacts to each ocean grid cell from hypothetical oil spill trajectories initiated over entire planning areas of the U.S. GOM OCS is tracked and tabulated. For a given wind and current data set, this information can be loaded into a database and statistics on any launch areas within GOM OCS planning areas can be retrieved later for use in the estimates of conditional probability of future lease sales. Using this method, conditional probability maps can be generated, thus allowing a quantitative evaluation of the effects of hindcast surface winds and ocean currents on conditional probability estimates. Instead of estimating conditional probability of contact at three fixed time intervals, this study calculates these probabilities each day from day 1 to day 30.

The conditional probabilities are calculated in this study using two sets of relatively high resolution hindcast surface wind (six-hourly) and ocean current (three-hourly) data from two time periods, 1993–1999 and 2000–2007. Both sets of surface ocean current data were generated by the Princeton Regional Ocean Forecast System (PROFS), which is described in detail by Oey and Lee (2002) [25], Oey et al. (2003) [26], Oey (2005) [27], and Chang et al. (2011) [28]. The model results were used to study the Loop Current, eddies, and related circulation in the GOM, and they were extensively verified with a variety of surface and subsurface observations including in situ and shipboard acoustic Doppler current profiler measurements, National Data Buoy Center data, and satellite and drifter data [29–40]. Chang et al. (2011) [28] demonstrated that trajectories generated by a long-term hindcast current data (2000–2007) from the PROFS can reasonably simulate the spread of the Deepwater Horizon oil spill in 2010. An in-depth description of these data sets is in Section 2.3. These data sets were chosen to match that used in the recent OSRA application in BOEM's Eastern Planning Area in the GOM [12]. BOEM also provided these data sets to NOAA for use in its analysis of long-term outlook of oil spill transport during the Deepwater Horizon oil spill incident [41].

Similar approach was used in the European Commission's NEREIDs project for assessing shoreline and offshore susceptibility to the hypothetical large oil spills around Suez Canal and nearby oil and gas fields in the Eastern Mediterranean Sea [42–44]. The model in NEREIDs project considers various factors that mitigate the impacts of oil spills and is designed for use in a small confined marine basin that needs a rapid response. The method proposed in this study aims for use in open sea with the focus on long-term contingency planning.

This paper begins with descriptions of the OSRA model domain and components, followed by comparisons of annual and monthly conditional probability calculated from two time periods, and an application of the new method to estimate probability of spill contact to a subset of environmental resources in the GOM OCS. The discussion focuses on the advantage of this method and explains how this method will improve BOEM's decision-making process in the future.

2. Methods

2.1. Study Area

As shown in Figure 2, the study area for OSRA extends from 98° W to 78° W and 18° N to 31° N, which includes portion of the western Atlantic Ocean. The study area was chosen to be large enough to allow hypothetical oil spill trajectories to develop without contacting the boundary at the east within 30 days (the maximum elapsed time considered). The OSRA model has a resolution of 0.1° latitude by 0.1° longitude and a total of 28,564 grid cells in the study area. The hypothetical oil spill locations, also referred to as the launch points, are shown as blue dots in Figure 2. There are 6044 launch points spaced at a resolution of 0.1° latitude by 0.1° longitude. These launch points are selected to represent the Western GOM, Central GOM, and Eastern GOM Planning Areas as displayed in Figure 3. Note that the launch points are located in the OCS waters, which are generally about three miles from the shore, except for the launch points off the western Florida Shelf and Texas coastline, which are about nine miles offshore. As of 3 December 2018, the GOM OCS planning areas comprised a total of 29,100 leasing blocks and 159,381,023 acres. The number of active leases was 2557, covering 13,540,330 acres of leased areas. About 86% of active leases are in the Central GOM Planning Area, and very few are in the Eastern GOM Planning Area.

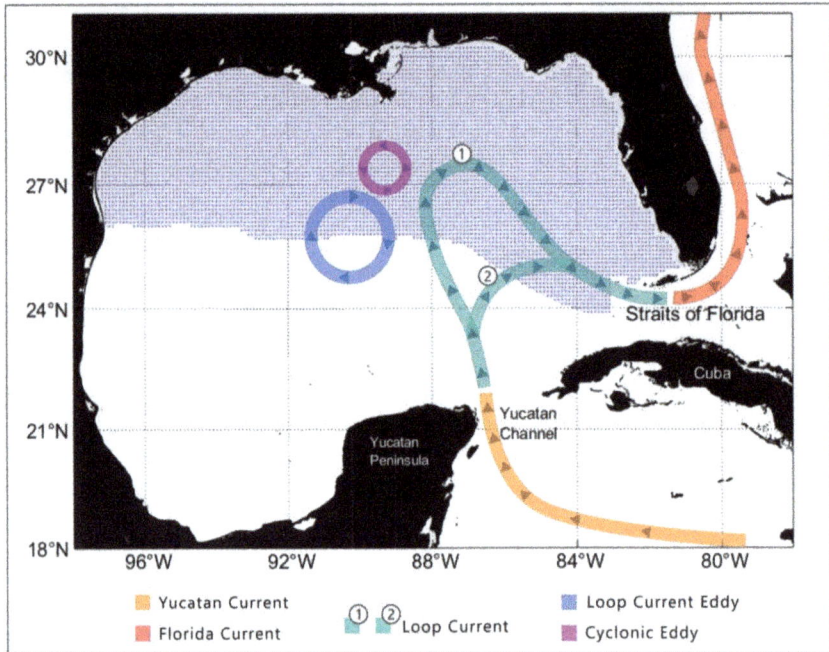

Figure 2. Map of the GOM used for the OSRA model simulation with schematic drawings showing the Loop Current and Loop Current Eddy. Blue dots denote hypothetical oil spill locations.

The dominant circulation features in the GOM are the Loop Current and Loop Current eddies (Figure 2). The Loop Current originates from the Yucatan Channel and loops inside the GOM before forming the Florida Current at the Straits of Florida around the Florida Peninsula. As part of the Gulf Stream System, the Florida Current flows from the Straits of Florida to Cape Hatteras along the U.S. southeastern coast. The Loop Current can reach a speed of 1.7 ms^{-1} inside the GOM [45] and extends deep into the GOM. The Loop Current exhibits a range of variations, which can be measured by how much farther north it penetrates into the GOM. The Loop Current eddies are large anti-cyclonic rings separating from the Loop Current when it becomes unstable as it extends farther north into the GOM, and these eddies subsequently drift to the west after separation [46]. The time interval of the separation events varies from 3 to 17 months with an average of about 9.5 months [46]. Cyclonic eddies, also referred to as the Loop Current Frontal eddies, are much smaller than the Loop Current eddies, and they originate from the boundary of the Loop Current and Loop Current eddies. These powerful currents can influence biological production, pollutant transport such as oil spills, design and operation of oil and gas facilities, fishery management in the GOM, and other processes and resources [47].

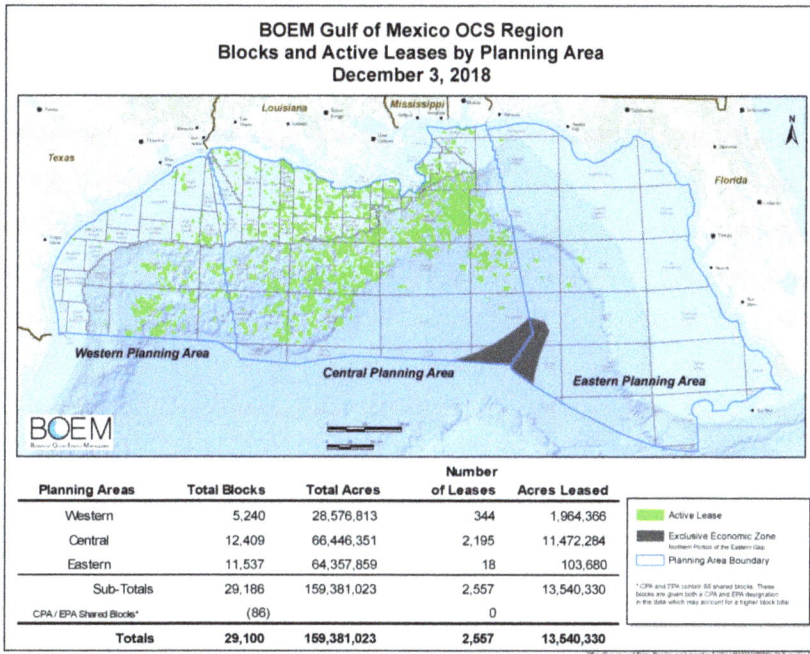

Figure 3. Active leases in the GOM planning areas (Western, Central, and Eastern) as of 3 December 2018.

2.2. Trajectory Simulations

One of the key components of the OSRA model is trajectory simulation. The path that a hypothetical oil spill moves under the forces of surface currents and winds is the modeled trajectory. The hypothetical oil spill trajectories are constructed using vector addition of a temporally and spatially varying ocean current field and an empirical wind-induced drift of the hypothetical oil spills [48]. The wind-drift factor was estimated to be 0.035, with a variable drift angle ranging from 0° to 25° clockwise that is inversely related to wind speed. The drift angle is computed as a function of wind speed according to the formula in Samuels et al. (1982) [48]. Collectively, the trajectories represent a statistical ensemble of simulated oil spill displacements produced by the fields of winds and ocean currents from numerical models with observations assimilated.

The existing OSRA model initiates the trajectories every day at the same time from hypothetical spill locations in areas of prospective drilling and production and along projected pipeline and tanker routes. The trajectories are advected every hour using the instantaneous surface current and wind data and are allowed to continue for as long as 30 days. The maximum travel time of 30 days is chosen because a typical GOM oil slick of 1000 barrels (bbl) or greater, when exposed to typical winds and currents, would not persist on the water surface beyond 30 days [49]. Considering the diurnal cycle of surface winds, initiating trajectories at random time during the day would be a better approach and will be pursued in the future.

It is worth noting that the trajectories simulated by the OSRA model represent only hypothetical pathways of oil slicks, and they do not consider cleanup, dispersion, or weathering processes that could alter the quantity or properties of oil that might eventually contact the environmental resources. However, an implicit analysis of weathering and decay can be considered by choosing a travel time that represents the likely persistence of the oil slick on the water surface.

2.3. Surface Wind and Ocean Current Data

The trajectory simulations in this study used two sets of six-hourly surface wind and three-hourly ocean current data, one from 1993 to 1999 and the other from 2000 to 2007. These wind and current data were further interpolated in the OSRA model to an hourly interval to calculate the trajectories, and seasonal statistics on conditional probability were derived from a huge ensemble (millions) of individual trajectories simulated on an hourly basis. These data were chosen because they were used in the recent OSRA report for lease sales in the Eastern GOM Planning Area [12], and therefore the results generated from this study can be validated before applying this method to BOEM's lease sales.

Surface wind data for 1993–1999 are taken from the European Centre for Medium-Range Weather Forecasts [27]. Surface wind data for 2000–2007 are from the National Centers for Environmental Prediction (NCEP) QuikSCAT blended (https://rda.ucar.edu/datasets/ds744.4/) [50]. Both sets of data are at six-hourly time intervals. The NCEP QuikSCAT blended wind data are derived from merging of high-resolution satellite data (SeaWinds instrument on the QuikSCAT satellite) and NCEP reanalysis. The NCEP–QuikSCAT blended winds were corrected using high-resolution wind fields from NOAA's Hurricane Research Division, which includes hurricane strength winds [39].

Surface winds are primarily northeasterly in the winter, becoming easterly or southeasterly in the summer. During the spring and fall time, the surface winds are mostly easterly or southeasterly. Winds at the west Florida shelf are primarily offshore, where strong winds (northeasterly in winter, easterly in spring and southeasterly to easterly in fall) push the particles offshore. The weak onshore winds tend to occur in the summer months.

As mentioned earlier, PROFS produced the ocean current data at three-hourly time intervals. PROFS is a version of Princeton Ocean Model (POM), which is a three-dimensional, time-dependent, primitive equation model using orthogonal curvilinear coordinates in the horizontal dimension and a topographically conformal coordinate in the vertical dimension [51]. These coordinates more realistically represent coastline and bottom topography in the model simulation. There are some similarities and differences in model configuration for these two time periods. Both simulations cover a large domain that includes the northwest Atlantic Ocean, extending to 55° W in the east, 50° N in the north, and the Caribbean Sea in the south. Monthly climatology of temperature and salinity is obtained from the World Ocean Atlas at NOAA's National Oceanographic Data Center and is used for initial conditions and boundary conditions for eastern boundary in the Atlantic Ocean [28]. The model simulation incorporates daily river discharges from 34 rivers in the northern GOM obtained from the U.S. Geological Survey [28]. Both simulations assimilate satellite-derived sea surface height and sea surface temperature. Major differences between these two runs are data assimilation scheme and resolution. The 2000–2007 simulation uses a nested grid in the GOM with a resolution of 3.5 km and adopts a more advanced Ensemble Kalman Filter data assimilation scheme. Simulation for 1993–1999 uses an optimal interpolation and has a resolution of 5 km. The temporal and spatial resolution of the PROFS is typical of the state-of-the-art ocean model used in the GOM for oil spill modelling. For example, during the Deepwater Horizon oil spill, MacFadyen et al. (2011) [23] applied ocean current output from six hydrodynamic models with spatial resolution ranging from ~3 to 14 km to NOAA's GNOME ensemble forecasting (daily 72 h) of surface oil transport.

The model simulations were extensively verified with many observations, from satellite-borne instrument to in situ measurements including moored current meters and drifters in the GOM [29,31–35,39,40]. These extensive observations afford a rigorous assessment of the POM's ability to reproduce ocean transport and prominent features in the GOM, such as the Loop Current and large, energetic eddies that spin off from the Loop Current. The POM reproduced realistic surface currents both on and off the continental shelf.

2.4. Environmental Resources

The environmental resources consist of biological, physical, and socioeconomic resources located in any onshore and offshore areas that could be potentially affected by OCS oil spills. BOEM analysts defined these resources in the GOM region with additional input from the National Marine Fisheries Service and the U.S. Fish and Wildlife Service. BOEM analysts also used information from the results of the Bureau's funded research projects, literature reviews, and consultations with other scientists to define resources. Typically, the OSRA model for the GOM OCS incorporates 184 offshore and 102 onshore environmental resources [12]. Those resources are not limited to environmental sensitive areas such as fish habitats; they also include the state offshore waters that are defined by each of five costal states (Texas, Louisiana, Mississippi, Alabama, and Florida) that border the GOM and seafloors of nearshore, shelf, and deepwater. Moreover, international waters of Cayman Islands, Bahamas, and Jamaica are considered. The onshore environmental resources include the U.S. coastline (grouped into counties or parishes, resource habitats, recreational beaches) and coastline of Mexico, Belize (country), and Cuba. For a comprehensive list of these resources, see Ji et al. (2013) [12].

The geographic locations of environmental resources are displayed as maps that can be digitalized onto the OSRA model grid. Each environmental resource has a seasonal vulnerability, defined as a time period when resources are present or susceptible to damage from an oil spill. The offshore environmental resources—the focus of this study—are delineated as the areas of surface waters overlying their locations.

In the existing OSRA model, the environmental resources are treated as inputs to the OSRA model. The spatially and temporally varying environmental resources are digitalized and hard-coded onto the OSRA model grid prior to OSRA model simulation. If the environmental resources are changed later, a new OSRA model run has to be performed. The new method proposed in this study tabulates the trajectory contacts to every ocean grid cell in the OSRA model and archives the number of counts to compute the conditional probability to a specific environmental resource later without re-running the OSRA model.

The OSRA model does not assess the susceptibility of these environment resources to oil spill, such as the 'Environmental Susceptibility Index' defined in Adler and Inbar (2007) [52]. The details of how and why these resources could be negatively impacted by oil spills due to BOEM's leasing activities are discussed thoroughly in the EIS.

2.5. Conditional Probability

The OSRA model geographically tracks the contacts of each hypothetical spill trajectory to the environmental resources. A contact occurs when a trajectory touches an environmental resource. At every hour, the OSRA model calculates the locations of the simulated spills and counts the number of oil spill contacts to the environmental resources. The OSRA model only tabulates the counts during the months when the environmental resources are vulnerable. For a given hypothetical launch point, the OSRA model divides the total number of contacts to the environmental resources by the total number of hypothetical spills initiated in the model after specific periods of time. These ratios are the estimated conditional probabilities of oil spill contact from a given hypothetical launch point at designated oil spill travel times, which are 3, 10, and 30 days in the GOM OCS.

2.6. Conditional Probability from Two Launch Points in 1998

Two launch points of the same latitude are selected to demonstrate how conditional probability is calculated using the new method. One launch point is east of Mississippi delta at 91.9° W, 29° N and the other launch point is west of Mississippi delta at 88.1° W, 29° N. The trajectories were launched from each launch point every day for year 1998 and were driven by the corresponding six-hourly surface winds and three-hourly ocean currents. Price et al. (2006) [15] compared the OSRA model generated trajectories using the 1993–1999 wind and current data described earlier with drifter data collected during five hydrographic survey from 1997 through 1999 and found that the cumulated errors in the input fields led to an average discrepancy of 78 km after 3 days. Nevertheless, this wind and current data were shown to be able to reproduce the similar oil spill patterns when used in the ORSA model to simulate the 2010 Deepwater Horizon oil spill [16].

The trajectories are calculated every hour and are allowed to continue for 30 days. The model tabulated contacts of trajectories to each ocean grid cell and estimated the conditional probability of contact at each ocean grid cell using the number of contacts divided by the total number of trajectories launched, i.e., 365. Figures 4 and 5 show the conditional probability of contact from these two launch points at 3, 10, and 30 days.

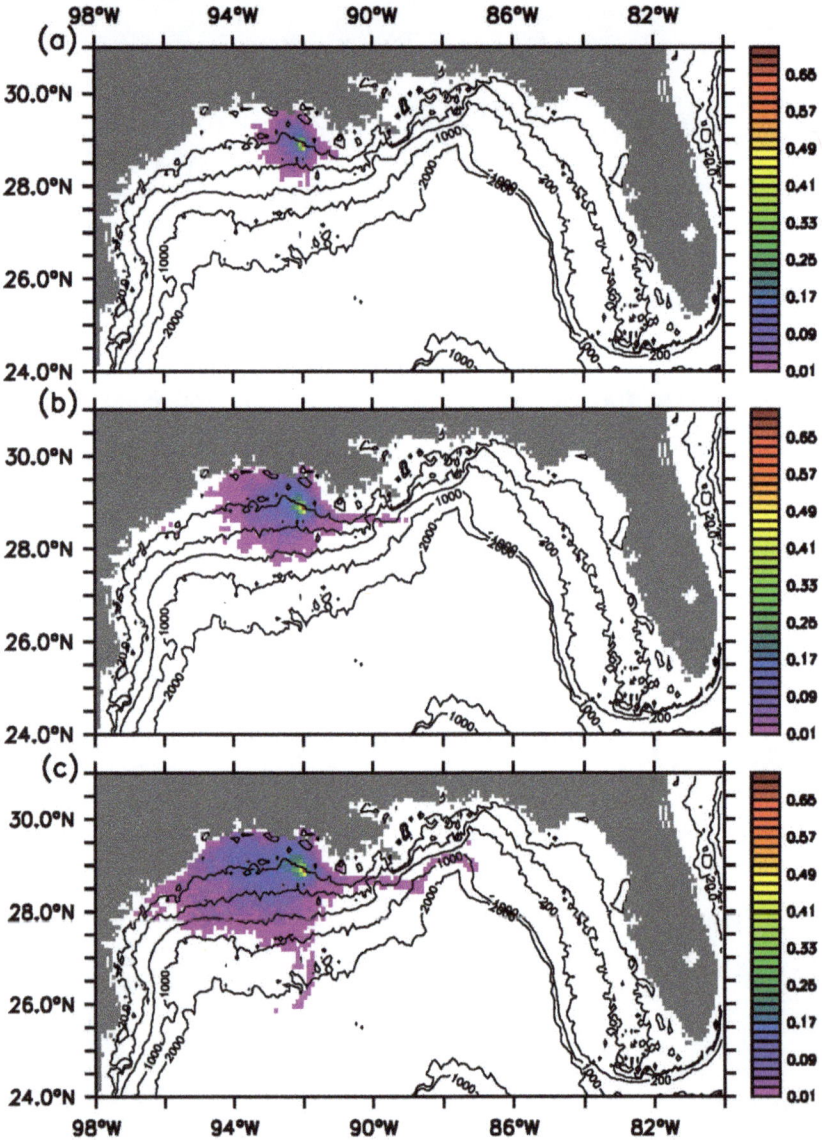

Figure 4. Conditional probability of contact from the launch point at 91.9° W, 29° N for year 1998: (**a**) 3 days; (**b**) 10 days; (**c**) 30 days. Isobaths of 20, 50, 200, 1000, and 2000 m are shown.

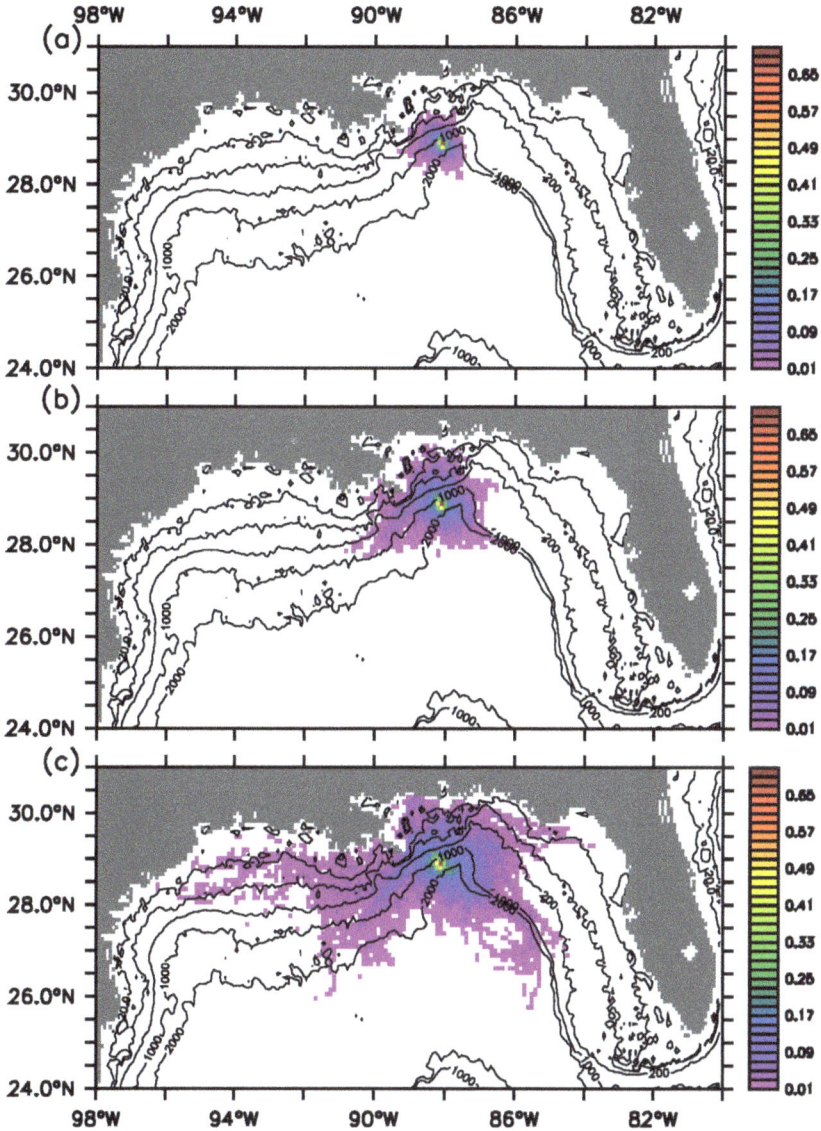

Figure 5. Conditional probability of contact from the launch point at 88.1° W, 29° N for year 1998:
(**a**) 3 days; (**b**) 10 days; (**c**) 30 days. Isobaths of 20, 50, 200, 1000, and 2000 m are shown.

Conditional probability maps for two launch points that are separated by the Mississippi River
(MR) Delta show a completely different behavior. For the launch point west of the MR Delta, the
conditional probability of contact remains mostly in areas west of the MR Delta; for the launch point
east of the MR Delta, the conditional probability of contact spreads both westward and eastward.

2.7. Sensitivity Studies

Significant amounts of computer time are required to calculate the trajectories from over
6000 launch points on an hourly basis and tabulate contacts to each of 20,615 ocean grid cells on

a daily basis for 30 days. Since trajectories are initiated every day over a combined time period of 13 years, over 28.6 million trajectories are generated. A sensitivity test was performed by reducing the number of launch points to 3022, i.e., every other point in Figure 2, to assess the differences in conditional probability fields between these two simulations. No significant differences were found in the conditional probability at day 30 when reducing the number of launch points by half. With the reduced number of launch points, over 1.1 million trajectories are initialized each year.

3. Results

3.1. Annual Conditional Probability

Annual conditional probability refers to the conditional probability calculated over an entire year for the environmental resources that are vulnerable all year round. Each year OSRA model launches over 1.1 million trajectories for estimating the annual conditional probability, as each launch point initiates one trajectory every day for a period of 365 days. Two sets of wind and current data, i.e., 1993–1999 and 2000–2007, are used to estimate annual conditional probability on a daily basis from day 1 to day 30. Because the OSRA model simulates the trajectories for 30 days, a trajectory launched at 31 December 1998 (or 31 December 2006) will need the first 30 days of data from year 1999 (or 2007) to complete a 30-day trajectory analysis. Thus, the annual conditional probability calculated is from 1993 to 1998 and from 2000 to 2006. Figure 6 through Figure 7 show the annual conditional probability at day 30 for each year for these time periods. The annual conditional probability displays a strong variability from year to year, with different spatial distributions. The distribution of the annual conditional probability reflects the convergence of the trajectory paths, which depends on convergence of surface ocean currents and drifting effects of surface winds.

Areas of highest annual conditional probability tend to occur near the Loop Current and Loop Current eddies. From 1993 to 1998, areas of highest annual conditional probability appear around the Loop Current near the western entrance of Florida Strait. Annual conditional probability in the Loop Current–Florida Current in 1997 and 1998 stands out as the highest among all. Another area of relatively high annual conditional probability for 1993–1998 is at the Texas shelf. A relatively high annual conditional probability occurs near the Texas shelf at the 20- to 50-m isobaths in 1994.

For 2000–2006, areas of large annual conditional probability are located in the interior of the GOM, where cyclonic and anti-cyclonic eddies dominate. It is not surprising that locations of largest annual conditional probability coincide with the most energetic portion of circulation in the GOM. The distribution pattern varies, with highest annual conditional probability occurring in 2001, 2003, and 2005.

The west Florida shelf remains one of the areas with lowest probability of contact despite the fact there are launch points adjacent to it. The low annual conditional probability in the west Florida shelf coincides with the so-called 'Forbidden Zone' described by Yang et al. (1999) [53]. In this zone, drifters do not enter the shallow waters off the coast of southwest Florida and Florida Bay (i.e., south of Tampa Bay and west of Florida Bay); it suggests that currents, winds, bathymetry, or all three combined, keep the drifters offshore. The drifters presented by Yang et al. (1999) [53] were launched by the Surface Current and Lagrangian-Drift Program (SCULP) II during February 1996 to June 1997 [54]. Over 300 passive drifters were launched at various locations in the northeastern GOM, from the Mississippi-Alabama border on the east to Cedar Key in Florida on the west; drifters were tracked via satellite throughout the GOM and along the Florida Current. Although drifters in the SCULP II were deployed mostly north of the Tampa Bay at 28°N, the launch points in this study were located inside the Forbidden Zone, but the trajectories initiated from these launch points were mostly driven away from the shore by the combination effects of currents, winds, and bathymetry.

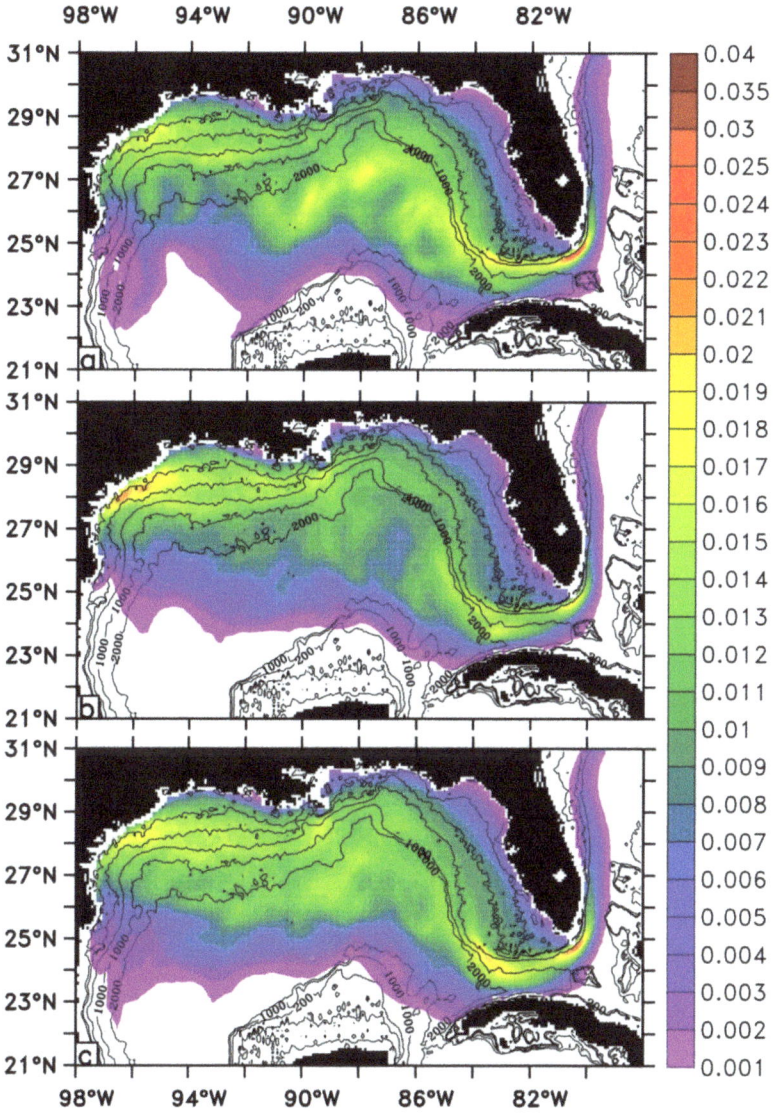

Figure 6. *Cont.*

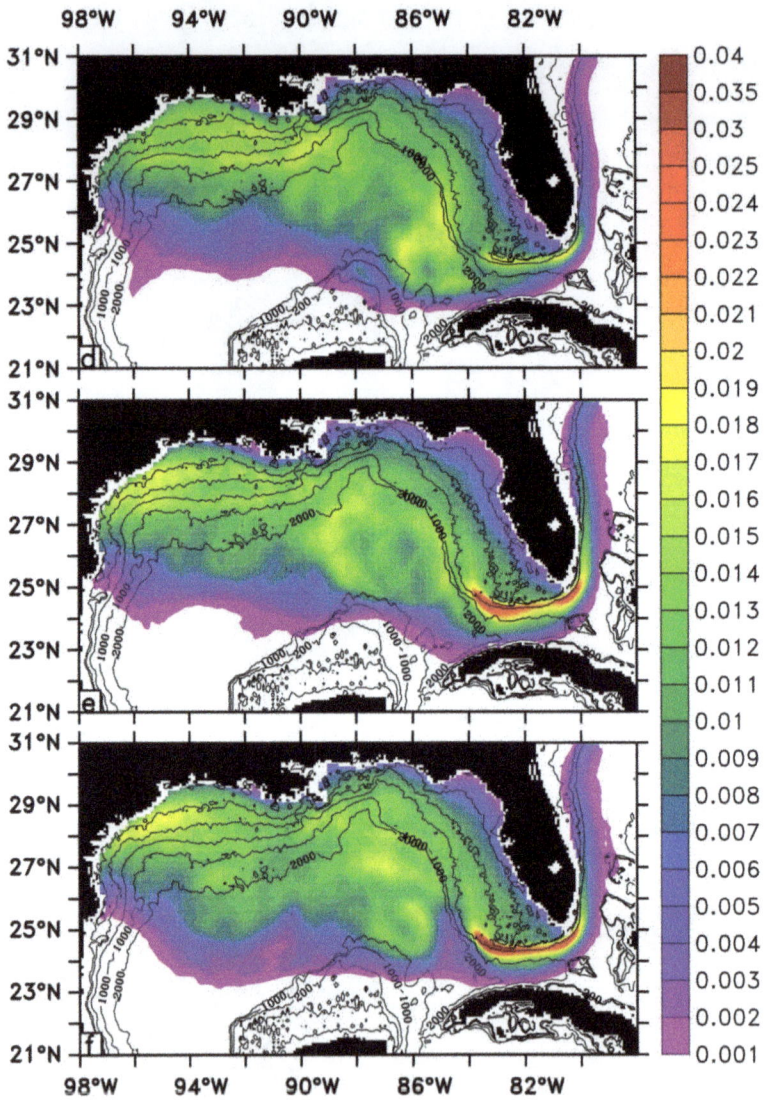

Figure 6. Annual conditional probability at day 30 for year: (**a**) 1993; (**b**) 1994; (**c**) 1995; (**d**) 1996; (**e**) 1997; (**f**) 1998. The color bar has an interval of 0.001 from 0.001 to 0.025 and an interval of 0.005 from 0.025 to 0.004. Isobaths of 20, 50, 200, 1000, and 2000 m are shown as black lines.

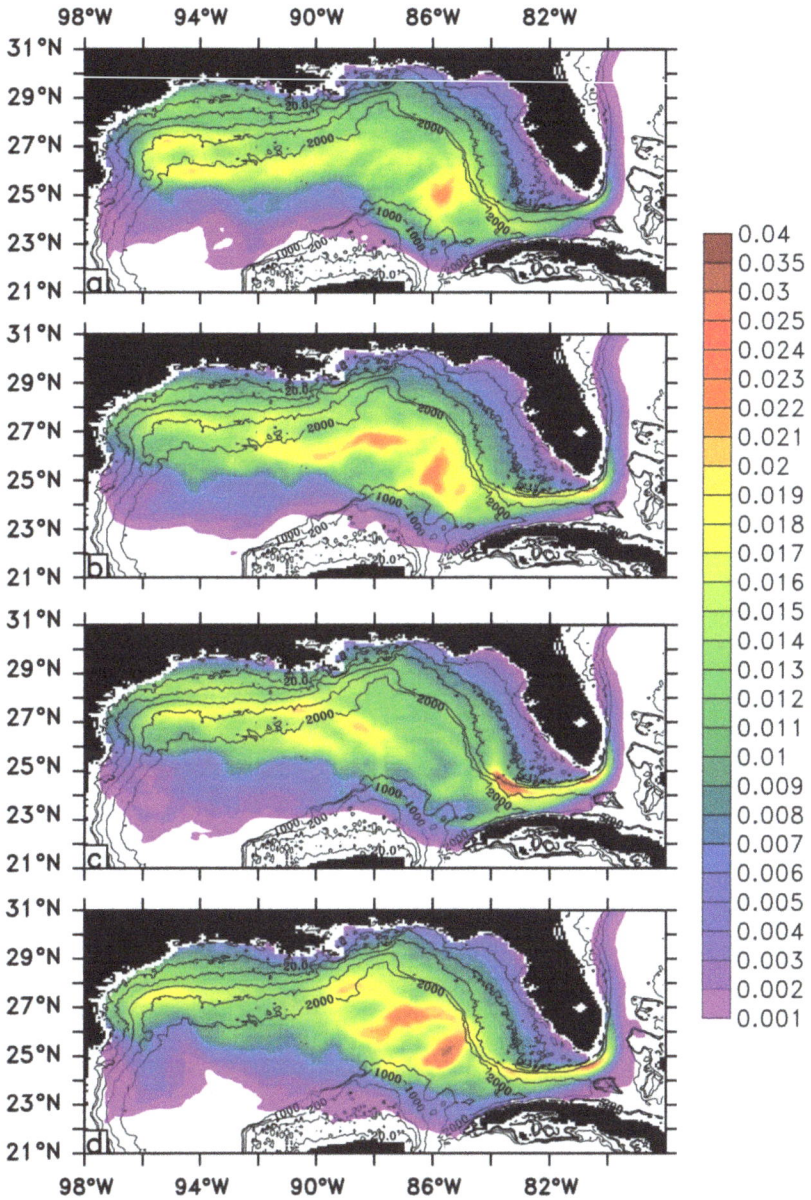

Figure 7. *Cont.*

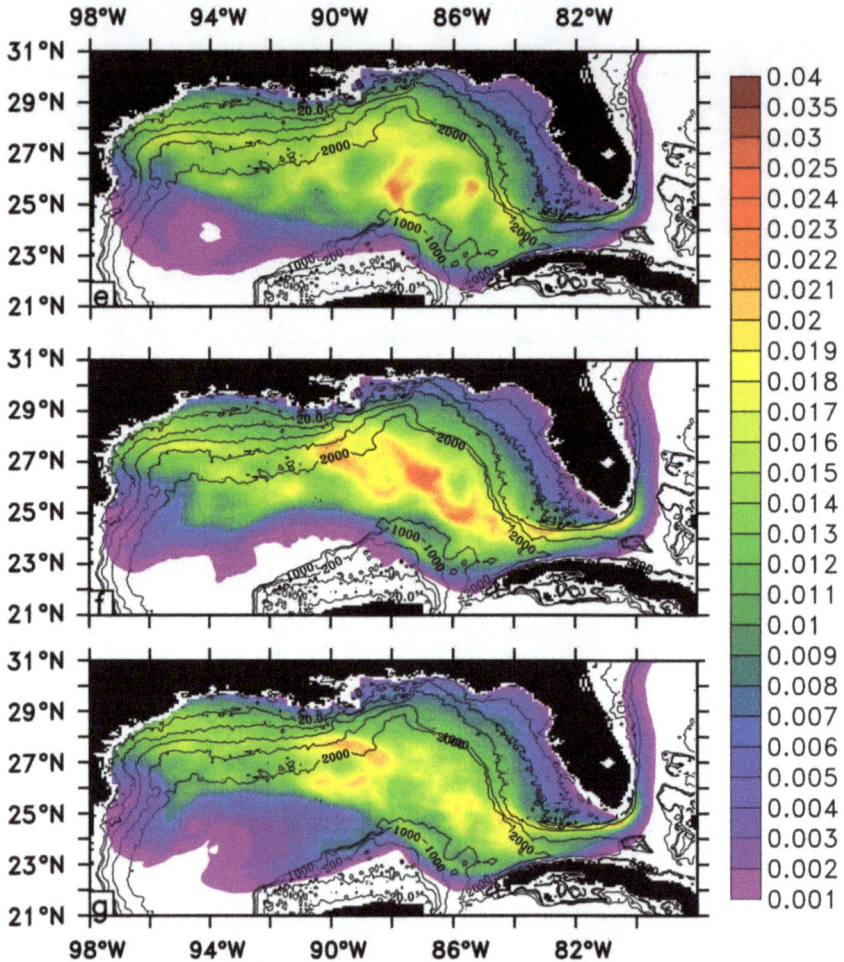

Figure 7. Annual conditional probability at day 30 for year: (**a**) 2000; (**b**) 2001; (**c**) 2002; (**d**) 2003; (**e**) 2004; (**f**) 2005; (**g**) 2006. The color bar has an interval of 0.001 from 0.001 to 0.025 and an interval of 0.005 from 0.025 to 0.004. Isobaths of 20, 50, 200, 1000, and 2000 m are shown as black lines.

3.2. Multi-Year Mean Annual Conditional Probability

This section analyzes multi-year mean annual conditional probability and standard deviations for these two time periods. The OSRA model launches 6,527,520 trajectories for 1993–1998, and 7,615,440 trajectories for 2000–2006. The model tabulates the contacts of these trajectories to each of 20,615 ocean grid cells to estimate the multi-year mean annual conditional probability and the standard deviations of annual conditional probability. These are calculated every day from day 1 to day 30.

As shown in Figure 8, maps of multi-year mean annual conditional probability at day 30 for these two time periods show different patterns. For 2000–2006, the multi-year mean annual conditional probability reaches a maximum value in areas of Loop Current eddies and has relatively large values in areas of Loop Current eddies; for 1993–1998, its counterpart has relatively large value in the Loop Current. For 1993–1998, the multi-year mean annual conditional probability is relatively larger near the Texas shelf, ranging from 0.016 to 0.018. Areas of maximum variations in the annual conditional probability coincide with the most energetic part of the circulation in the GOM as indicated by large standard deviations.

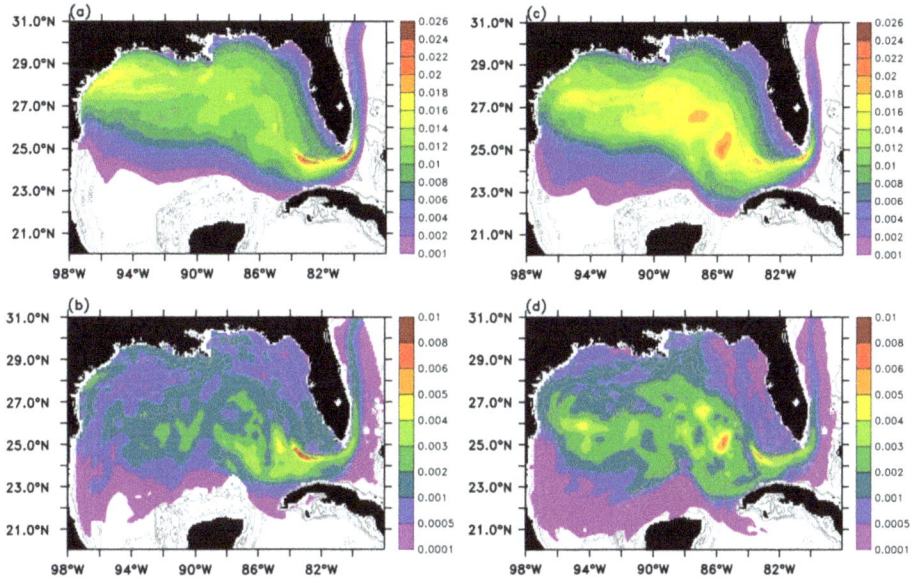

Figure 8. (**a**) Multi-year mean of annual conditional probability for 1993–1998 at day 30; (**b**) Standard deviation of annual conditional probability for 1993–1998 at day 30; (**c**) Multi-year mean conditional probability for 2000–2006 at day 30; (**d**) Standard deviation of annual conditional probability for 2000–2006 at day 30. Isobaths of 20, 50, 200, 1000, and 2000 m are shown as grey lines.

3.3. Monthly Conditional Probability

The monthly conditional probability at day 30 is calculated every month from January to December for each year from 1993 to 1998, and a multi-year averaged conditional probability for each month is generated. Figure 9 shows the monthly conditional probability averaged from 1993 to 1998 at day 30 in January, March, May, July, September, and November. Relatively low monthly conditional probability occurs at the Texas shelf from May to August, and the monthly conditional probability starts to increase during the fall and winter months, as would be expected from the seasonal difference in the Texas–Louisiana coastal circulation [55].

Figure 10 shows the monthly conditional probability averaged from 2000 to 2006 at day 30 in January, March, May, July, September, and November. Very high monthly conditional probability occurs at the Loop Current in January and November. The monthly conditional probability at Loop Current–Florida Current near the Florida Strait in January and March is among the highest of all months for this time period. The monthly conditional probability in May remains the lowest of all months. Compared to 1993–1998, the monthly conditional probability for 2000–2006 is much higher in areas around the Loop Current and Loop Current eddies, reflecting the presence of stronger Loop Current and Loop Current eddies for this time period.

Standard deviations of monthly conditional probability for these two time periods are calculated (not shown). Generally speaking, large variability tends to occur in the Florida Current in 1993–1998, versus in areas of Loop Current and Loop Current eddies in 2000–2006. For 1993–1998, largest standard deviation occurs in May and July near the Loop Current centered at about 83.5° W and overlaid on top of the 200-m isobaths. For 2000–2006, largest standard deviation occurs in September and November in regions dominated by the Loop Current and Loop Current eddies.

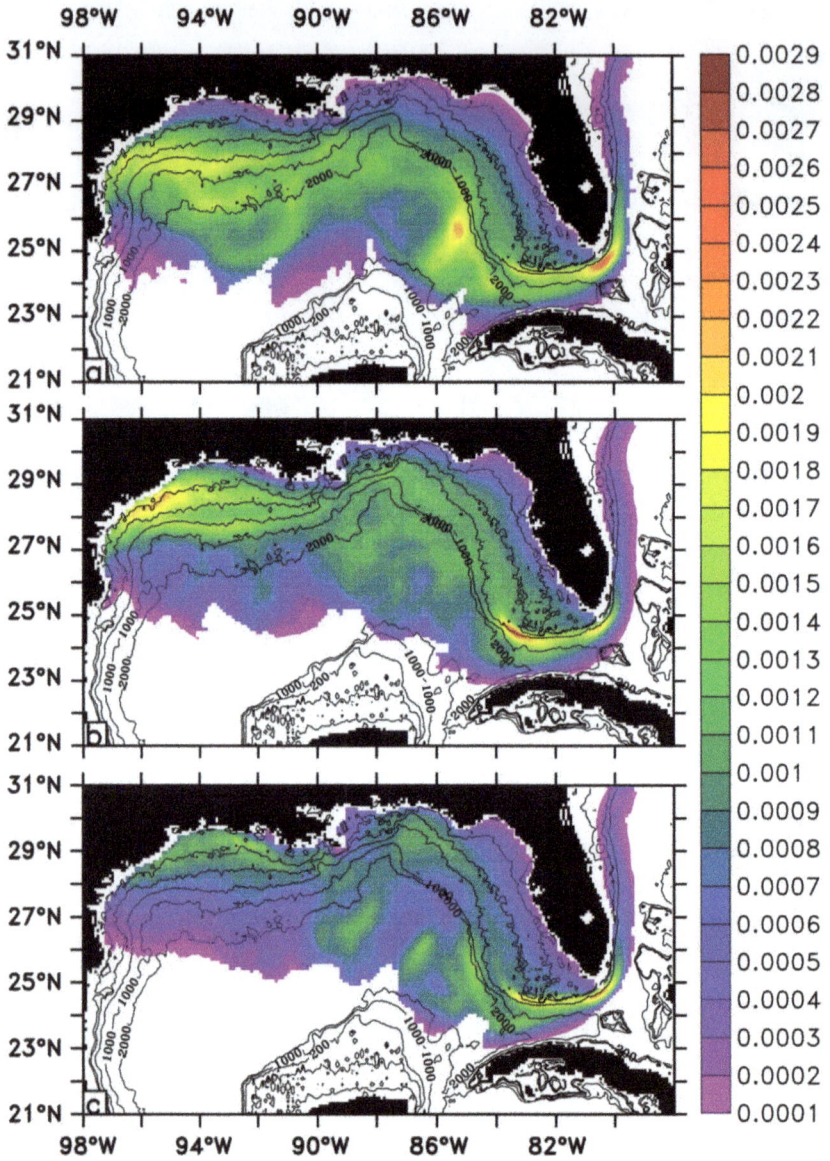

Figure 9. *Cont.*

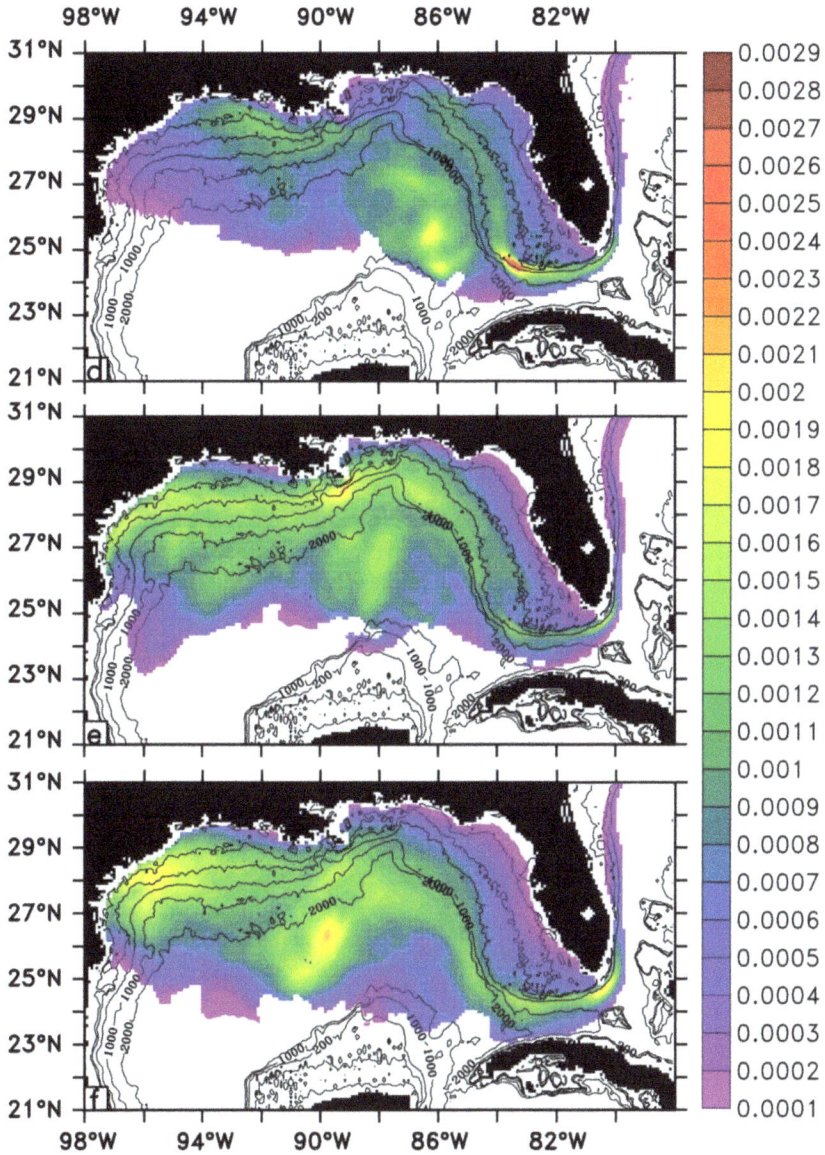

Figure 9. Monthly conditional probability averaged from 1993 to 1998 at day 30 in (**a**) January; (**b**) March; (**c**) May; (**d**) July; (**e**) September; (**f**) November. Isobaths of 20, 50, 200, 1000, and 2000 m are shown as black lines.

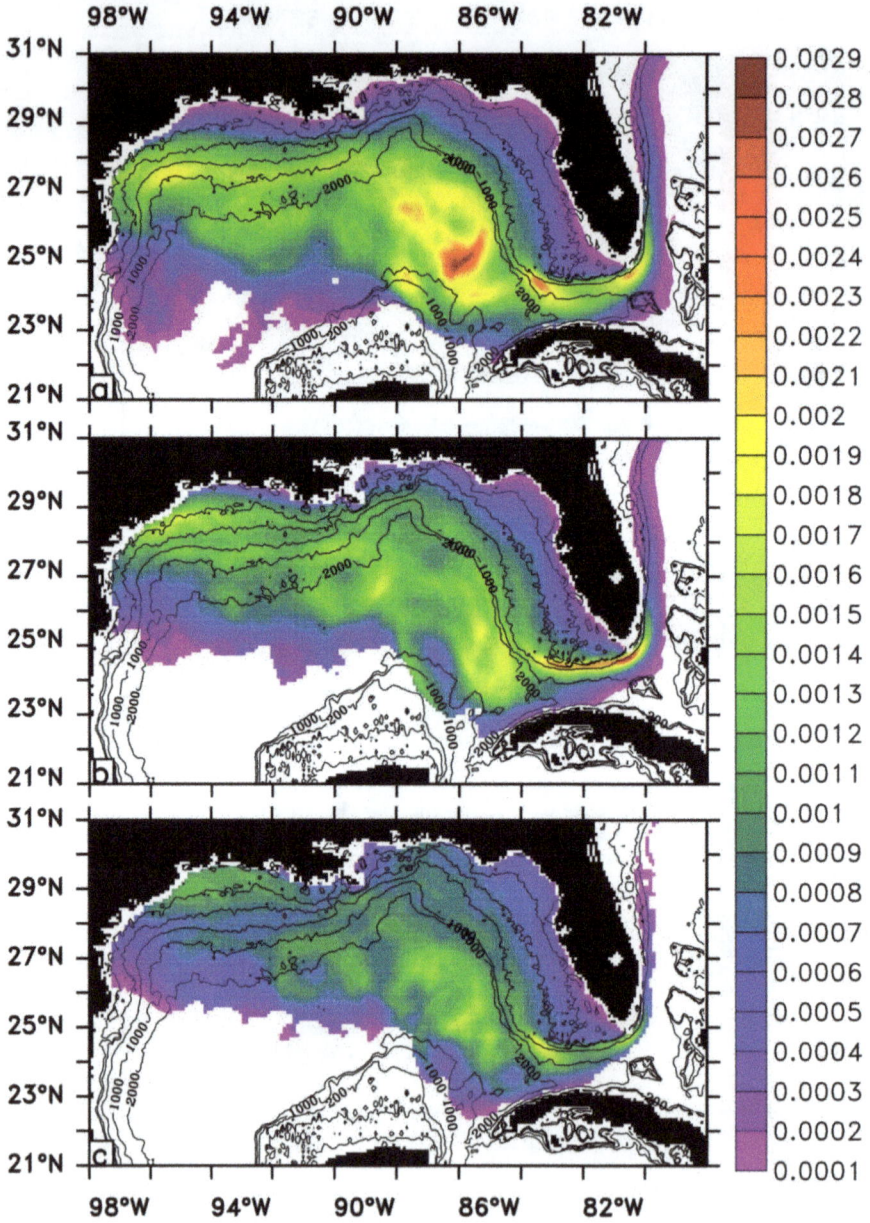

Figure 10. *Cont.*

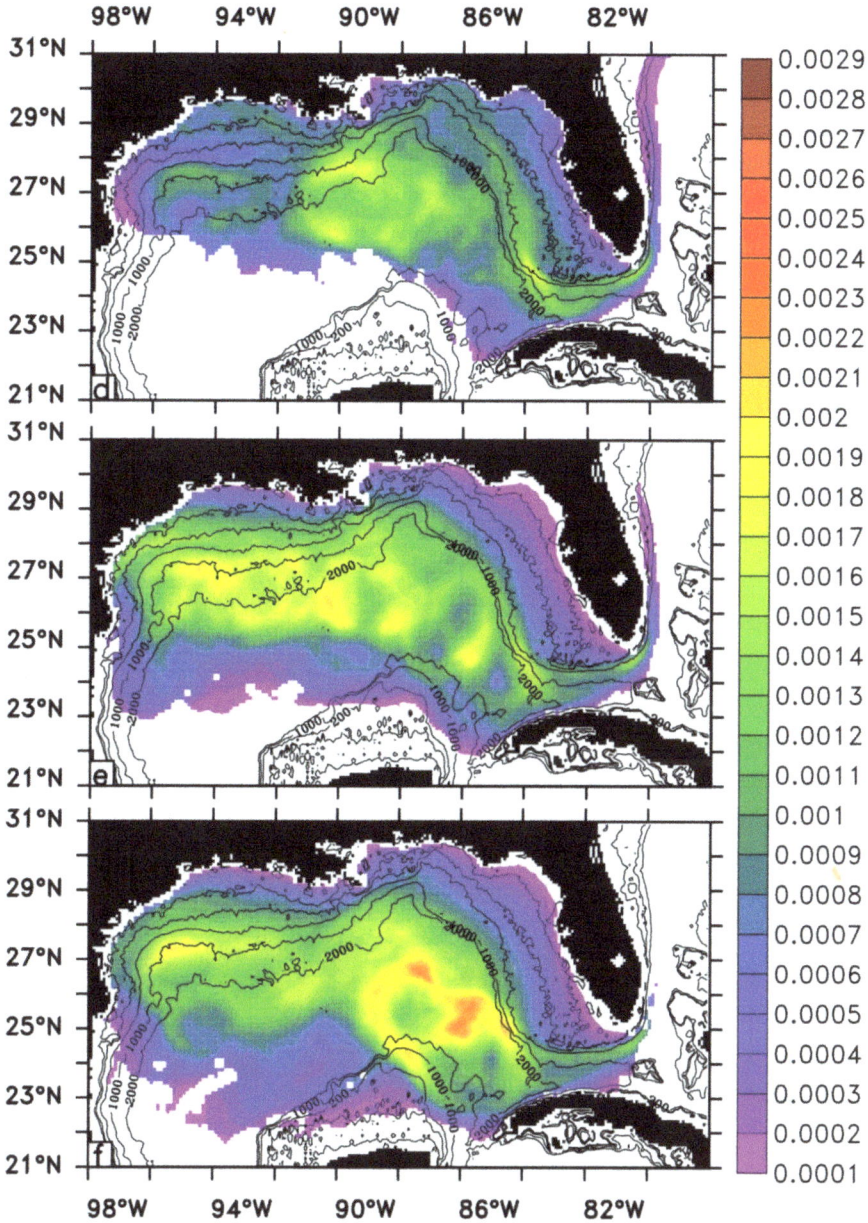

Figure 10. Monthly conditional probability averaged from 2000 to 2006 at day 30 in (**a**) January; (**b**) March; (**c**) May; (**d**) July; (**e**) September; (**f**) November. Isobaths of 20, 50, 200, 1000, and 2000 m are shown as black lines.

3.4. Estimation of Annual Conditional Probability for a Subset of Environmental Resources

To demonstrate the advantage of the new method, the annual conditional probability of a few selected offshore environmental resources in the GOM OCS is estimated. Note that these estimates are for demonstration purpose only; the hypothetical oil spills are assumed to occur over the entire planning areas, and therefore these estimates do not represent real lease sales. Because annual and monthly conditional probability are estimated from each launch point in the planning areas to every ocean grid cell in the OSRA model and are estimated on a daily basis from 1 to 30 days, a database could be created to archive these results. For any future lease sales, conditional probability for any offshore environmental resources can be calculated later from the database without re-running the OSRA model, and they can be estimated at any designated oil spill travel time from 1 to 30 days if wind and current data remain the same.

The resources shown in Figure 11 are examples of offshore environmental resources included in the EISs prior to a lease sale in the GOM. These particular resources are part of the group of resources described as Habitat Areas of Particular Concern (HAPC). HAPCs are defined by the NOAA's National Marine Fisheries Service, and regional Fishery Management Councils identify habitats that fall within HAPCs. These areas provide important ecological functions and/or are especially vulnerable to degradation. HAPCs are discreet subsets of Essential Fish Habitat (EFH). HAPCs are considered high priority areas for conservation, management, or research because they are rare, sensitive, stressed by development, or important to ecosystem function. The HAPC designation does not necessarily mean additional protections or restrictions are placed upon an area, but it helps to prioritize and focus on conservation efforts. Although these habitats are particularly important for healthy fish populations, other EFH areas that provide suitable habitat functions are also necessary to support and maintain sustainable fisheries and a healthy ecosystem. Most of the resources areas in this group are topographic features, meaning that the ocean water over the feature is shallower than much of the surrounding sea floor. These areas are frequently habitats for a variety of fish species, including commercial, recreational, and non-commercial fish.

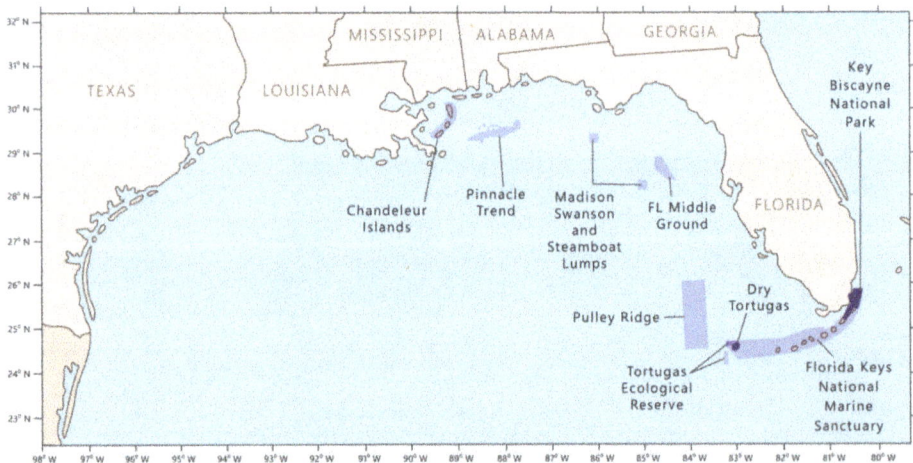

Figure 11. Locations of selected offshore environmental resources in the GOM OCS.

Table 1 lists the estimated mean annual conditional probability and standard deviations for selected environmental resources at day 30, which are calculated from OSRA model output shown in Figures 6 and 7. The environmental resources located near the western edge of the Florida Current, such as the North and South Tortugas Ecological Reserve, have relatively high annual conditional probability and are often associated with high standard deviations due to the interannual and seasonality variability of Loop Current positions. The mean annual conditional probability is slightly higher for 1993–1998 for all selected environmental resources than that of 2000–2006.

Table 1. Mean and standard deviations (SD) of annual conditional probability for a few selected environmental resources in the GOM OCS at day 30 calculated from two data sets.

Environmental Resource	1993–1998		2000–2006	
	Mean	SD	Mean	SD
Chandeleur Islands	0.0061	0.0009	0.0054	0.0010
Madison Swanson and Steamboat Lumps Marine Reserve	0.0100	0.0009	0.0084	0.0005
Florida Middle Ground	0.0084	0.0008	0.0072	0.0007
Pulley Ridge	0.0123	0.0022	0.0118	0.0017
Pinnacle Trend	0.0107	0.0008	0.0098	0.0013
Tortugas Ecological Reserve (North & South)	0.0178	0.0050	0.0143	0.0031
Key Biscayne National Park	0.0088	0.0018	0.0061	0.0011
Dry Tortugas	0.0100	0.0009	0.0090	0.0011
Florida Keys National Marine Sanctuary	0.0112	0.0015	0.0093	0.0014 [1]

[1] Note that these estimates are based on the assumption that the hypothetical oil spills occur in the entire planning areas and do not represent a real leasing scenario. They are for demonstration purposes only with a very small subset of environmental resources considered.

4. Discussion

The OSRA model plays an important role in BOEM's decision-making process as it provides critical information for BOEM's NEPA documents and oil spill response planning. BOEM continues to improve the OSRA model on several fronts, including updating ocean circulation model on a recurring schedule, improving the pre-processing and post-processing of OSRA model input and output files, and developing tools for visualization of OSRA results.

However, utilizing the OSRA model to meet the needs for NEPA analyses still proves to be a challenge, especially when a lease sale is announced without adequate time to perform an updated OSRA model run. This study attempts to provide a solution that will speed up the OSRA process by 'extracting' conditional probability from an existing database without re-running the model. For each leasing scenario, the conditional probability field for the hypothetical oil spills at a specific travel time can be generated for BOEM's contingency planning. Typically, BOEM updates the ocean model hindcast data every five to seven years depending on the availability of funds from BOEM's Environmental Study Program. With this method, the conditional probability can be derived from conditional probability database and used in NEPA analyses for any lease sales, as long as the wind and current data are not updated within the five- to seven-year time frame.

There are many other applications for this method. Because the OSRA model typically uses a long-term, hindcast wind and current data to generate a large ensemble of trajectories for statistical analyses, the variations in conditional probability can be estimated by calculating the standard deviations to reflect the annual and seasonal variability in the forcing fields. Second, this method can be used to analyze the environmental resources that are not distributed evenly in the area represented by the polygon. For this type of environmental resource, a spatial- and/or temporal-dependent density function can be created, and this function can be used in combination with the number of counts tabulated to generate a more accurate estimate of conditional probability. Third, because the database will store the information of the hypothetical oil spill locations, the 'source of pollution' may be identified from convergence of the conditional probability. This is equivalent to running the OSRA model in a reverse mode. These locations can be presented in the form of probability map showing the hypothetical oil spill locations with a certain travel time. Fourth, this method can be used to answer questions; for example, what is the likelihood of the potential oil spills from proposed leasing areas exiting the GOM via the Florida Straits and into the Atlantic Ocean, and at what travel time?

It is important to note that the OSRA model is not designed for use as an oil spill response tool. It was developed for assessing the spill risk prior to a lease sale, without knowing the oil properties. The trajectories are calculated without any approximations of weathering or intervention. It is a conservative approach specifically for the pre-sale process. The use of the term 'contact' to the environmental resources was chosen by USDOI before 1982 to allow a calculation of the probabilities while leaving out the estimation of the 'impact' of the possible oiling. The impact estimation was based on the probability of contact, with the subject matter expert on the particular resource making the assessment. Barker (2011) [41] discussed the similarities and differences between the OSRA model and GNOME when used for longer-term planning purposes, such as the Deepwater Horizon oil spill. The OSRA model is capable of considering the probability that a spill may occur, while NOAA's GNOME and Trajectory Analysis Planner approaches do not have this capability. The OSRA model only considers the surface release of the oil spills because much of the oil released at depth was shown to surface within a few hours and at a radius of a few kilometers [56]. Though the OSRA model includes the sub-surface environmental resources, the model only calculates the probability of contact at the surface and the subject matter experts will use the OSRA model results to assess the impacts to these resources in BOEM's EIS documents.

This method can be improved by conducting more sensitivity tests to derive an optimal number of launch points and model resolutions corresponding to a specific set of wind and current data. Spacing between the launch points may depend on the geographic areas, whether mesoscale or sub-mesoscale eddies dominate. It is constrained by the Rossby radius of deformation, which varies from 10 km in the shelf to 40 km in the interior of the GOM [57]. This study identifies the 'least contacted areas' (i.e., the west Florida shelf, east coast of Florida), and the 'most contacted areas' (i.e., Loop Current and eddies) in the GOM on seasonal and interannual time scales, based on the assumptions that there are hypothetical spills from BOEM's entire planning areas. The analysis can be further grouped into each planning area, such as the Western GOM, Central GOM, and Eastern GOM Planning Areas, with field of conditional probability generated for each planning area. Other OCS regions can use this method to speed up the spill risk assessment process.

Author Contributions: Z.L. conceived the ideas, performed the model runs, analyzed the model output, made the plots, and drafted and revised the paper. W.J. provided mentorship and critical review.

Funding: This research received no external funding.

Acknowledgments: We thank BOEM for covering the costs for publication of this article in open access. The opinions presented here are of authors and may not represent the viewpoints or policy of BOEM. We appreciate the editorial support from Paulina Chen and graphic support from Russell Yerkes. Thanks to Guillermo Auad for reviewing our manuscript.

Conflicts of Interest: The authors declare no conflict of interest.

References

1. Zeringue, B.A.; Yu, C.W.; Riches, T.J., Jr.; De Cort, T.M.; Maclay, D.M.; Wilson, M.G. *U.S. Outer Continental Shelf Gulf of Mexico Region Oil and Gas Production Forecast: 2018–2027*; OCS Report BOEM 2017–082; Department of the Interior, Bureau of Ocean Energy Management, Gulf of Mexico OCS Region: New Orleans, LA, USA, 2017; p. 32.
2. Smith, R.A.; Slack, J.R.; Davis, R.K. *An Oil Spill Risk Analysis for the North Atlantic Outer Continental Shelf Lease Area*; Geological Survey Open-File Report 76-620; U.S. Geological Survey: Reston, VA, USA, 1976; p. 38.
3. Smith, R.A.; Slack, J.R.; Wyant, T.; Lanfear, K.J. *The Oilspill Risk Analysis Model of the U.S. Geological Survey*; Geological Survey Professional Paper 1227; United States Government Printing Office: Washington, DC, USA, 1982; p. 40.
4. LaBelle, R.P.; Samuels, W.B.; Amstutz, D.E. An Examination of the Argo Merchant Oil Spill Incident Using a Probabilistic Oil Spill Model. Presented at the 47th Annual Meeting of the American Society of Limnology and Oceanography, Vancouver, BC, Canada, 11–14 June 1984.

5. Johnson, W.R.; Marshall, C.F.; Lear, E.M. *Oil-Spill Risk Analysis: Pacific Outer Continental Shelf Program, Minerals Management Service*; OCS Report 2000-057; OCS Report: Herndon, VA, USA, 2000; p. 290.

6. Anderson, C.M.; LaBelle, R.P. Comparative Occurrence Rates for Offshore Oil Spills. *Spill Sci. Technol. Bull.* **1994**, *1*, 131–141. [CrossRef]

7. Anderson, C.M.; LaBelle, R.P. Update of Comparative Occurrence Rates for Offshore Oil Spills. *Spill Sci. Technol. Bull.* **2000**, *6*, 303–321. [CrossRef]

8. Anderson, C.M.; Mayes, M.; LaBelle, R.P. *Update of Occurrence Rates for Offshore Oil Spills*; OCS Report 2012-069; Bureau of Ocean Energy Management Division of Environmental Assessment, and Bureau of Safety and Environmental Enforcement: Herndon, VA, USA, 2012. Available online: http://www.boem.gov/uploadedFiles/BOEM/Environmental_Stewardship/Environmental_Assessment/Oil_Spill_Modeling/AndersonMayesLabelle2012.pdf (accessed on 21 December 2018).

9. Price, J.M.; Johnson, W.R.; Marshall, C.F.; Ji, Z.-G.; Rainey, G.B. Overview of the Oil Spill Risk Analysis (OSRA) Model for Environmental Impact Assessment. *Spill Sci. Technol. Bull.* **2003**, *8*, 529–533. [CrossRef]

10. Price, J.M.; Johnson, W.R.; Ji, Z.-G.; Marshall, C.F.; Rainey, G.B. Sensitivity Testing for Improved Efficiency of a Statistical Oil-Spill Risk Analysis Model. *Environ. Model. Softw.* **2004**, *19*, 671–679. [CrossRef]

11. Guillen, G.; Rainey, G.; Morin, M. A Simple Rapid Approach Using Coupled Multivariate Statistical Methods, GIS and Trajectory Models to Delineate Areas of Common Oil Spill Risk. *J. Mar. Syst.* **2004**, *45*, 221–235. [CrossRef]

12. Ji, Z.-G.; Johnson, W.R.; Li, Z.; Green, R.E.; O'Reilly, S.E.; Gravois, M.P. *Oil-Spill Risk Analysis: Gulf of Mexico Outer Continental Shelf (OCS) Lease Sales, Eastern Planning Area, 2012–2017, and Eastern Planning Area OCS Program, 2012–2051*; OCS Report BOEM 2013–0110; Department of the Interior, Bureau of Ocean Energy Management: Herndon, VA, USA, 2013; p. 61.

13. Johnson, W.R.; Ji, Z.-G.; Marshall, C.F. Statistical Estimates of Shoreline Oil Contact in the Gulf of Mexico. In Proceedings of the International Oil Spill Conference, Miami Beach, FL, USA, 15–19 May 2005; pp. 547–551.

14. Lugo-Fernandez, A.; Morin, M.V.; Ebesmeyer, C.C.; Marshall, C.F. Gulf of Mexico Historic (1955–1987) Surface Drifter Data Analysis. *J. Coast. Res.* **2001**, *17*, 1–6.

15. Price, J.M.; Reed, M.; Howard, M.K.; Johnson, W.R.; Ji, Z.-G.; Marshall, C.F.; Guinasso, C.N., Jr.; Rainey, G.B. Preliminary Assessment of an Oil-Spill Trajectory Model using Satellite-Tracked, Oil-Spill-Simulating Drifters. *Environ. Model. Softw.* **2006**, *21*, 258–270. [CrossRef]

16. Ji, Z.-G.; Johnson, W.R.; Li, Z. Oil Spill Risk Analysis Model and its Application to the Deepwater Horizon Oil Spill Using Historical Current and Wind data. In *Monitoring and Modeling the Deepwater Horizon Oil Spill: A Record-Breaking Enterprise*; Liu, Y., MacFadyen, A., Ji, Z.-G., Weisberg, R.H., Eds.; American Geophysical Union: Washington, DC, USA, 2011; pp. 227–236.

17. Zheng, L.; Yapa, P.D.; Chen, F.H. A Model for Simulating Deepwater Oil and Gas Blowouts—Part I: Theory and Model Formulation. *J. Hydraul. Res.* **2003**, *41*, 339–351. [CrossRef]

18. Chen, F.; Yapa, P.D. A Model for Simulating Deep Water Oil and Gas Blowouts—Part II: Comparison of Numerical Simulations with "Deepspill" Field Experiments. *J. Hydraul. Res.* **2003**, *41*, 353–365. [CrossRef]

19. Yapa, P.D.; Chen, F.H. Behavior of Oil and Gas from Deepwater Blowouts. *J. Hydraul. Res.* **2004**, *130*, 540–553. [CrossRef]

20. LaBelle, R.P.; Lane, J.S. Meeting the Challenge of Deepwater Spill Response. In Proceedings of the International Oil Spill Conference, Tampa, FL, USA, 26–29 March 2001; pp. 705–708.

21. LaBelle, R.P. Overview of US Minerals Management Service Activities in Deepwater Research. *Mar. Pollut. Bull.* **2001**, *43*, 256–261. [CrossRef]

22. Beegle-Krause, C.J. General NOAA Oil Modeling Environment (GNOME): A New Spill Trajectory Model. In Proceedings of the International Oil Spill Conference, Tampa, FL, USA, 26–29 March 2001; pp. 865–871.

23. MacFadyen, A.; Watabayashi, G.Y.; Barker, C.H.; Beegle-Krause, C.J. Tactical Modeling of Surface Oil Transport during the Deepwater Horizon Spill Response. In *Monitoring and Modeling the Deepwater Horizon Oil Spill: A Record-Breaking Enterprise*; Liu, Y., MacFadyen, A., Ji, Z.-G., Weisberg, R.H., Eds.; American Geophysical Union: Washington, DC, USA, 2011; pp. 167–178.

24. Reed, M.; Singsaas, I.; Daling, P.S.; Faksness, L.-G.; Brakstad, O.G.; Hetland, B.A.; Hokstad, J.N. Modeling the Water-accommodated Fraction in OSCAR2000. In Proceedings of the International Oil Spill Conference, Tampa, FL, USA, 26–29 March 2001; pp. 1083–1091.

25. Oey, L.-Y.; Lee, H.-C. Deep Eddy Energy and Topographic Rossby Waves in the Gulf of Mexico. *J. Phys. Oceanogr.* **2002**, *32*, 3499–3527. [CrossRef]
26. Oey, L.-Y.; Lee, H.-C.; Schmitz, W.J., Jr. Effects of Winds and Caribbean Eddies on the Frequency of Loop Current Eddy Shedding: A Numerical Model Study. *J. Geophys. Res.* **2003**, *108*, 3324. [CrossRef]
27. Oey, L.-Y. *Circulation Model of the Gulf of Mexico and the Caribbean Sea: Development of the Princeton Regional Ocean Forecast (& Hindcast) System–PROFS, and Hindcast Experiment for 1992–1999*; OCS Study MMS 2005–049, Final Report; Department of the Interior, Minerals Management Service: Herndon, VA, USA, 2005; p. 174.
28. Chang, Y.-L.; Oey, L.; Xu, F.-H.; Lu, H.-F.; Fujisaki, A. Oil spill: Trajectory Projections Based on Ensemble Drifter Analyses. *Ocean Dyn.* **2011**, *61*, 829–839. [CrossRef]
29. Wang, D.-P.; Oey, L.-Y.; Ezer, T.; Hamilton, P. Near-Surface Currents in DeSoto Canyon (1997–1999): Comparison of Current Meters, Satellite Observations, and Model Simulation. *J. Phys. Oceanogr.* **2003**, *33*, 313–326. [CrossRef]
30. Ezer, T.; Oey, L.-Y.; Lee, H.-C.; Sturges, W. The Variability of Currents in the Yucatan Channel: Analysis of Results from a Numerical Ocean Model. *J. Geophys. Res.* **2003**. [CrossRef]
31. Fan, S.J.; Oey, L.-Y.; Hamilton, P. Assimilation of Drifter and Satellite Data in a Model of the Northeastern Gulf of Mexico. *Cont. Shelf Res.* **2004**, *24*, 1001–1013. [CrossRef]
32. Oey, L.-Y.; Ezer, T.; Lee, H.-C. Loop Current, Rings and Related Circulation in the Gulf of Mexico: A Review of Numerical Models and Future Challenges. In *Circulation in the Gulf of Mexico: Observations and Models*; Geophysical Union Geophysical Monograph Series; Wiley: Hoboken, NJ, USA, 2005; Volume 161, pp. 32–56.
33. Oey, L.-Y.; Ezer, T.; Forristall, G.; Cooper, C.; DiMarco, S.; Fan, S. An Exercise in Forecasting Loop Current and Eddy Frontal Positions in the Gulf of Mexico. *Geophys. Res. Lett.* **2005**, *32*, L12611. [CrossRef]
34. Oey, L.-Y.; Ezer, T.; Wang, D.-P.; Fan, S.; Yin, X.-Q. Loop Current Warming by Hurricane Wilma. *Geophys. Res. Lett.* **2006**, *33*, L08613. [CrossRef]
35. Oey, L.-Y.; Ezer, T.; Wang, D.-P.; Yin, X.-Q.; Fan, S.-J. Hurricane-induced Motions and Interaction with Ocean Currents. *Cont. Shelf Res.* **2007**, *27*, 1249–1263. [CrossRef]
36. Oey, L.-Y.; Inoue, M.; Lai, R.; Lin, X.-H.; Welsh, S.; Rouse, L., Jr. Stalling of Near-inertial Waves in a Cyclone. *Geophys. Res Lett.* **2008**, *35*, L12604. [CrossRef]
37. Oey, L.-Y. Loop Current and Deep Eddies. *J. Phys. Oceanogr.* **2008**, *38*, 1426–1449. [CrossRef]
38. Oey, L.-Y.; Chang, Y.-L.; Sun, Z.-B.; Lin, X.-H. Topocaustics. *Ocean Model.* **2009**, *29*, 277–286. [CrossRef]
39. Yin, X.-Q.; Oey, L.-Y. Bred-ensemble Ocean Forecast of Loop Current and Rings. *Ocean Model.* **2007**. [CrossRef]
40. Lin, X.-H.; Oey, L.-Y.; Wang, D.-P. Altimetry and Drifter Data Assimilations of Loop Current and Eddies. *J. Geophys. Res.* **2007**, *112*, C05046. [CrossRef]
41. Barker, C.H. A Statistical Outlook for the *Deepwater Horizon* Oil Spill. In *Monitoring and Modeling the Deepwater Horizon Oil Spill: A Record-Breaking Enterprise*; Liu, Y., MacFadyen, A., Ji, Z.-G., Weisberg, R.H., Eds.; American Geophysical Union: Washington, DC, USA, 2011; pp. 237–244.
42. Alves, T.M.; Kokinou, E.; Zodiatis, G.A. Three-step Model to Assess Shoreline and Offshore Susceptibility to Oil Spills: The South Aegean (Crete) as an Analogue for Confined Marine Basins. *Mar. Pollut. Bull.* **2014**, *86*, 443–457. [CrossRef] [PubMed]
43. Alves, T.M.; Kokinou, E.; Zodiatis, G.A.; Lardner, R.; Panagiotakis, C.; Radhakrishnan, H. Modelling of Oil Spills in Confined Maritime Basins: The Case for Early Response in the Eastern Mediterranean Sea. *Environ. Pollut.* **2015**, *206*, 390–399. [CrossRef]
44. Alves, T.M.; Kokinou, E.; Zodiatis, G.A.; Radhakrishnan, H.; Panagiotakis, C.; Lardner, R. Multidisciplinary Oil Spill Modeling to Protect Coastal Communities and the Environment of the Eastern Mediterranean Sea. *Sci. Rep.* **2016**, *6*. [CrossRef]
45. Forristal, G.Z.; Schaudt, K.J.; Cooper, C.K. Evolution and Kinematics of a Loop Current Eddy in the Gulf of Mexico during 1985. *J. Geophys. Res.* **1992**, *97*, 2173–2184. [CrossRef]
46. Sturges, W.; Leben, R. Frequency of Ring Separations from the Loop Current in the Gulf of Mexico: A Revised Estimate. *J. Phys. Oceanogr.* **2000**, *30*, 1814–1819. [CrossRef]
47. Sturges, W.; Lugo-Fernandez, A. (Eds.) *Circulation in the Gulf of Mexico: Observations and Models*; Geophysical Monograph Series; American Geophysical Union: Washington, DC, USA, 2005; Volume 161, p. 347.
48. Samuels, W.B.; Huang, N.E.; Amstutz, D.E. An Oilspill Trajectory Analysis Model with a Variable Wind Deflection Angle. *Ocean Eng.* **1982**, *9*, 347–360. [CrossRef]

49. Department of the Interior, Bureau of Ocean Energy Management. *Gulf of Mexico OCS oil and Gas Lease Sales: 2014 and 2016; Eastern Planning Area Lease Sales 225 and 226—Final Environmental Impact Statement*; OCS EIS/EA BOEM 2013-200; Department of the Interior, Bureau of Ocean Energy Management, Gulf of Mexico Region: New Orleans, LA, USA, 2013; Volume 1.

50. Oey, L.-Y. Extended Hindcast Calculation of Gulf of Mexico Circulation: Model Development, Comparison with Observations, and Application to the 2010 Oil Spill. Unpublished work.

51. Mellor, G.L. *User Guide for a Three-Dimensional, Primitive Equation, Numerical Ocean Model (Jul/2002 Version)*; Program in Atmospheric and Oceanic Sciences, Princeton University: Princeton, NJ, USA, 2002; p. 42. Available online: http://www.ccpo.odu.edu/POMWEB/UG.10-2002.pdf (accessed on 21 December 2018).

52. Adler, E.; Inbar, M. Shoreline Sensitivity to Oil Spills, the Mediterranean Coast of Israel:Assessment and Analysis. *Ocean Coast. Manag.* **2007**, *50*, 24–34. [CrossRef]

53. Yang, H.; Weisberg, R.H.; Niiler, P.P.; Sturges, W.; Johnson, W. Lagrangian Circulation and Forbidden Zone on the West Florida Shelf. *Cont. Shelf Res.* **1999**, *19*, 1221–1245. [CrossRef]

54. Sturges, W.; Niiler, P.P.; Weisberg, R.H. *Northeastern Gulf of Mexico Inner Shelf Circulation Study*; Final Report, MMS Cooperative Agreement 14-35-0001-30787, OCS Report MMS 2001–103; US Minerals Management Service: Herndon, VA, USA, 2001; p. 90.

55. Nowlin, W.D., Jr.; Jochens, A.E.; DiMarco, S.F.; Reid, R.O.; Howard, M.K. Low-frequency Circulation over the Texas-Louisiana Continental Shelf. In *Circulation in the Gulf of Mexico: Observations and Models*; Geophysical Monograph Series; Sturges, W., Lugo-Fernandez, A., Eds.; American Geophysical Union: Washington, DC, USA, 2005; Volume 161, pp. 219–240.

56. Johansen, O.; Rye, H.; Cooper, C. DeepSpill—Field Study of a Simulated Oil and Gas Blowout in Deep Water. *Spill Sci. Technol. Bull.* **2003**, *8*, 433–443. [CrossRef]

57. Chelton, D.B.; de Szoeke, R.A.; Schlax, M.G.; El Naggar, K.; Siwertz, N. Geographical Variability of the First Baroclinic Rossby Radius of Deformation. *J. Phys. Oceanogr.* **1998**, *28*, 433–460. [CrossRef]

Journal of
*Marine Science
and Engineering*

MDPI

Article

Oil Spill Scenarios in the Kotor Bay: Results from High Resolution Numerical Simulations

Giulia Zanier [1,*,†], **Massimiliano Palma** [1,†], **Andrea Petronio** [1,†], **Federico Roman** [1,†], **Vincenzo Armenio** [2,†]

1   Iefluids s.r.l., Piazzale Europa 1, 34127 Trieste, Italy; maxpalma@fastwebnet.it (M.P.);
    a.petronio@iefluids.com (A.P.); f.roman@iefluids.com (F.R.)
2   Department of Engineering and Architecture, University of Trieste, 34127 Trieste, Italy;
    VINCENZO.ARMENIO@dia.units.it
*   Correspondence: g.zanier@iefluids.com; Tel.: +39-040-558-3470
†   These authors contributed equally to this work.

Received: 2 January 2019; Accepted: 17 February 2019; Published: 25 February 2019

check for
updates

Abstract: A major threat for marine and coastal environment comes from oil spill accidents. Such events have a great impact on both the ecosystem and on the economy, and the risk increases over time due to increasing ship traffic in many sensitive areas. In recent years, numerical simulation of oil spills has become an affordable tool for the analysis of the risk and for the preparation of contingency plans. However, in coastal areas, the complexity of the bathymetry and of the orography requires an adequate resolution of sea and wind flows. For this reason, we present, to the best of the author's knowledge, the first study on the subject adopting Large Eddy Simulations for both the low-atmosphere and sea dynamics in order to provide highly-resolved marine surface current and wind stress to the oil slick model, within a one-way coupling procedure. Such approach is applied to the relevant case of Kotor Bay (UNESCO heritage since 1979), in Montenegro, which is a semi-closed basin surrounded by mountains that is subject to an intense ship traffic for touristic purposes. Oil spill spots are tracked along ship paths, in two wind scenarios.

Keywords: oil spill; numerical simulation; LES; low atmosphere; coastal flow; contingency plan; Kotor bay

1. Introduction

Oil spill accidents represent a major threat to marine and coastal environment, impacting both biological species and human health, as well as economic, touristic and commercial activities. For example, according to data collected from 1977 to 2003 about 304,700 tons of oil have been released in the Mediterranean Sea mainly due to extensive marine traffic of oil tankers and ships [1,2].

Weather conditions, oil physical and chemical characteristics determine oil fate and persistence at sea. Most kinds of oils spread on the sea surface as a thin film, the slick is then driven by the sea currents and wind stress; furthermore, if the oil temperature drops below the pour-point, oil can solidify and form tar balls. In case of wavy and turbulent seas, small oil drops can detach from the oil slick and then, depending on their density, particles can either sink, float on the surface or be transported along the water column. Moreover, oil interacts with the surrounding environment. Immediately after spills, oil can evaporate and, under the action of wind and waves, it can absorb water droplets producing emulsion. Such phenomena, called weathering processes, change oil physical and chemical properties in time, strongly affecting oil fate and persistence at sea [3–6].

Given the significant impact of oil spill on the environment and economy of the area, over the years, efforts have been devoted to the preparation of contingency plans, aimed at ensuring fast response and to facilitate clean-up operations after accidents. In this context, oil spill numerical models have been widely established as helpful tools, both for development of contingency plans and for guiding clean-up operations. Oil slick models are usually integrated with hydrodynamic and meteorological models that provide sea currents and wind data. The modeling approaches can be classified as Lagrangian, Eulerian and Lagrangian/Eulerian hybrid models [4,7]. In the Lagrangian models, e.g., [2], oil slick is treated as a multitude of finite size particles, which are advected by a mean drift velocity plus a fluctuating turbulent component, the latter usually parametrized by means of a random walk technique. In the Eulerian method, e.g., [8], oil slick dynamics are derived from mass and momentum conservation equations. Finally, in hybrid Lagrangian/Eulerian models (see, for example [9]), a large number of particles parametrizes the oil slick immediately after the spill, and, as far as the width of the slick reaches a terminal value, the computation switches to a Eulerian model.

In the present paper, we study the case of a hypothetical oil spill accident due to ship collision in Boka Kotorska Bay, a long and tortuous fjord situated in the Adriatic Sea. The study is aimed at preparation of contingency plans for the area under investigation. This area is under the UNESCO protection since 1979, for its own important natural and historical heritage. The prevention from possible hazardous situations is becoming urgent in light of the increased maritime traffic over the recent years. To make the study as realistic as possible, the typical ship path is considered within the bay [10,11] in conjunction with oil spill spots identified as dangerous from an environmental point of view.

We use a novel approach to simulate oil slick dispersion in coastal areas characterized by surface currents and low atmosphere circulations governed by complex bathymetry, coastline and orography. We use a two-dimensional Eulerian model derived by Nihoul's theory [12–14] for the oil slick. The oil slick simulation is coupled with LESCOAST [15,16], a high resolution hydrodynamic model used to simulate water circulation in coastal areas. Given the complex orography surrounding the bay, a preliminary low-atmosphere wind simulation is required to take into account the horizontal variability of wind stress. This latter is the main forcing item driving both the oil slick and the sea current in the upper layers, and it has to be properly modeled, for example as suggested in [17].

The paper is organized as follows: in Section 2, we provide a brief overview of the hydrodynamic models for water and air domains; then, we introduce the oil spill model and finally we briefly describe the features of the area under investigation along with the boundary and initial condition for the simulations. Results of the most significant scenarios are reported and examined in Section 3. Finally, the discussion is provided in Section 4.

2. Materials and Methods

In this section, we describe the methodology used for the study: in Section 2.1, we provide a brief description of the LESCOAST/LESAIR model used for the marine and low-atmosphere simulations; in Section 2.2, we present the oil spill model for the analysis of pollutant dispersion; in Section 2.3, we give a description of the Boka Kotorska Bay; in Section 2.4, we discuss the set-up of the simulations.

2.1. Hydrodynamical Model

LESCOAST/LESAIR model [18,19] solves the filtered form of three-dimensional, non-hydrostatic Navier–Stokes equations under the Boussinesq approximation along with the transport equations for scalar quantities, i.e., salinity and temperature/humidity and temperature in marine/atmosphere simulations, respectively.

The LESCOAST/LESAIR model uses a Large Eddy Simulation approach to parametrize turbulence, and the variables are filtered by a low-pass filter function represented by the size of the cells. The subgrid-scale fluxes (SGS), which come out from the filtering operation, are parametrized by a two-eddy viscosity anisotropic Smagorinsky model developed in [18]. Such method is effective in simulating coastal flows on sheet-like anisotropic computational grids.

The complex geometry, which usually characterizes harbor and coastal areas, is treated using an Immersed Boundary Method (IBM), based on a direct forcing approach, as described in [20]; the technique is used to reproduce coastline, anthropogenic structures, bathymetry and topography.

The filtered Boussinesq form of the Cartesian Navier–Stokes equations reads as follows:

Continuity equation:

$$\frac{\partial \overline{u}_j}{\partial x_j} = 0, \tag{1}$$

Momentum equation:

$$\frac{\partial \overline{u}_i}{\partial t} + \frac{\partial \overline{u}_i \overline{u}_j}{\partial x_j} = -\frac{1}{\rho_0}\frac{\partial \overline{p}}{\partial x_i} + v\frac{\partial^2 \overline{u}_i}{\partial x_j \partial x_j} - 2\epsilon_{ijk}\Omega_j \overline{u}_k - \frac{\Delta\rho}{\rho_0}g_i\delta_{i3} - \frac{\partial \tau_{ij}}{\partial x_j}, \tag{2}$$

Scalar transport equation:

$$\frac{\partial \overline{s}}{\partial t} + \frac{\partial \overline{u}_j \overline{s}}{\partial x_j} = k^s\frac{\partial^2 \overline{s}}{\partial x_j \partial x_j} - \frac{\partial \lambda_j^s}{\partial x_j}, \tag{3}$$

where '$-$' represents the filtering operation, u_i represents the ith-component of the Cartesian velocity vector (u, v, w), x_i represents the ith-component of the Cartesian coordinates (x, y, z), t is time, ρ_0 is the reference density, $\overline{p}$ is the hydrodynamic pressure, v is the kinematic viscosity, ϵ_{ijk} is the Levi–Civita tensor, Ω_i is the ith-component of the Earth rotation vector, $\Delta\rho$ is the density anomaly, g_i is the ith-component of the gravity vector, and τ_{ij} is the SGS stress tensor which comes from the nonlinearity of the advective term, $\overline{s}$ is a scalar quantity (e.g., temperature and salinity/humidity), k^s is scalar diffusivity and λ_j^s is the SGS scalar flux.

In the present low-atmosphere simulations, density variations are small and therefore buoyancy effects are neglected. In the marine simulations, we solve the transport equation of the scalar quantities, temperature T and salinity S, respectively. The fluid density is computed through the state equation:

$$\frac{\Delta\rho}{\rho_0} = \frac{\rho - \rho_0}{\rho_0} = -\beta^T(T - T_0) + \beta^S(S - S_0), \tag{4}$$

where ρ_0 is the reference density at the temperature T_0 and salinity S_0; β^T and β^S are respectively the coefficient of temperature expansion and haline contraction.

At immersed boundaries, we apply the wall-layer model (IBWLM) presented in [21]; and at the open boundaries the Orlanski boundary condition is enforced [22], and it reads as:

$$\frac{\partial \overline{u}_i}{\partial t} + C_i\frac{\overline{u}_i}{\partial x_i} = 0, \tag{5}$$

where C_i is the phase velocity, calculated as the flux at the cell face.

The effect of the wind imposed over the free surface is taken into account by means of the formula proposed in [23]. It calculates the stress at the sea surface as:

$$\tau_w = \rho_a C_{10} U_{10}^2, \tag{6}$$

where ρ_a is the density of air and C_{10} is the drag coefficient which is a function of wind speed as:

$$C_{10} = (0.8 + 0.065 U_{10}) 10^{-3}, \tag{7}$$

where U_{10} is the wind velocity 10 m above the mean sea level, which is provided by the low-atmosphere simulations.

For scalar quantities, we apply a no-flux condition at solid walls and, at the surface; at the open boundaries, the Dirichlet condition is enforced with values interpolated from measured data.

Equations (1)–(3) are integrated using the finite difference semi-implicit fractional step method of [24]; it is second-order accurate in both space and time. The model adopts a non-staggered grid, meaning that the primitive variables, like velocity, pressure and scalars are located at the cell centroids, while the fluxes are defined at the cell faces. More details about the numerical model can be found in [18]. In [15], the LESCOAST model is validated against measured and numerical results.

2.2. Oil Spill Model

LESOIL [13,25] is a two-dimensional Eulerian numerical model which simulates the main physical processes governing oil behavior at sea from the start of the release for a time period of the order of 24 h. The processes are: transport and spreading under gravity, friction and Coriolis forces, and the short-term weathering processes, namely evaporation and emulsification.

The model, derived from Nihoul's theory [12], considers oil slick as a thin-film, whose evolution is driven by gravity and friction forces. The equation that is solved for the thickness of the oil slick h reads as:

$$\frac{\partial h}{\partial t} + \frac{\partial v_j h}{\partial x_j} = Q + \frac{\partial}{\partial x_j}\left(\frac{\alpha \partial h}{\partial x_j}\right),$$ (8)

where v is the transport velocity induced by sea currents and wind stress; Q is the source/sink term that takes into account a continuous release of oil, or oil loss due to dispersion of particles or evaporation. The term $\alpha = gh^2(\rho_w - \rho_0)\rho_0/0.02\rho_w$ can be interpreted as the oil slick diffusion coefficient and it depends on gravitational acceleration g, oil and water densities (respectively ρ_0 and ρ_w). Based on literature findings [8], the contribution of sea current and wind stress on the transport velocity of oil slick is given by:

$$v = u_c + k_w U_{10},$$ (9)

where u_c is the velocity induced by current which is provided at each time step by the LESCOAST model; k_w is the wind drift factor, set equal to 0.03 in agreement with relevant literature [8,9,26–28]; U_{10} is supplied by the low-atmosphere simulation.

Immediately after the spill, weathering processes can take place, changing oil density and volume, and eventually influencing oil slick fate and persistence. The model can take into account the main processes, namely evaporation and emulsification, by means of established literature models [29,30]. Parametrization of these effects, in Eulerian models, requires an average of quantities, such as wind velocity, over the whole surface/volume of the slick. This approach is suitable for simulating oil slick in an open ocean scenario where wind stress is almost constant, but it is not properly suited for a slick undergoing a highly varying wind-sea current flow conditions. For these reasons, in this study, the weathering processes are not considered. Although this assumption may appear less realistic, it is still reasonable over a time scale of the order of 12 h and it allows for underlining the spreading mechanism due to the combined wind and sea currents' actions.

In order to facilitate coupling with LESCOAST, the oil model Equation (8) is run on a surface mesh that perfectly matches the horizontal discretization of the first cells layer at the free surface of the marine model. We use a second-order Adams–Bashfort scheme to integrate numerically Equation (8), the diffusion terms are treated using a centered second-order finite differences method, while the advective terms are discretized using SMART, a third order accurate, monotonic scheme [31]. Compared to literature results [8], the method is proved to be accurate without appreciable numerical diffusion [13,25]. At the immersed bodies, we apply a no-flux condition, neglecting the phenomenon of oil deposition over coastlines.

2.3. Kotor Bay Area Description

The area under investigation, illustrated in Figure 1, is Boka Kotorska Bay, or Kotor Bay, a semi-enclosed karstic basin situated in the southeastern side of the Adriatic Sea (Montenegro). The bay covers an area of about 87 km². The fjord consists of three different sub-bays: Kotor-Risan

Bay, Tivat Bay and Hergeg-Novi Bay. The inner one, Kotor-Risan Bay, is further divided into two smaller basins: Kotor Bay (southeast) and Risan Bay (northwest). Two straits connect the three bays: the Kumbor Strait links Tivat Bay with Herceg-Novi Bay, and a narrow canyon, Verige Strait, connects Tivat with Kotor-Risan Bay. The bathymetry decreases rapidly up to a depth of 40 m in the largest part of the basin, the average depth and the maximum depth are 27.3 m and 60 m, respectively. The largest and narrowest widths of the bay are respectively 7 km and 0.3 km [32–34].

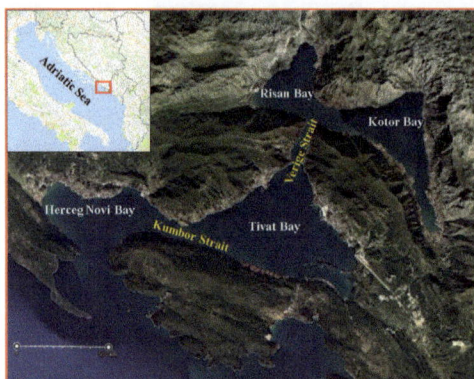

Figure 1. Kotor Bay area and its location in the Adriatic Sea—image from Google Maps (online, Italy, 2017).

Numerous fresh water inputs characterize the bay, such as submarine springs, precipitations and rivers. In the upper layers, the water circulation is driven mainly by wind, tide, river runoff, and density gradients; denser water from the Adriatic Sea flows into the bay in the bottom layer. The fresh water inflow mainly impacts the inner bay and, as a matter of fact, salinity at the sea surface increases moving from Kotor-Risan Bay towards the outer basin. In the winter season, characterized by higher precipitations and river runoff, the saline vertical stratification can be more pronounced. One of the most frequent wind conditions for this area is Bora, a strong katabatic wind which flows from the first quadrant. Bora is more frequent in winter time, and it can last for 3–7 days [33,34]. Because of the mountains surrounding the bay, the fetch is small, preventing the generation of significant waves [32]; for this reason, we neglect the effect of waves on sea current and oil slick simulations.

2.4. Case Set-Up

The computational domain for sea circulation has overall dimensions of $L_x = 18,435$ m, $L_y = 21,575$ m, and $L_z = 72$ m. The computational domain is Cartesian and it is discretized uniformly by $640 \times 1024 \times 24$ grid cells, respectively, with dimensions of about $\Delta x \approx \Delta y \approx 25$ m and $\Delta z \approx 3$ m. The coastline, the anthropogenic structures and bathymetry, as well as topography for the low-atmosphere simulation, are reproduced by the IBM [20]. The computational grid is illustrated in Figure 2; the black rectangle identifies grid borders, and its four corners are labelled with Cardinal points according to grid orientation. The immersed body used to reproduce the coastline is shaded in dark-gray, while the contour colors represent the depth of sea bottom as reproduced in the computational domain. On the southwest side, the basin exchanges water with the Adriatic Sea. Here, we apply the Orlanski boundary condition (Equation (5), [22]), which allows for the simulation of the inversion of the currents over the vertical. At the sea surface, currents are forced by wind stress, whose intensity and direction are obtained by LESAIR Simulations (LAS).

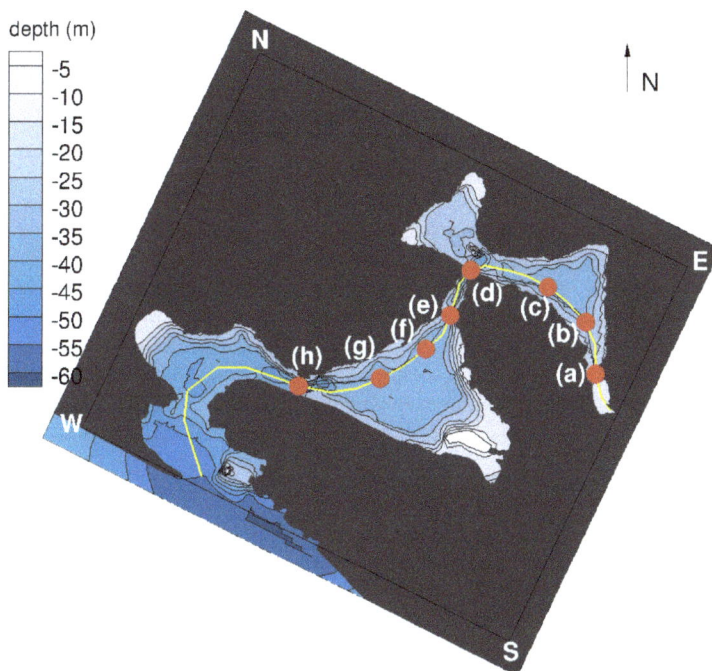

Figure 2. Computational domain: the black lines define grid borders, the Immersed Boundary Method is used to reproduce the coastline (gray shade) and bathymetry (contour plot) of the fjord. The yellow line indicates the main ship route in the bay as reported in literature [10,11], red dots indicate the position where we consider oil spill occurrence in our simulations.

For the low-atmosphere, first, a simulation was run on a large and coarse grid (hereafter referred to as LASc); then, the obtained velocity data were nested as initial and boundary conditions of a second simulation on a smaller and finer grid (hereafter labeled as LASf). The height of the domain ($L_v = 2000$ m) is the same for both low-atmosphere grids, and it is discretized by 24 grid cells, stretched in order to obtain a finer resolution near the surface. For LASf, the horizontal domain dimension and discretization are the same as those adopted for the marine domain, while, for the LASc, the horizontal dimensions of the domain are three times larger than in LASf, retaining the same number of grid points.

For the atmosphere simulations, two summer-wind conditions are chosen: the first is the Bora wind case, which is the most frequent wind in the area; the second is the Libeccio (southwest) wind condition. Although it is not the most frequent wind in the area, it is chosen because it represents the worst scenario in case of oil spills, especially in low river runoff conditions. Libeccio, blowing inland from the Adriatic Sea, prevents oil slicks from flowing out of the bay.

The boundary conditions applied for the LASc are: a logarithm inflow velocity with intensity $U_{10} = 6$ m/s, set at the wind side (northeast for Bora wind simulation $\theta = 65°$ and southwest for Libeccio wind simulation $\theta = 206°$), named Log Inflow in Table 1. On the opposite sides, southwest and northeast for Bora and Libeccio cases, respectively, the Orlanski boundary condition is adopted. On the northwest and southeast sides, periodicity is enforced. On the top boundary, we use a free slip condition. On the bottom boundary, at the interface with the sea, a free slip is used, while, at the immersed boundaries surface, a wall model is applied. The LASf boundaries and initial conditions are provided by a nesting procedure, interpolating data from the coarser simulation. Finally, from the computed wind velocity 10 m above the surface, the stress at the free surface of the hydrodynamic model is calculated using Equation (6). The summary of boundary conditions applied for air and sea simulations is reported in Table 1.

For the sea simulation, temperature and salinity fields are initialized according to literature field data of Boka Kotorska Bay [33,35], while, at the southwest boundary, Copernicus database values [36] are adopted. Density stratification is principally driven by temperature gradients in Kotor-Risan Bay. In the inner bay, the temperature varies from 24° (C) at the surface to 17° (C) at the bottom, while the salinity varies from 33 (PSU) close to the surface to 36 (PSU) at the bottom. In the other bays, the vertical distribution of the two quantities is almost constant with values approximately of $T = 24°$ (C) and $S = 38$ (PSU).

Oil spill simulations are set considering a ship collision that releases a light oil (CPC-BLEND), with density $\rho_0 = 809$ kg/m^3 and a pour point of $-36°$ (C), low enough to prevent formation of tars. In Figure 2, the yellow line indicates the ship navigation route as reported in [10], obtained from [11]. The accidents are assumed to happen along the path in Verige Strait, the narrowest region of the bay. The red dots in Figure 2, labeled with letters (a)–(h), indicate the oil release points. The spill rate reproduces a tub emptying process, considering a total amount of spilled oil of $V = 1400$ m^3, in 4 h.

Table 1. Summary of boundary conditions applied.

	LASc Bora	LASc Libeccio	LASf	Sea sim.
Top	Free Slip	Free Slip	Free Slip	Wind stress Equation (6)
Bottom	Free Slip/IBWLM	Free Slip/IBWLM	Free Slip/IBWLM	IBWLM
Side N-E	Log Inflow: $U_{10} = 6$ m/s, $\theta = 65°$N	Orlanski Outflow Equation (5)	Nesting	Ibm
Side S-E	Periodic	Periodic	Nesting	Ibm
Side W-S	Orlanski Outflow Equation (5)	Log Inflow:, $U_{10} = 6$ m/s $\theta = 206°$N	Nesting	Ibm/Orlanski Outflow
Side N-W	Periodic	Periodic	Nesting	Ibm

BM = Immersed boundary method; IBWLM = Wall Layer Model for Immersed boundary.

3. Results

In this section, we show and describe the dynamics of low-atmosphere, sea and oil spill, first for the Bora wind scenario and successively for the Libeccio one. Finally, we analyze the effect of wind stress on oil slick dynamics.

3.1. Low-Atmosphere Simulations for Bora Wind Scenario

Figure 3 shows the wind velocity magnitude and vectors obtained from low-atmosphere simulations. Panel (a) shows the contour plot of wind velocity 10 m above the sea level in the LASc. The immersed bodies adopted to reproduce mountains surrounding the bay, are represented by green-brown contour plot. Panel (b) shows the inhomogeneous map of wind velocity obtained over the bay in simulation LASf; there are regions where the magnitude of horizontal velocity is close to zero, especially by the leeward side of the surrounding mountains. For example, in panel (c), a recirculation zone leeward a hill is well visible at a vertical cross-section over Herceg-Novi Bay. The position and direction of the section is marked by a red line and rectangle in Figure 3b.

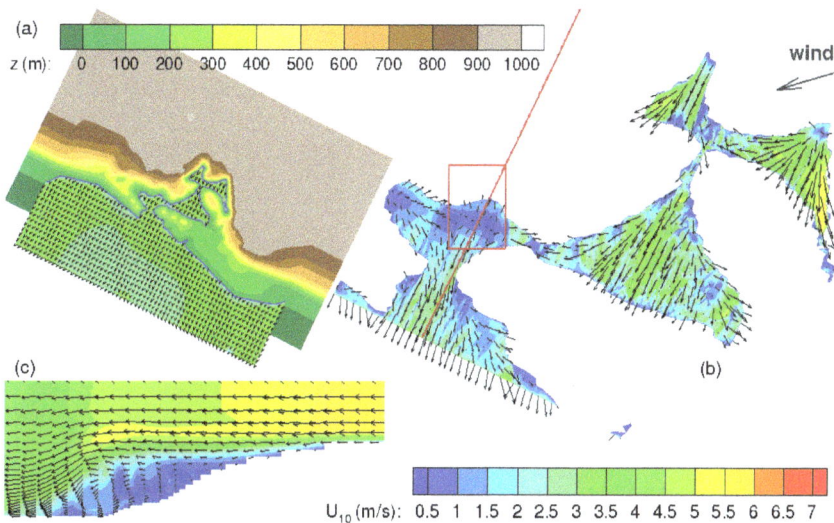

Figure 3. Low-atmosphere simulation of Bora wind case: horizontal instantaneous velocity 10 m above the sea surface for the LASc (**a**) and for the LASf (**b**); (**c**) a cross section along wind direction shows the development of recirculation zones close to mountains slopes. The location of the section is shown with the red line and rectangle in (**b**). Velocity vectors are plotted every 16 nodes in both horizontal directions. The green-brown contour plot in (**a**) represents the immersed body that reproduces the terrain elevation z. The velocity vector on the top-right shows wind direction, set up as inflow boundary condition in LASc.

3.2. Sea Current Simulation for the Bora Wind Scenario

Figure 4a shows the surface contour plot of the horizontal instantaneous velocity field. The horizontal space variability of wind stress directly affects the sea current, which flows from the inner bay towards Adriatic Sea. The red lines indicate the position of two vertical sections (W-E Figure 4b and N-S Figure 4c). The flow features within the fjord are made evident by streamlines: at the surface, the current follows the wind direction, while, at the bottom, it flows in the opposite direction. These up-welling and down-welling phenomena are typical in coastal regions. On the vertical section, horizontal axis vortices, spanning all over the water column, are visible. Such structures generated by wind stress at the sea surface have also been observed previously in [15,17].

Horizontal and vertical distributions of water density are shown in Figure 5. Panel (a) illustrates density variations at the surface layer; we notice that in the Kotor-Risan Bay, more than in other bays, sea currents are driven by both wind stress and density differences. Panels (b) and (c) show vertical distribution of density in two cross-sections indicated by red lines in panel (a). In Kotor-Risan Bay (panel (b)), analysis of the contour plot and streamlines suggest that wind stress is able to break stratification, thus enhancing water mixing along the vertical depth: lighter surface water is transported towards the denser region along the wind direction. This phenomenon appears more pronounced in panel (c), which shows that low-density water is pushed by wind from Kotor-Risan Bay towards Tivat Bay. The horizontal axis vortex developing in Verige Strait is promoted by both bathymetry and the horizontal stratification effect.

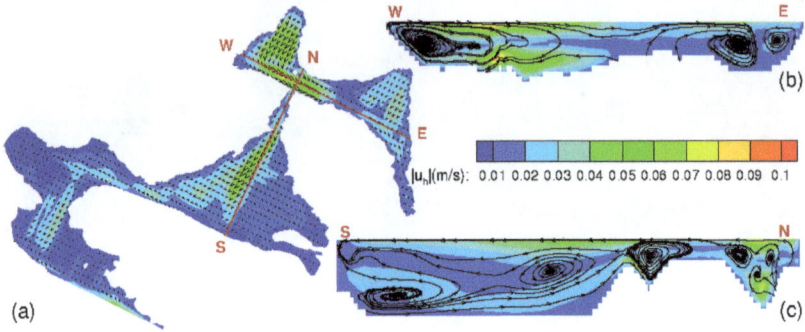

Figure 4. Sea water horizontal instantaneous velocity $|u_h| = \sqrt{u^2 + v^2}$ in Bora wind scenario. (**a**) contour plot and velocity vectors close to the surface Vectors are skipped every 12 node cells in both directions. (**b**) contour plot in a vertical section along Kotor-Risan Bays; (**c**) contour plot in a vertical section along Kotor-Tivat Bays. The location of sections is shown in (**a**) with red lines. Streamlines indicate fluid flow direction.

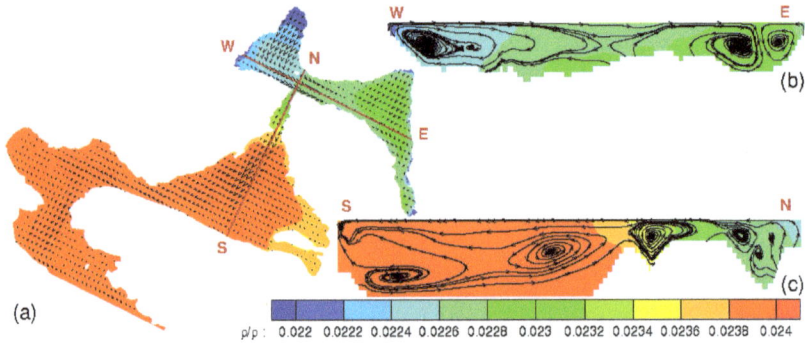

Figure 5. Contour plot of instantaneous density in Bora wind scenario: (**a**) at the surface; (**b**) in a vertical plane in Kotor-Risan Bays; (**c**) in a vertical plane in Kotor-Tivat Bays. The location of sections is shown in (**a**) with red lines. Streamlines and vectors indicate fluid flow direction; vectors are skipped every 12 node cells in both directions.

3.3. Oil Spill Simulation for Bora Wind Scenario

Figure 6 shows different oil spill scenarios, simulated for the Bora wind case. Following a typical ship's path, from Kotor harbor to Kumbor Strait (yellow line in Figure 2), oil is released at different points (identified by panels from (a) to (h)) along the route. Film thickness contour lines of $h = 10^{-3}$ mm show the oil slick position at different simulation times after the initial release, t_r: $t_r = 5$ min (red line), $t_r = 1$ (green line), $t_r = 5$ h (purple line) and $t_r = 12$ h (blue line). Vectors indicate the direction and intensity of currents at the sea surface (dark gray), and wind velocity 10 m above sea level (orange); for clarity, vectors are shown every 20 grid cells in both horizontal directions and wind velocity vectors are scaled by a factor of 0.01. For each scenario, the area of interest for the oil spill event is shown on the left; on the right, we show a zoom over the oil slick, in order to better visualize oil spreading and transport dynamics.

In the first scenario (Figure 6a), oil is released in the narrow bay (700 m wide) close to Kotor harbor. Here, both sea currents and wind are weak and the slick spreading is driven by gravity force rather than friction. One hour after the spill, the slick reaches the eastern coastline, and, later on, the western one. Five hours after the spill, the slick spreads over the entire narrow strait and, after twelve hours, the oil slick occupies an area of about $6 \cdot 10^5$ m², still moving towards Kotor harbor and slipping along the coastlines.

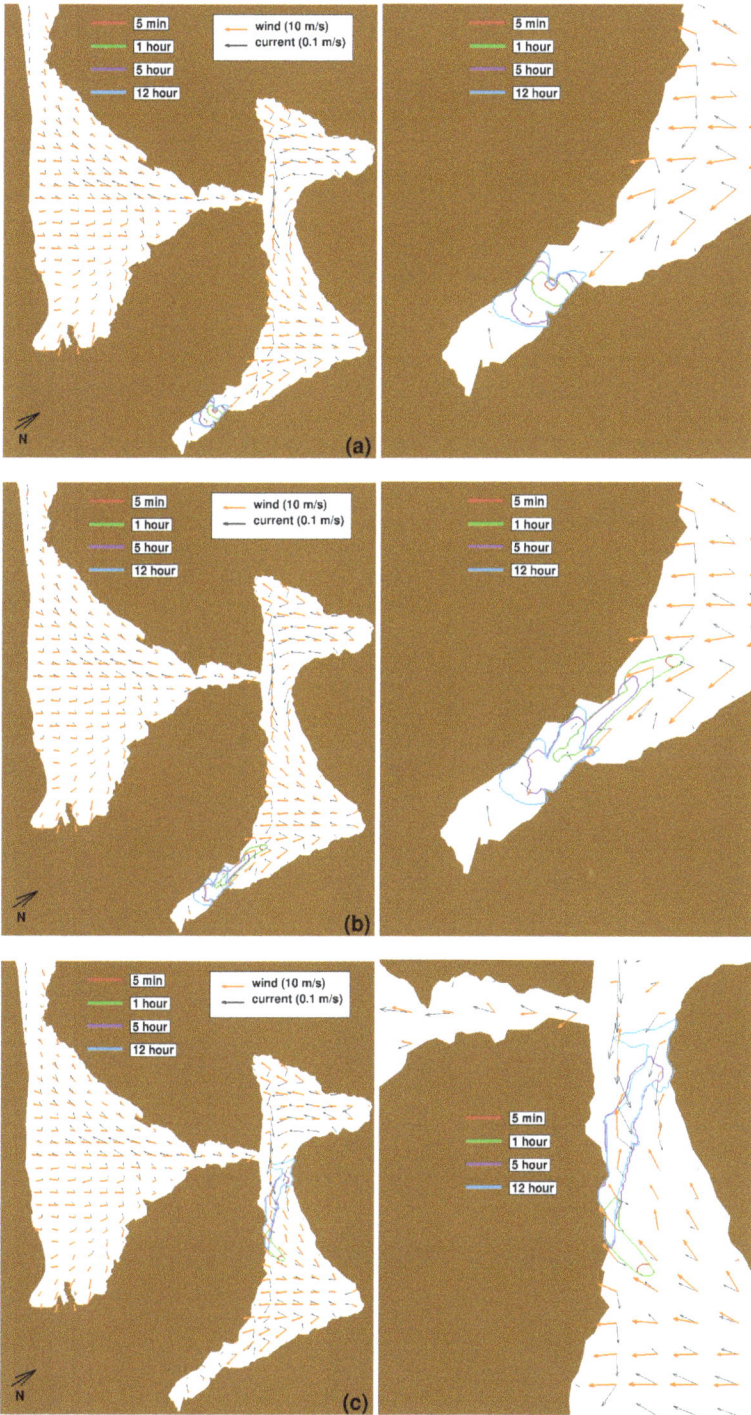

Figure 6. *Cont.*

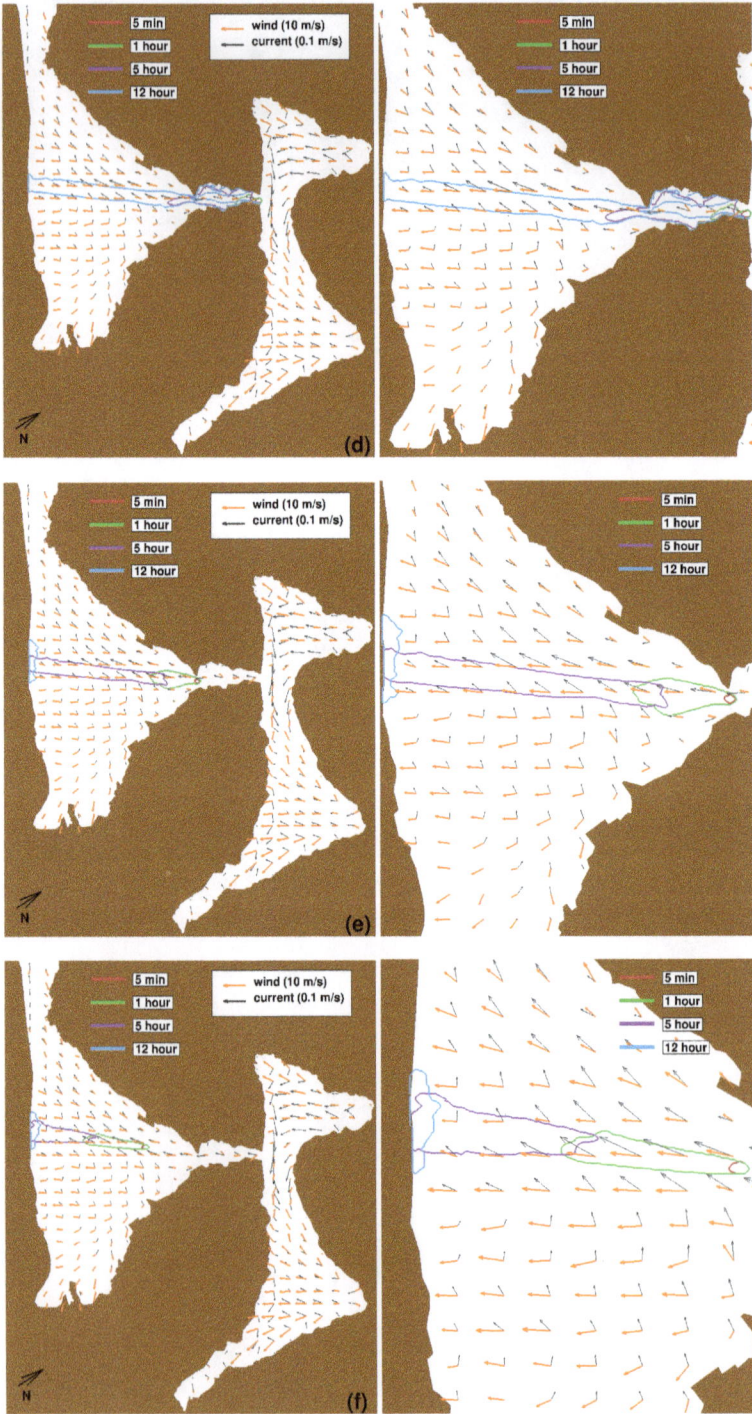

Figure 6. *Cont.*

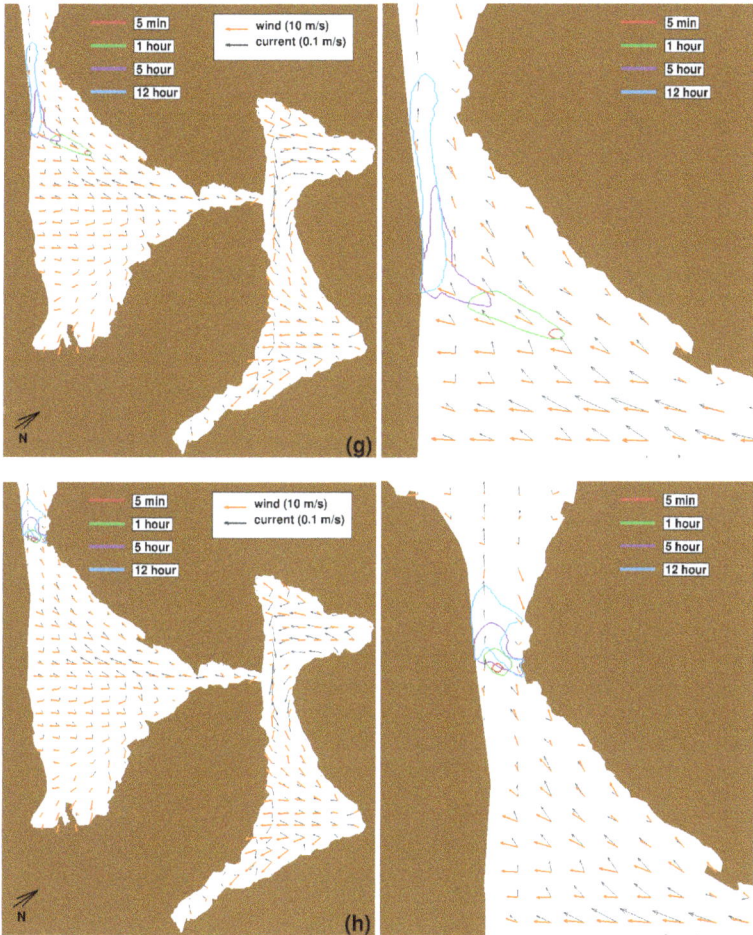

Figure 6. Oil slick, released at different points, at different simulation times in the Bora wind scenario; the film thickness represented is 10^{-3} mm. Vectors show the direction and intensity of sea currents (dark grey) and wind (orange), vectors are shown every 20 grid cells in both horizontal directions and wind vectors are scaled by a factor 0.01. Zoom factors of the right panels are: (**a**) 4; (**b**) 4; (**c**) 4; (**d**) 2.5; (**e**) 3.25; (**f**) 4.5; (**g**) 3.75; (**h**) 3.25.

At the second release point, wind and sea currents are relevant (Figure 6b). During the first hour after the spill, oil starts spreading in the direction of the wind and is transported southward by the sea current towards Kotor harbor. Five hours after the spill, the slick reaches approximately the position described in the previous scenario. Here, wind and sea currents are weaker and the oil slick starts spreading radially driven by gravity force, similarly to the case of scenario (a).

In panel (c), the release spot is in the central part of Kotor Bay (Figure 6c), where wind and current point westward. Oil impacts on the southern coastline and is then transported by currents northward, close to Verige Strait, where sea currents flow in the direction opposite to the wind. Once oil reaches the northern coastline, it starts accumulating and then spreads driven by friction and gravity forces. The flow pattern of recirculation wind in this area makes the evolution of the oil slick more complex. Wind is weak because of the presence of mountains and the opposite sea currents (from Risan and from Kotor Bays, respectively) drive oil towards Verige Strait.

Panels (d) to (f) show the evolution of oil spill taking place in the Verige Strait. In each scenario, oil is transported mainly along wind and currents' directions. In the scenario shown in Figure 6d, oil reaches the eastern shore of Verige Strait in less than one hour. Later on, because of lateral spreading, the slick also reaches the other side of the strait, covering a distance of about 150 m. Five hours after the spill, the slick occupies the whole strait area. After 12 h, the slick is transported inside Tivat Bay, where it finally reaches the southern coastline and starts accumulating because of the overall weaker transport velocity. In scenario (e), oil reaches the southern coastline about five hours after the spill and then it starts accumulating along the shore. Scenario (f) is similar to the aforementioned one; oil spot is 3 km south with respect to the previous scenario and the slick impacts the shore sooner.

In the scenario illustrated in Figure 6g, the oil slick spreads along wind and sea currents and it moves towards Kumbor Strait. Five hours after the spill, the oil slick reaches the southern shoreline where it starts accumulating and to be transported towards the strait along the coastline.

Finally, for the scenario with oil release spot in the Kumboir Strait (Figure 6h), where wind stress is weaker, the slick is spread mainly by friction and gravity force. The slick advection velocity is slower than in previous scenarios because of the lower wind friction in the area. In five hours, the slick impacts both shores of the strait and it moves towards Herget-Novi Bay.

3.4. Low-Atmosphere Simulations for Libeccio Wind Scenario

In Figure 7, we show wind horizontal velocity 10 m above the sea surface obtained in the two low-atmosphere simulations of Libeccio wind case. In panels (a) and (b), we show contour plot of wind velocity obtained in the LASc and LASf, respectively (note that the green-brown contour plot in panel (a) illustrates the immersed body used to reproduce the topography around the bay). For clarity, velocity vectors are plotted every 16 nodes in the horizontal directions. Similar to the Bora wind scenario, regions with different wind stress intensities can be identified. In Tivat Bay, at the entrance of Verige Strait, the wind accelerates and reaches a peak value of about $U_{10} = 10$ m/s.

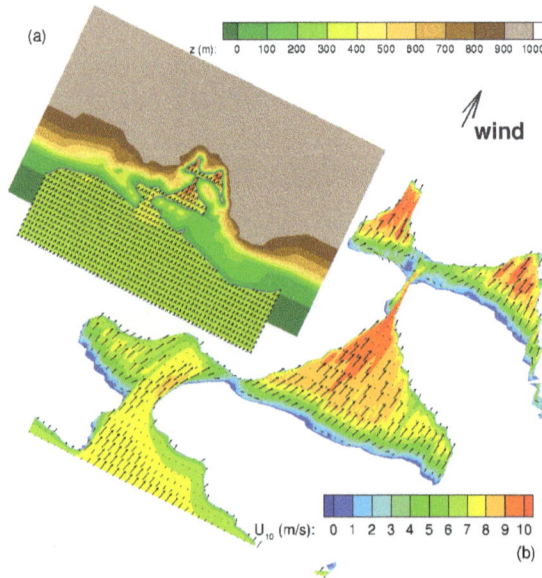

Figure 7. Contour plot of instantaneous horizontal wind velocity 10 m above the sea level for Libeccio wind case: (**a**) LASc; (**b**) LASf. Velocity vectors are plotted every 16 nodes in both the horizontal directions. The green-brown contour plot in (**a**) represents the immersed body reproducing terrain elevation z. The velocity vector in top-right shows wind direction set as inflow boundary condition in LASc.

3.5. Sea Current Simulation for Libeccio Wind Scenario

Contour plots of the horizontal velocity and vectors in Figure 8a indicate the intensity and the direction of the surface current. It moves along the wind direction and become more intense in correspondence of wind peaks. In the vertical sections W-E (Figure 8b) and N–S (Figure 8c), the streamlines show the behavior of fluid flow along the vertical direction. At the surface, the current is aligned with wind direction and flows from Tivat Bay to the inner one, while, at the bottom, the current moves in the opposite direction. Close to the coastline or in correspondence of rapidly varying bathymetry, up-welling and down-welling phenomena are visible.

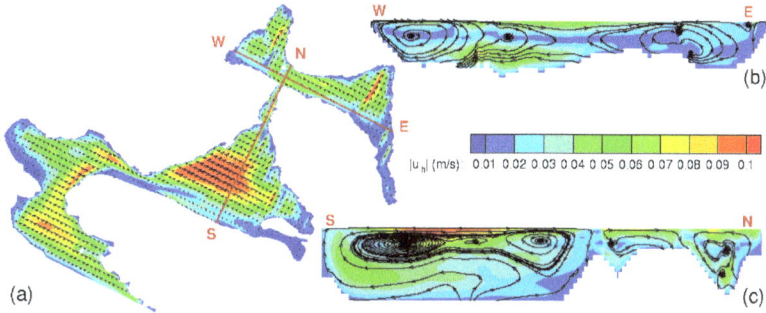

Figure 8. Sea water instantaneous horizontal velocity $|u_h| = \sqrt{u^2 + v^2}$ for Libeccio wind scenario. (a) contour plot and velocity vectors close to the surface, vectors are skipped every 12 nodes for cells in both horizontal directions. (b) contour plot in a vertical section of Kotor-Risan Bays; (c) contour plot in a vertical section in Kotor-Tivat Bays. The location of sections is shown in the (a) with red lines. Streamlines indicate fluid flow direction.

3.6. Oil Spill Simulation for Libeccio Wind Scenario

Figure 9 illustrates oil spills events for Libeccio wind case. In the following, all notation, colors, resolution and scaling factors are the same as for the Bora wind case, if not explicitly reported.

The oil spill dynamics, in all the distinct scenarios under Libeccio wind conditions, seem to indicate Kotor Bay as the most sensitive area in case of an oil spill accident. In fact, in all cases shown in Figure 9a–h, the oil slick is transported towards the northeast in Kotor Bay, where both wind and sea currents entrap the oil.

In the first scenario, shown in Figure 9a, oil starts spreading immediately after the spill under the action of gravity force rather than friction, since wind and sea currents are weak in this region, as also observed for Bora wind scenario. One hour after the spill, the slick impacts the western shore and from here it starts spreading northward along the coastline; then, it starts to be drifted eastward under the action of the more intense wind. Finally, it impacts the eastern coastline about twelve hours after the spill.

In the second scenario, panel (b), wind and currents are stronger at the oil spill location: the slick is immediately transported toward the eastern coastline. The impact takes place at about $t_r = 1$ h. Then, the slick is driven northward along the shore and eventually gets trapped in the bay.

The oil spill scenario shown in Figure 9c is similar to the one described in (b): oil is firstly transported towards the eastern coastline in less than one hour, and then it moves along the coastline toward the bay.

In the scenario in Figure 9d, oil is released immediately after the Verige Strait. The slick impacts the portion of coastline situated in front of the strait, in less than one hour after the spill, and then it travels along the coastline towards the inlet.

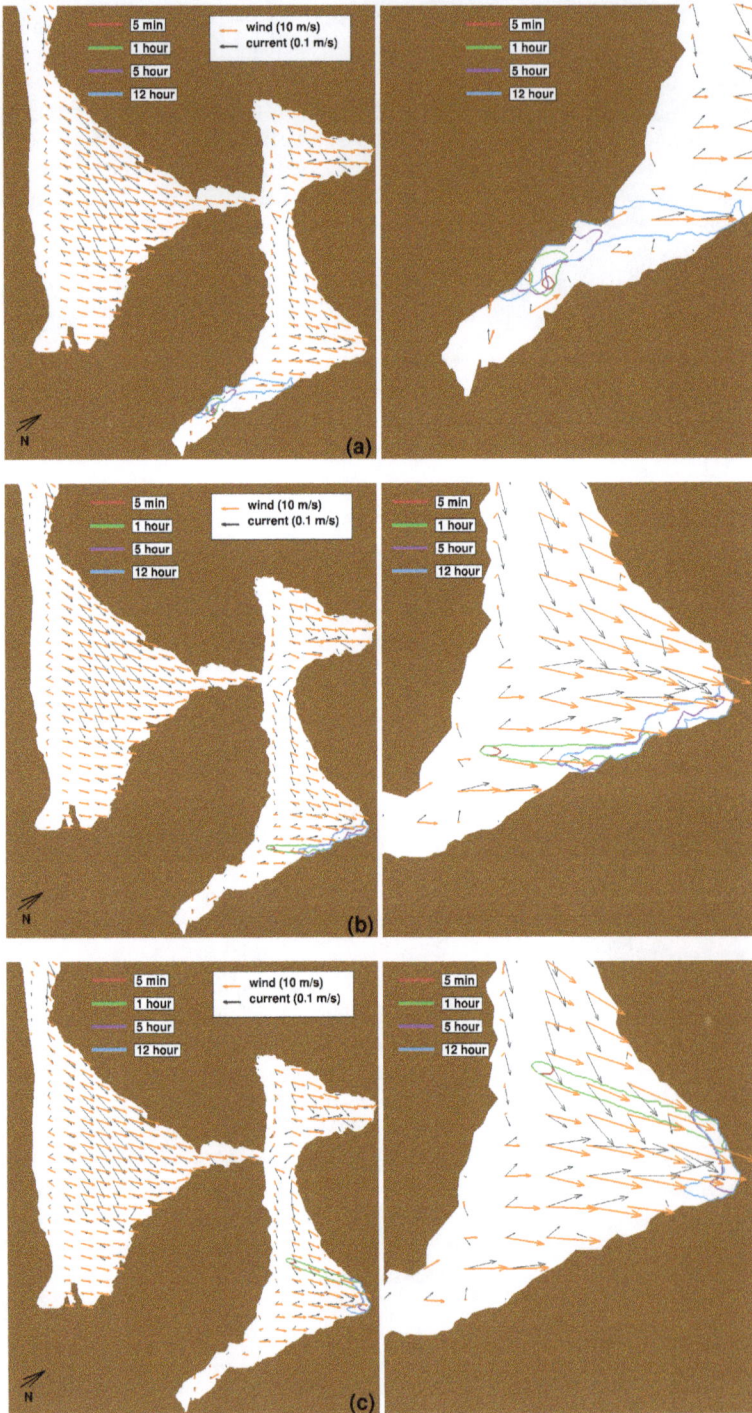

Figure 9. *Cont.*

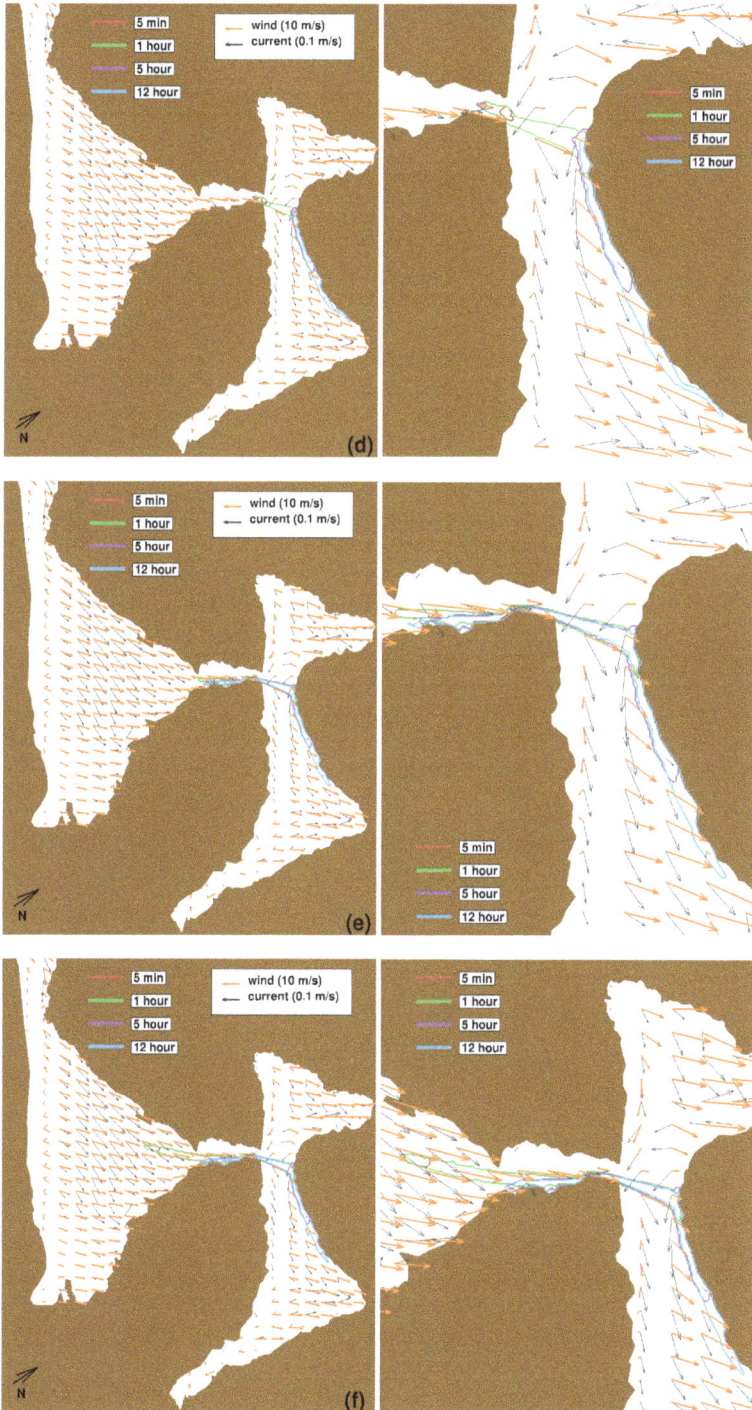

Figure 9. *Cont.*

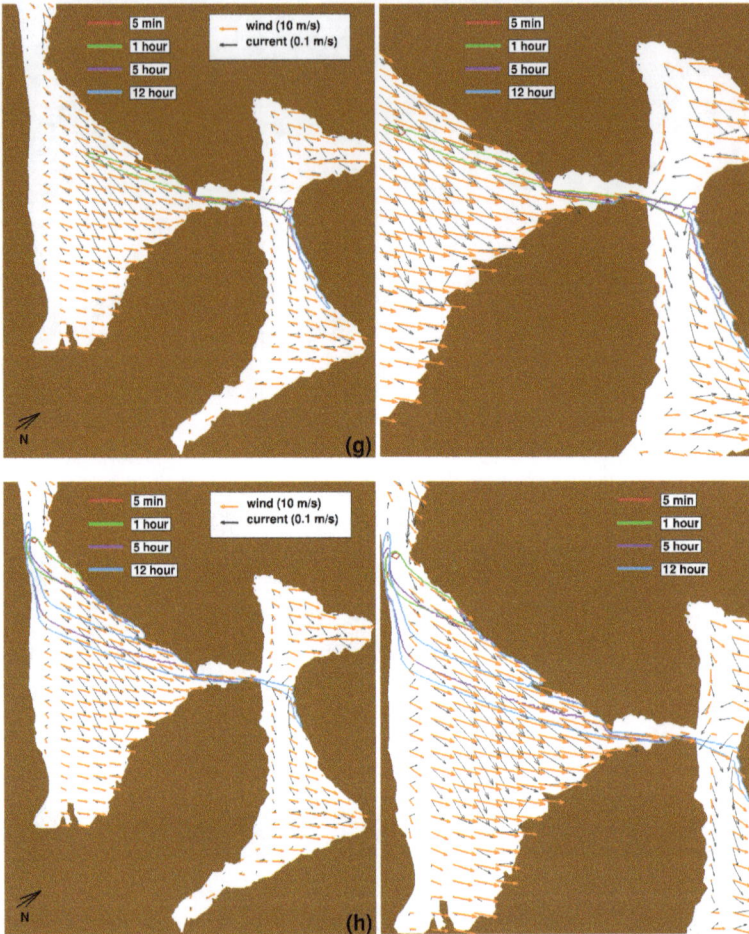

Figure 9. Oil slick, released at different points, at different simulation times for Libeccio wind scenario; the film thickness represented is 10^{-3} mm. Vectors show the direction and intensity of sea currents (dark grey) and wind (orange), vectors are shown every 20 grid cells in both horizontal directions and wind vectors are scaled by a factor 0.01. Zoom factors of the right panels are, respectively: (**a**) 4; (**b**) 4; (**c**) 4; (**d**) 4; (**e**) 4; (**f**) 3; (**g**) 2.4; (**h**) 2.1.

In the scenario illustrated in Figure 9e, oil is released at the entrance of Verige Strait. In less than one hour, the slick impacts the eastern coastline of the strait first, and, later on, the coastline in front of the mouth of the strait. From here, oil is further transported along the coastline. Inside the strait, a fraction of the oil slick slows down and starts accumulating on the strait shore, slowly releasing more oil in the bay.

In scenario (f) (Figure 9f), the oil slick moves towards the Verige Strait and impacts its eastern shoreline in less than one hour. At about $t_r = 5$ h, it reaches the coastline in front of the strait and then it slips along the shore eastward. As in scenario (e), a part of the slick slows down, impacting the shore, releasing an elongated tail, departing from the main slick.

In scenario (g) (Figure 9g), oil spreads along the wind direction towards the Verige Strait. In about one hour, it impacts the western coastline at the entrance of the strait, and then it is transported by wind and currents on the other side of the strait. As for scenarios (e) and (f), when part of the slick starts slowing down approaching the shore, an elongated tail departs from the main slick. At $t_r = 12$ h, part of the oil is still located along the eastern strait's shore.

The last scenario is illustrated in Figure 9h. The spill occurs in an area where both wind and sea currents are weak; therefore, as mentioned in scenario (a), oil spreads radially mainly due to gravity force. A part of the slick reaches the southwestern coastline where wind and sea currents are almost absent; here, oil starts to accumulate along the shore. Another part of the slick meets stronger currents and starts to spread along the flow direction. At the $t_r = 1$ h, the slick has also reached the northwestern coastline. The slick is then transported through Tivat Bay along the coastline, and, in the meantime, part of the oil from the slick accumulated along the southwestern shore starts detaching. As a consequence of this continuous release of oil along the coastline, the oil slick is much larger than in the other scenarios analyzed. Five hours after the spill, the oil slick arrives at the entrance of the Verige Strait. Twelve hours after the spill, the slick crosses the Tivat Bay and Verige Strait. Part of the oil is accumulated at the entrance along the western coast and at the eastern shore of the strait. The slick also reaches the coastline in front of the straits and it is transported eastward.

3.7. Effect of Wind Stress Parametrization on Oil Slick Transport

In most models present in literature, oil slick advection velocity is given by the sum of two contributions proportional to sea current and wind velocity, respectively (see Equation (9)). Usually, for sea current, the drift coefficient is set equal to 1.0, while the wind drift factor adopted is $k_w = 0.03$ [8,9,28]. More recently, some authors proposed different values for the parametrization, as, for example [2]. They calibrated the drift coefficients of their oil spill Lagrangian model using trajectories of buoys; the wind drift coefficient was found to be $k_w = 0.005$, a value one order of magnitude smaller than the standard literature one. Such difference can completely modify the trajectory of the spill predicted by the model. At the moment, we cannot draw any conclusion on the correct value to be used in simulations, especially in the presence of Eulerian models. However, assessing the difference in results considering the two values can shed light upon the necessity of more fundamental work on the topic. For this reason, we apply the newly proposed parametrization value in the scenarios studied in Section 3.3, for the Bora wind case, considering the release points (d) and (e). In such cases, the relative importance of wind and sea current can lead to completely different predicted trajectories. Both points are located in Verige Strait, which is surrounded by mountains rising up to 600 m. In this canyon, the wind blows from Kotor-Risan Bay towards Tivat Bay. With the standard k_w value, the slick spreads along the wind direction while lateral diffusion is inhibited. The wind stress is dominating over the other forces. The oil slick moves following the path where wind and currents reach the maximum velocity as can be deduced by an analysis of the wind and current. Figure 10 shows a vertical cross-section downwind the Verige Strait, the position being indicated by the red line in the black top view of the bay. Contour plots indicate the plane-normal velocities for air and water flows, in the upper and lower panels of the figure, respectively. The dark green line indicates the position and the extension of the oil slick at $t_r = 12$ h, for scenario (d). We can notice that, leeward, the Verige Strait, the velocity of wind and sea current are higher and move from the strait towards the bay; oil slick is located in the area where both wind and sea currents are stronger.

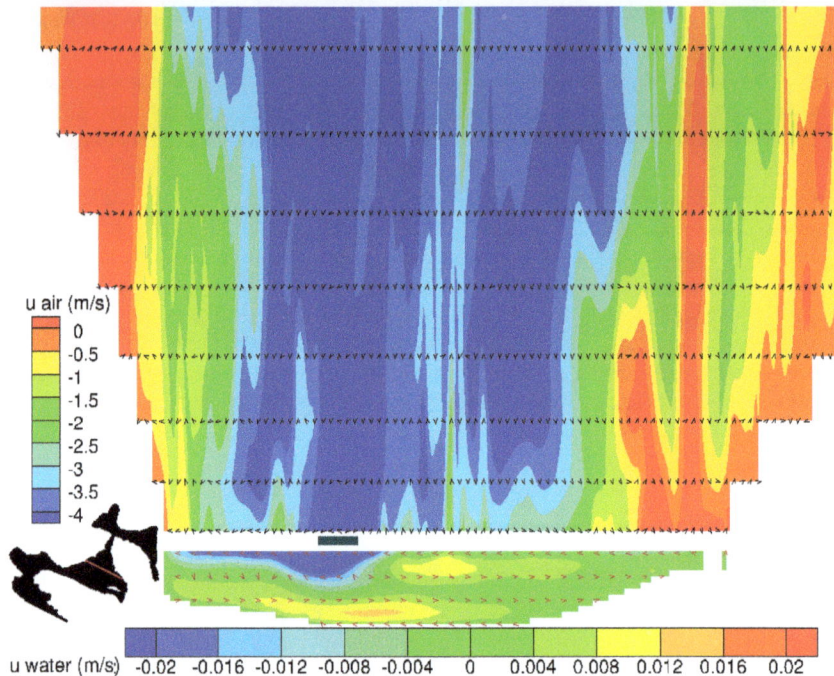

Figure 10. Plot of cross-shore plane leeward the Verige Strait. The contour plot shows the plane-normal instantaneous velocities of air (top panel) and water (bottom panel) simulations, the vectors indicate the plane-tangential velocity components. The dark green line between the panels shows the position of the oil slick at $t_r = 12$ h for scenario (d).

On the other hand, the use of $k_w = 0.005$ produces a variation of the trajectories as depicted in Figure 11 for scenarios (d) and (e). In scenario (d), oil is transported slower and it spreads laterally and backward in the Kotor Bay. At $t_r = 12$ h, the oil slick has moved along the wind direction inside the strait, but also against wind, following the sea current close to the southern coast of the Kotor Bay.

In scenario (e), the slick spreads laterally following the sea current. After $t_r = 12$ h, the slick is still in the middle of the bay.

This brief analysis highlights the importance of a correct wind stress parametrization in order to correctly predict oil slick trajectory and spreading, especially in the presence of complex air and sea flows as in coastal areas. The new parametrization seems to have a minor impact in case of a strong wind, while, in case of light wind, it overestimates the effects of sea currents in oil slick transport, as we can observe in Figure 11, for the scenarios (e) (on the right panel) and (d) (on the left panel), respectively.

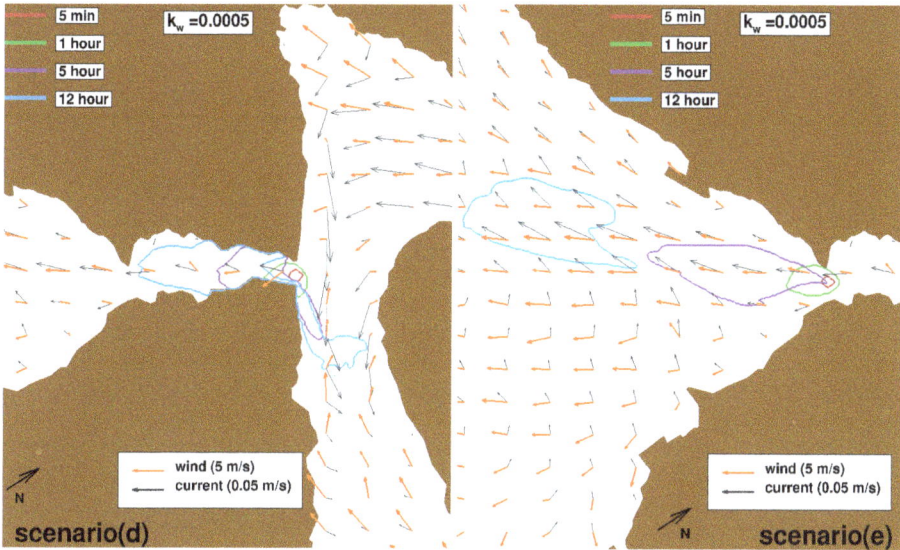

Figure 11. Contour plot of oil slick position at different simulation times. Oil is released windward Verige Strait (**left figure**) and leeward Verige Strait (**right figure**). Here, we use the wind drift factor proposed in [2] $k_w = 0.005$. Vectors show the direction and intensity of sea currents (dark grey) and wind (orange), vectors are shown every 20 grid cells in both horizontal directions and wind vectors are scaled by a factor 0.01.

4. Discussion

In this study, we analyze the dynamics of a hypothetical oil spill accident in the Boka Kotorska Bay, a fjord characterized by a rapidly varying orography, a complex bathymetry and sinuous coastline. A high resolution model is used to reproduce this complex dynamics and to obtain reliable oil slick predictions. In this study, the oil model does not take into account the weathering processes, in order to underline the role played by wind and sea current on the oil fate. Its inclusion will be considered in future works.

As suggested in [17], the complex orography can affect the horizontal distribution of wind stress in the bay, making it highly inhomogeneous. For this reason, two low-atmosphere simulations are run: LASc uses a large domain with a coarse grid, LASf uses a small domain with a fine grid. The latter provides an accurate distribution of the wind stress, used for both marine and oil simulations. The analysis of low-atmosphere results highlights the presence of zones where wind stress is strong, spots where wind is almost absent or even recirculating in the opposite direction. The wind stress is adopted as surface boundary conditions for marine simulations. We investigate different oil spill scenarios, characterized by different locations of oil release within the bay, along the ship route. We consider two meteorological conditions: Bora wind (NE), which is the most frequent wind in the area; Libeccio wind (SW). We consider a summer situation, in which river runoff is almost absent and river discharge in the bay can be neglected. The absence of river runoff and hence lower outflowing currents turn out to be pejorative since the pollutant remains longer in the bay.

In both cases, we find that oil slick approaches the coastline in few hours and then it spreads along it, constrained by wind forcing and by the alongshore surface sea current. This small time scale underlines the need of contingency plan for this type of situation, allowing for an effective pollutant mitigation action.

The complex orography has a strong impact on the oil slick because wind flow is influenced by the presence of the mountains. This aspect and its interaction with the coastline generate complex flow patterns at the sea surface, eventually influencing pollutant dispersion.

On leeward coastline, because of the mountains, the wind exhibits separation, with low stress regions and recirculations with respect to the wind direction. On the windward coastline, down-welling takes place, but, due to the mountains on the same side, the wind is then forced to blow parallel to the coastline, and the sea current adjusts accordingly. This behaviour is accentuated under strongly stratified water column, since the energy transfer from the wind to the sea, is confined in the upper water layer; this results in an intensification of the horizontal component of the surface current, rather than in an increase of down-welling.

Depending on the wind direction, the gulch between Tivat Bay and Risan-Kotor Bay behaves as a bottleneck (see Libeccio case), where the wind, constrained by converging mountains, accelerates increasing the stress at the sea surface. Moreover, the wind, constrained by lateral and bottom boundaries in the fjord, forms secondary flows characterized by vortexes elongated in the streamwise direction. The resulting stress at the sea surface reduces lateral oil spreading.

The complex orography and coastline also determine areas where the sea current is opposite to the wind direction. Considering Bora case, scenario (c), such effect is evident and could trigger counterintuitive oil spreading: sea current flows in one direction while the oil spreads on the opposite one. In this kind of analysis, the wind parametrization adopted in the oil model becomes crucial for cases where air and surface water move in opposite directions. As underlined by the numerical simulations, variations in the wind parametrization can bring to completely different spreading scenarios. In this study, we adopted the literature's widely accepted value and successively, for two spill scenarios, we re-run some cases with a smaller value recently proposed in literature. The differences are significant.

These aspects show the difficulties in the prediction of oil fate in such a complex situation, and that a proper wind pattern representation, together with a robust wind stress parametrization for the oil model, are of crucial importance.

Author Contributions: Conceptualization, F.R.; Formal analysis, M.P. and A.P.; Investigation, G.Z.; Supervision, V.A.

Funding: This study was partially supported by the Know-How Exchange Program, KEP-Italy 2016, Central European Initiative No1206.005-16, and by Ocean Montenegro d.o.o. and Logicar d.o.o.

Acknowledgments: The authors thanks Radmilla Gacić and Danilo Nikolic from the University of Montenegro for help in data collection and analysis.

Conflicts of Interest: The authors declare no conflict of interest.

References

1. IMO/UNEP. *Regional Information System, Part C, Databanks, Forecasting Models and Decision Support Systems, Section 2: List of Alerts and Accidents in the Mediterranean*; Technical Report; RAMPEC: Valletta, Malta, 2002.
2. Cucco, A.; Sinerchia, M.; Ribotti, A.; Olita, A.; Fazioli, L.; Perilli, A.; Sorgente, B.; Schroeder, K.; Sorgente, R. A high-resolution real-time forecasting system for predicting the fate of oil spills in the Strait of Bonifacio (western Mediterranean Sea). *Mar. Pollut. Bull.* **2012**, *64*, 1186–1200. [CrossRef] [PubMed]
3. ITOPF. *TIP2 Fate of Marine Oil Spills*; Technical Report; The International Tanker Owners Pollution Federation Limited: London, UK, 2011.
4. National Research Council. *Oil in the Sea III: Inputs, Fates, and Effects*; The National Academies Press: Washington, DC, USA, 2003.
5. De Dominicis, M.; Pinardi, N.; Zodiatis, G.; Lardner, R. MEDSLIK-II, a Lagrangian marine surface oil spill model for short-term forecasting—Part 1: Theory. *Geosci. Model Dev.* **2013**, *6*, 1851–1869. [CrossRef]
6. Zodiatis, G.; Lardner, R.; Alves, T.M.; Krestenidis, Y.; Perivoliotis, L.; Sofianos, S.; Spanoudaki, K. Oil spill forecasting (prediction). *Sea Sci. Ocean Predict. J. Mar. Res.* **2017**, *75*, 923–953. [CrossRef]

7.   Yapa, P.D. Modeling Oil Spills to Mitigate Coastal Pollution. In *Handbook of Environmental Fluid Dynamics, Volume Two*; Taylor and Francis Group: Abingdon, UK, 2013; pp. 243–255.

8.   Tkalich, P. A CFD solution of oil spill problems. *Environ. Model. Softw.* **2006**, *21*, 271–282. [CrossRef]

9.   Guo, W.; Wang, Y. A numerical oil spill model based on a hybrid method. *Mar. Pollut. Bull.* **2009**, *58*, 726–734. [CrossRef] [PubMed]

10.   Nikolić, D.; Gagić, R.; Ivošević, S. Estimation of Air Pollution from Ships in the Boka Kotorska Bay. In *The Boka Kotorska Bay Environment*; Springer International Publishing: Cham, Switzerland, 2017; pp. 117–128.

11.   Traffic, M. "Marine Traffic", 2007. Available online: http://www.marinetraffic.com/ (accessed on 11 April 2016).

12.   Nihoul, J. A non-linear mathematical model for the transport and spreading of oil slicks. *Ecol. Model.* **1984**, *22*, 325–339. [CrossRef]

13.   Zanier, G.; Petronio, A.; Armenio, V. The Effect of Coriolis force on oil slick transport and dispersion at sea. *J. Hydraul. Res.* **2017**, *55*, 409–422. [CrossRef]

14.   Zanier, G.; Giunto, F.; Petronio, A.; Roman, F.; Gagić, R.; Nikolic, D. Kotor Bay Area Hydrodynamics and Pollutant Dispersion Simulations: A Tool for Contingency Plans. In Proceedings of the Technology and Science for the Ships of the Future, Trieste, Italy, 20–22 June 2018; pp. 604–611.

15.   Galea, A.; Grifoll, M.; Roman, F.; Mestres, M.; Armenio, V.; Sanchez-Arcilla, A.; Zammit Mangion, L. Numerical Simulation of Water Mixing and Renewals in the Barcelona Harbour Area: The winter season. *Environ. Fluid Mech.* **2014**, *14*, 1405–1425. [CrossRef]

16.   Petronio, A.; Roman, F.; Nasello, C.; Armenio, V. Large-Eddy Simulation model for wind driven sea circulation in coastal areas. *Nonlinear Process. Geophys.* **2013**, *20*, 1095–1112. [CrossRef]

17.   Santo, M.; Toffolon, M.; Zanier, G.; Giovannini, L.; Armenio, V. Large eddy simulation (LES) of wind-driven circulation in aperi-alpine lake: Detection of turbulent structures andimplications of a complex surrounding orography. *J. Geophys. Res. Oceans* **2017**, *122*, 4704–4722. [CrossRef]

18.   Roman, F.; Stipcich, G.; Armenio, V.; Inghilesi, R.; Corsini, S. Large eddy simulation of mixing in coastal areas. *Int. J. Heat Fluid Flow* **2010**, *31*, 327–341. [CrossRef]

19.   Petronio, A.; Roman, F.; Armenio, V.; Stel, F.; Giaiotti, D. Large eddy simulation model for urban areas with thermal and humid stratification effects. In *Direct and Large-Eddy Simulation 9*; Springer: Cham, Switzerland, 2013.

20.   Roman, F.; Napoli, E.; Milici, B.; Armenio, V. An improved Immersed Boundary Method for curvilinear grids. *Comput. Fluids* **2009**, *38*, 1510–1527. [CrossRef]

21.   Roman, F.; Armenio, V.; Frohlich, J. A simple wall layer model for LES with IBM. *Phys. Fluids* **2009**, *21*, 101701. [CrossRef]

22.   Orlanski, I. A Simple Boundary Condition for Unbounded Hyperbolic Flows. *J. Comput. Phys.* **1976**, *21*, 251–269. [CrossRef]

23.   Wu, J. Wind-stress coefficient over sea surface from breeze to hurricane. *J. Geophys. Res.* **1982**, *87*, 9704–9706. [CrossRef]

24.   Zang, J.; Street, R.; Koseff, J. A non-staggered grid, fractional step method for time-dependent incompressible Navier–Stokes equations in curvilinear coordinates. *J. Comput. Phys.* **1994**, *114*, 18–33. [CrossRef]

25.   Zanier, G.; Petronio, A.; Roman, F.; Armenio, V. High Resolution Oil Spill Model for Harbour and Coastal Areas. In Proceedings of the 3rd IAHR Europe Congress, Porto, Portugal, 14–16 April 2014.

26.   Smith, J. (Ed.) *Torrey Canyon Pollution and Marine Life*; Cambridge University Press: Cambridge, UK, 1968; Chapter 8, pp. 150–162.

27.   ASCE. The state-of-the-art of modeling oil spills. *J. Hydraul. Eng.* **1996**, *122*, 594–609. [CrossRef]

28.   Goeury, C.; Hervouet, J.M.; Baudin-Bizien, I.; Thouvenel, F. A Lagrangian/Eulerian oil spill model for continental waters. *J. Hydraul. Res.* **2014**, *51*, 36–48. [CrossRef]

29.   Mackay, D.; Paterson, S.; Nadeau, S. Calculation of the Evaporation Rate of Volatile Liquids. In Proceedings of the National Conference on Control of Hazardous Material Spills, Louisville, KY, USA, 13–15 May 1980; pp. 364–368.

30.   Mackay, D.; Buist, I.; Mascarenhas, R.; Petersen, S. *Oil Spill Processes and Models*; Report EE-8; Environmental Protection Service: Gatineau, QC, Canada, 1980.

31.   Zhu, J. On the higher-order bounded discretization schemes for finite volume computations of incompressible flows. *Comput. Methods Appl. Mech. Eng.* **1992**, *98*, 345–360. [CrossRef]

32. Joksimović, A.; Đurović, M.; Semenov, A.V.; Zonn, I.S.; Kostianoy, A.G. Introduction. In *The Boka Kotorska Bay Environment*; Springer International Publishing: Cham, Switzerland, 2017; pp. 1–17.

33. Mandić, M.; Regner, S.; Gačić, Z.; Durović, M. Marković, O.; Ikica, Z. Composition and diversity of ichthyplankton in the Boka Kotorska Bay (South Adriatic Sea). *Acta Adriat.* **2014**, *55*, 229–244.

34. Mandić, S.; Radović, I.; Radović, D. Physical and Geographical Description of the Boka Kotorska Bay. In *The Boka Kotorska Bay Environment*; Springer International Publishing: Cham, Switzerland, 2017; pp. 43–67.

35. Dautović, J.; Strmečki, S.; Pestorić, B.; Vojvodić, V.; Plašvić, M.; Krivokapić, S.; Ćosović, B. Organic matter in the karstic enclosed bay (Boka Kotorska Bay, South Adriatic Sea). Influence of freshwater input. *Fresenius Environ. Bull.* **2012**, *21*, 995–1006.

36. Copernicus Database. Available onine: http://marine.copernicus.eu/ (accessed on 15 December 2016).

MDPI

St. Alban-Anlage 66

4052 Basel

Switzerland

Tel. +41 61 683 77 34

Fax +41 61 302 89 18

www.mdpi.com

Journal of Marine Science and Engineering Editorial Office

E-mail: jmse@mdpi.com

www.mdpi.com/journal/jmse